# CLASSIFIED INDEX

OF

# SUBJECTS OF INVENTION

ADOPTED IN

THE U. S. PATENT OFFICE,

MARCH 1, 1872.

WASHINGTON:
GOVERNMENT PRINTING OFFICE.
1872.

# INDEX OF CLASSES.

| No. | Class. | Page. |
|---|---|---|
| 1 | Aëration and Bottling | 15 |
| 2 | Apparel | 16 |
| 3 | Artificial Limbs | 18 |
| 4 | Baths and Closets | 19 |
| 5 | Beds | 20 |
| 6 | Bee-hives | 21 |
| 7 | Beer and Wine | 22 |
| 8 | Bleaching and Dyeing | 23 |
| 9 | Boats | 24 |
| 10 | Bolts, Nuts, and Rivets | 25 |
| 11 | Book-binding | 26 |
| 12 | Boots and Shoes | 27 |
| 13 | Brakes and Gins | 29 |
| 14 | Bridges | 30 |
| 15 | Brushes and Brooms | 31 |
| 16 | Builders' Hardware | 32 |
| 17 | Butchering | 34 |
| 18 | Caoutchouc | 35 |
| 19 | Carding | 36 |
| 20 | Carpentry | 37 |
| 21 | Carriages and Wagons | 38 |
| 22 | Casting | 40 |
| 23 | Chemical, Miscellaneous | 41 |
| 24 | Clasps and Buckles | 43 |
| 25 | Clay | 44 |
| 26 | Cloth | 45 |
| 27 | Coffins | 46 |
| 28 | Cordage | 47 |
| 29 | Crinoline and Corsets | 48 |
| 30 | Cutlery | 49 |
| 31 | Dairy | 50 |
| 32 | Dental | 51 |
| 33 | Drafting | 52 |
| 34 | Dryers and Kilns | 53 |
| 35 | Educational | 54 |
| 36 | Electricity | 55 |
| 37 | Excavators | 57 |
| 38 | Felting and Hats | 58 |
| 39 | Fences | 59 |
| 40 | Files | 60 |
| 41 | Fine Arts | 61 |
| 42 | Fire-arms | 62 |
| 43 | Fishing | 63 |
| 44 | Fuel | 64 |
| 45 | Furniture | 65 |
| 46 | Games and Toys | 67 |
| 47 | Garden and Orchard | 68 |
| 48 | Gas | 70 |
| 49 | Glass | 71 |
| 50 | Governors | 72 |
| 51 | Grinding and Polishing | 73 |
| 52 | Gunpowder | 74 |
| 53 | Hardware Manufacture | 75 |
| 54 | Harness | 78 |
| 55 | Harrows | 80 |
| 56 | Harvesters | 81 |
| 57 | Hoisting | 83 |
| 58 | Horology | 84 |
| 59 | Horse-shoes | 85 |
| 60 | Hose and Belting | 86 |
| 61 | Hydraulic Engineering | 87 |
| 62 | Ice | 89 |
| 63 | Jewelry | 90 |
| 64 | Journals and Bearings | 91 |
| 65 | Kitchen Utensils | 92 |
| 66 | Knitting and Netting | 94 |
| 67 | Lamps and Gas-fittings | 95 |
| 68 | Laundry | 96 |
| 69 | Leather | 97 |
| 70 | Locks and Latches | 98 |
| 71 | Manures | 99 |
| 72 | Masonry | 100 |
| 73 | Measuring Instruments | 101 |
| 74 | Mechanical Powers | 103 |
| 75 | Metallurgy | 104 |
| 76 | Metal Working: 1. Bending and Straightening | 106 |
| 77 | Metal Working: 2. Boring and Drilling. | 107 |
| 78 | Metal Working: 3. Forging, Swaging, and Riveting | 108 |
| 79 | Metal Working: 4 Punching, Cutting, and Shearing | 109 |
| 80 | Metal Working: 5. Rolling | 110 |
| 81 | Metal Working: 6. Tools | 111 |
| 82 | Metal Working: 7. Turning, Planing, and Milling | 113 |
| 83 | Mills | 114 |
| 84 | Music | 115 |
| 85 | Nails | 116 |
| 86 | Needles and Pins | 117 |
| 87 | Oils, Fats, and Glue | 118 |
| 88 | Optics | 119 |
| 89 | Ordnance | 120 |
| 90 | Ore | 121 |
| 91 | Paint | 122 |
| 92 | Paper-making | 123 |
| 93 | Paper Manufactures | 124 |
| 94 | Paving | 125 |
| 95 | Photography | 126 |
| 96 | Plating | 127 |
| 97 | Plows | 128 |
| 98 | Pneumatics | 129 |
| 99 | Preserving Food | 130 |
| 100 | Presses | 131 |
| 101 | Printing | 132 |
| 102 | Projectiles | 134 |
| 103 | Pumps | 136 |
| 104 | Railways: 1. The "Way" | 137 |
| 105 | Railways: 2. Cars and Interior Fittings | 138 |
| 106 | Railways: 3. Exterior Mountings and Fittings | 139 |
| 107 | Railways: 4. Track and Car Irons and Fittings, Manufacture of | 140 |
| 108 | Roofing | 141 |
| 109 | Safes | 142 |
| 110 | Saws | 143 |
| 111 | Seeders and Planters | 144 |
| 112 | Sewing-machines | 145 |
| 113 | Sheet-metal | 146 |
| 114 | Ships: 1. Construction | 147 |
| 115 | Ships: 2. Propulsion | 149 |
| 116 | Signals | 150 |
| 117 | Silk | 151 |
| 118 | Spinning | 152 |
| 119 | Stabling | 153 |
| 120 | Stationery | 154 |
| 121 | Steam: 1. Engines | 156 |
| 122 | Steam: 2. Boilers | 157 |
| 123 | Steam: 3. Locomotives | 159 |
| 124 | Stills | 160 |
| 125 | Stone, Lime, and Cement | 161 |
| 126 | Stoves and Furnaces | 163 |
| 127 | Sugar | 165 |
| 128 | Surgery | 166 |
| 129 | Tanning | 168 |
| 130 | Thrashing | 169 |
| 131 | Tobacco | 170 |
| 132 | Toilet | 171 |
| 133 | Trunks | 172 |
| 134 | Tubing and Wire | 173 |
| 135 | Umbrellas and Fans | 174 |
| 136 | Valves | 175 |
| 137 | Water Distribution | 176 |
| 138 | Water-wheels | 177 |
| 139 | Weaving | 178 |
| 140 | Wire-working | 180 |
| 141 | Wood-screws | 181 |
| 142 | Wood-working: 1. Lathes | 182 |
| 143 | Wood-working: 2. General-work Machinery | 183 |
| 144 | Wood-working: 3. Special-work Machinery | 184 |
| 145 | Wood-working: 4. Tools | 187 |

# CLASSES.

| No. | Class. | Page. | Examiner. |
|---|---|---|---|
| 1 | Aëration and Bottling.... | 15 | |
| 2 | Apparel.................. | 16 | |
| 3 | Artificial Limbs.......... | 18 | |
| 4 | Baths and Closets ........ | 19 | |
| 5 | Beds..................... | 20 | |
| 6 | Bee-hives................ | 21 | |
| 7 | Beer and Wine ............ | 22 | |
| 8 | Bleaching and Dyeing..... | 23 | |
| 9 | Boats.................... | 24 | |
| 10 | Bolts, Nuts, and Rivets.... | 25 | |
| 11 | Book-binding ............. | 26 | |
| 12 | Boots and Shoes.......... | 27 | |
| 13 | Brakes and Gins....... ... | 29 | |
| 14 | Bridges .................. | 30 | |

| No. | Class. | Page. | Examiner. |
|---|---|---|---|
| 15 | BRUSHES AND BROOMS....... | 31 | |
| 16 | BUILDERS' HARDWARE....... | 32 | |
| 17 | BUTCHERING ................ | 34 | |
| 18 | CAOUTCHOUC ..... .......... | 35 | |
| 19 | CARDING .......... ........ | 36 | |
| 20 | CARPENTRY ................ | 37 | |
| 21 | CARRIAGES AND WAGONS .... | 38 | |
| 22 | CASTING.................... | 40 | |
| 23 | CHEMICAL, MISCELLANEOUS .. | 41 | |
| 24 | CLASPS AND BUCKLES ....... | 43 | |
| 25 | CLAY ...................... | 44 | |
| 26 | CLOTH ...... .............. | 45 | |
| 27 | COFFINS ...... ............ | 46 | |
| 28 | CORDAGE.................... | 47 | |
| 29 | CRINOLINE AND CORSETS .... | 48 | |
| 30 | CUTLERY.................... | 49 | |

| No. | Class. | Page. | Examiner. |
|---|---|---|---|
| 31 | DAIRY .................... | 50 | |
| 32 | DENTAL .................... | 51 | |
| 33 | DRAFTING .................... | 52 | |
| 34 | DRYERS AND KILNS ......... | 53 | |
| 35 | EDUCATIONAL .............. | 54 | |
| 36 | ELECTRICITY ............... | 55 | |
| 37 | EXCAVATORS ................ | 57 | |
| 38 | FELTING AND HATS ........ | 58 | |
| 39 | FENCES .................... | 59 | |
| 40 | FILES .................... | 60 | |
| 41 | FINE ARTS .................. | 61 | |
| 42 | FIRE-ARMS ................ | 62 | |
| 43 | FISHING .................... | 63 | |
| 44 | FUEL .................... | 64 | |
| 45 | FURNITURE ................ | 65 | |
| 46 | GAMES AND TOYS ........... | 67 | |

| No. | Class. | Page. | Examiner. |
|---|---|---|---|
| 47 | Garden and Orchard...... | 68 | |
| 48 | Gas ...... .................... | 70 | |
| 49 | Glass.......................... | 71 | |
| 50 | Governors ................. | 72 | |
| 51 | Grinding and Polishing.... | 73 | |
| 52 | Gunpowder...... .......... | 74 | |
| 53 | Hardware Manufacture .. | 75 | |
| 54 | Harness...................... | 78 | |
| 55 | Harrows ................... | 80 | |
| 56 | Harvesters ................ | 81 | |
| 57 | Hoisting ..... ............ | 83 | |
| 58 | Horology ................... | 84 | |
| 59 | Horse-shoes ... ............ | 85 | |
| 60 | Hose and Belting ......... | 86 | |
| 61 | Hydraulic Engineering ... | 87 | |
| 62 | Ice .. . ..................... | 89 | |

| No. | Class. | Page. | Examiner. |
|---|---|---|---|
| 63 | JEWELRY .......... | 90 | |
| 64 | JOURNALS AND BEARINGS.... | 91 | |
| 65 | KITCHEN UTENSILS .......... | 92 | |
| 66 | KNITTING AND NETTING...... | 94 | |
| 67 | LAMPS AND GAS-FITTINGS.... | 95 | |
| 68 | LAUNDRY .................. | 96 | |
| 69 | LEATHER.................. | 97 | |
| 70 | LOCKS AND LATCHES ........ | 98 | |
| 71 | MANURES................ | 99 | |
| 72 | MASONRY.................. | 100 | |
| 73 | MEASURING INSTRUMENTS.... | 101 | |
| 74 | MECHANICAL POWERS ....... | 103 | |
| 75 | METALLURGY .............. | 104 | |
| 76 | METAL WORKING: 1. BENDING AND STRAIGHTENING. | 106 | |
| 77 | METAL WORKING: 2. BORING AND DRILLING. | 107 | |

| No. | Class. | Page. | Examiner. |
|---|---|---|---|
| 78 | METAL WORKING: 3. FORGING, SWAGING, AND RIVETING. | 108 | |
| 79 | METAL WORKING: 4. PUNCHING, CUTTING, AND SHEARING. | 109 | |
| 80 | METAL WORKING: 5. ROLLING. | 110 | |
| 81 | METAL WORKING: 6. TOOLS.. | 111 | |
| 82 | METAL WORKING: 7. TURNING, PLANING, AND MILLING. | 113 | |
| 83 | MILLS | 114 | |
| 84 | MUSIC | 115 | |
| 85 | NAILS | 116 | |
| 86 | NEEDLES AND PINS | 117 | |
| 87 | OILS, FATS, AND GLUE | 118 | |
| 88 | OPTICS | 119 | |
| 89 | ORDNANCE | 120 | |
| 90 | ORE | 121 | |

| No. | Class. | Page. | Examiner. |
|---|---|---|---|
| 91 | PAINT ...................... | 122 | |
| 92 | PAPER-MAKING .............. | 123 | |
| 93 | PAPER MANUFACTURES ...... | 124 | |
| 94 | PAVING ...................... | 125 | |
| 95 | PHOTOGRAPHY.............. | 126 | |
| 96 | PLATING .................... | 127 | |
| 7 | PLOWS ....................... | 128 | |
| 98 | PNEUMATICS................. | 129 | |
| 99 | PRESERVING FOOD........... | 130 | |
| 100 | PRESSES ..................... | 131 | |
| 101 | PRINTING .................... | 132 | |
| 102 | PROJECTILES ................ | 134 | |
| 103 | PUMPS ....................... | 136 | |

| No. | Class. | Page. | Examiner. |
|---|---|---|---|
| 104 | RAILWAYS: 1. THE "WAY".. | 137 | |
| 105 | RAILWAYS: 2. CARS AND INTERIOR FITTINGS. | 138 | |
| 106 | RAILWAYS: 3. EXTERIOR MOUNTINGS AND FITTINGS. | 139 | |
| 107 | RAILWAYS: 4. TRACK AND CAR IRONS AND FITTINGS, MANUFACTURE OF. | 140 | |
| 108 | ROOFING .................. | 141 | |
| 109 | SAFES... .................. | 142 | |
| 110 | SAWS ...................... | 143 | |
| 111 | SEEDERS AND PLANTERS. ... | 144 | |
| 112 | SEWING-MACHINES ........... | 145 | |
| 113 | SHEET-METAL..... .......... | 146 | |
| 114 | SHIPS: 1. CONSTRUCTION..... | 147 | |
| 115 | SHIPS: 2. PROPULSION ....... | 149 | |
| 116 | SIGNALS ..... .............. | 150 | |
| 117 | SILK ...................... | 151 | |

| No. | Class. | Page. | Examiner. |
|---|---|---|---|
| 118 | Spinning | 152 | |
| 119 | Stabling | 153 | |
| 120 | Stationery | 154 | |
| 121 | Steam: 1. Engines | 156 | |
| 122 | Steam: 2. Boilers | 157 | |
| 123 | Steam: 3. Locomotives | 159 | |
| 124 | Stills | 160 | |
| 125 | Stone, Lime, and Cement | 161 | |
| 126 | Stoves and Furnaces | 163 | |
| 127 | Sugar | 165 | |
| 128 | Surgery | 166 | |
| 129 | Tanning | 168 | |
| 130 | Thrashing | 169 | |
| 131 | Tobacco | 170 | |

| No. | Class. | Page. | Examiner. |
|---|---|---|---|
| 132 | Toilet | 171 | |
| 133 | Trunks | 172 | |
| 134 | Tubing and Wire | 173 | |
| 135 | Umbrellas and Fans | 174 | |
| 136 | Valves | 175 | |
| 137 | Water Distribution | 176 | |
| 138 | Water-wheels | 177 | |
| 139 | Weaving | 178 | |
| 140 | Wire-working | 180 | |
| 141 | Wood Screws | 181 | |
| 142 | Wood-working: 1. Lathes | 182 | |
| 143 | Wood-working: 2. General-work Machinery. | 183 | |
| 144 | Wood-working: 3. Special-work Machinery. | 184 | |
| 145 | Wood-working: 4. Tools | 187 | |

# CLASSIFICATION

OF

# SUBJECTS OF INVENTION.

---

## CLASS 1.

### AËRATION AND BOTTLING.

Aërated Liquor Apparatus and Processes, Soda-fountains and Fire-annihilators, Barrel-filling and Bungs, Bottling and Bottle-stoppers.

Aërated liquids.
Aërated liquor apparatus.
Aërating dough, Apparatus for
Barrel-cleaners.
Barrel-fillers.
Barrel-lining. (*See Class* 91.)
Barrel-pitching. (*See Class* 91.)
Barrel-rollers.
Barrel-stands and skids.
Barrel-tilters.
Barrel-washers.
Bottle-cleaners.
Bottle envelopes.
Bottle-fillers.
Bottle-racks.
Bottles, Boxes for transporting
Bottle-stoppers.
Bottle-washers.
Bottling-implements.
Bottling-machines.
Bungs.
Bungs, Bushes for
Bungs, Extractors for
Bushings for barrels.
Collapsible casks for beer, &c.
Corking-machines.
Cork-making machines. (*See Class* 142.)
Cork-presses.
Corks, Artificial
Effervescing preparations for beverages.
Fire-annihilators, Gas-generating
Fire-engines, Chemical
Gazogenes.
Mineral-waters, Compositions for
Siphon-bottles.
Sirup-gages.
Sirup-fountains.
Soda-fountains.
Soda-fountains, Apparatus for charging
Soda-water apparatus.
Vents for barrels.

# CLASS 2.

## APPAREL.

Clothing and Clothes-making, Work-table appliances, Patterns for Garments.

[*Excepting Boots and Shoes* ........ *Class* 12.
*Head-coverings* ......... *Class* 38.
*Hoop-skirts and Corsets* .. *Class* 29.
*Knitting-machines* ....... *Class* 66.
*Paper-garments* .... .... *Class* 93.
*Sewing-machines* ........ *Class* 112.]

Balmorals.
Belts, Body
Boas.
Bodkins.
Body-conformators.
Bonnet-stands.
Bosom-folders.
Bosoms.
Boxes for spools and thread.
Braces.
Capes.
Cloaks.
Cloth crimping machines.
Cloth-cutting machines.
Cloth-plaiting machines.
Clothes-pressing machines.
Coats.
Collars, (*except of Paper*, *Class* 93.)
Cravat-holders.
Cravats.
Crochet.
Cuffs.
Cuff-supporters.
Cushions for pins, needles, and sewing-work.
Cutting out garments.
Darning-lasts.
Darning-machines.
Diapers.
Drawers.
Dress-protectors.
Dress-supporters.
Dummies for displaying clothing.
Eyeleting-machines. (*See Class* 24.)
Feather dressers.
Finger-guards.
Fitting garments.
Fluting-machines.
Frames for measuring and fitting garments.
Frames for tambour-work and quilting.
Frilling.
Fur, Articles of
Gages for cutting garments.
Gaiterettes, (of cloth.)
Garments, (*except of Paper*, *Class* 93.)
Garments, Gages for cutting
Garments, Machines for cutting out
Garments, Patterns for
Garments, Systems of measuring and cutting
Garters.
Glove-making.
Gloves.
Glove-stretching machines.
Glove-turning machines.
Hangers for clothing.
Hemmers, Hand
Hosiery. (*See Class* 66.)
Hosiery-ironing machines.
Housewifes.
Ironing-machines, Tailors' and hosiery
Ladies' figure-stands.
Mittens.
Muffs.
Neck-tie retainers.
Neck-ties.
Needle-boxes.
Needle-cases.
Needle-guards.
Needles.
Needle-threaders.
Needle-wrappers.

Pantalets.
Pantaloon-protectors.
Pantaloons.
Pantaloon-straps.
Pantaloon-stretchers.
Patterns for garments.
Pin-boxes.
Pin-cushions.
Pinking-machines.
Plaiting-machines.
Pockets.
Pockets, Safety
Pressing garments.
Quilting-frames.
Ripping-tool.
Ruffles.
Ruffles, Machines for making
Safety-guards for watches and pocket-books.
Sewing-birds.
Sewing-boxes.
Sewing-guards.
Sewing-pins.
Sewing-stands.
Sewing-work holders.
Shawls.
Shirt-bosoms.
Shirts.
Shoulder-braces.
Shoulder-straps.
Shoulder-supporters.
Shuttles, Tatting
Skirt-covers.
Skirt-elevators.
Skirts, Binding for
Skirts, (*except Crinoline, Class* 29.)
Skirt-supporters.
Sleeves.
Spool-cases.
Spool-holders.
Spool-stands.
Stockings, Toe and heel protectors for
Stocking-supporters.
Stocks, Neck
Suspenders.
Tailoring.
Tailors' ironing-machines.
Tatting-shuttles.
Thimbles.
Thread and spool boxes.
Thread-cutters.
Thread-envelopes.
Threaders.
Tippets.
Trimmings for garments.
Under-garments, (*except Knitted, Class* 66.)
Under-sleeves.
Veils.
Vests.
Work-baskets.
Work boxes, Sewing-
Work holders, Sewing-
Work-table, Appliances for the
Wrappers.

# CLASS 3.

## ARTIFICIAL LIMBS.

Including all Prosthetic Parts, Splints, Fracture, and Orthopedic Appliances.

[*Except Artificial Teeth and Dentures, Class* 32.]

---

Arms, Artificial
Artificial arms.
Artificial ears.
Artificial eyes.
Artificial feet.
Artificial hands.
Artificial legs.
Artificial palates.
Auricles.
Club feet, Apparatus for
Counter-extension apparatus.
Crutches.
Dislocation apparatus.
Ears, Artificial
Ear-trumpets.
Ear-tubes.
Eyes, Artificial
Extension-apparatus for fractures.
Feet, Artificial
Fractures, Apparatus for reducing
Hands, Artificial
Legs, Artificial
Orthopedic apparatus.
Palates, Artificial
Spinal curvatures, Instruments for treating
Splints.

# CLASS 4.

## BATHS AND CLOSETS.

Baths, Water and Earth Closets, Urinals, Wash Stands and Basins, Sinks, and Stench-traps.

[*Except Faucets, Class* 136.]

[*For Manufacture of Water-pipe Fittings and Faucets, see Class* 53, *Section K.*]

---

Air-traps for sinks and sewers.
Bath-heaters. (*See Class* 126.)
Bathing apparatus.
Bathing tubs.
Baths.
Baths, Douche
Baths, Medicated
Baths, Shower
Baths, Vapor
Baths, Voltaic. (*See Class* 36.)
Bidets.
Chamber-pots.
Commodes.
Deodorizing materials and processes. (*See Manures, Class* 71.)
Earth closets.
Faucets. (*See Class* 136.)
Kitchen-sinks.
Sinks.
Stench-traps.
Traps, Stench
Traps, Water
Urinals.
Urinals, Portable
Valves, Water-closet. (*See Class* 136.)
Wash-basins.
Wash-stands.
Water-closets.
Water-closets, Appliances of
Water-traps.

# CLASS 5.

## BEDS.

Bedsteads, Bedding, and Bed-furniture.

---

Bed-bottoms.
Bed-clothes clasps.
Bed-bugs, Traps for
Bedding.
Beds.
Bedstead canopies.
Bedstead-cording.
Bedstead-fastenings.
Bedstead-keys.
Bedsteads.
Bedsteads and other furniture combined.
Bedsteads, Invalid
Bedsteads, Sofa
Bedsteads, Spring-slats for
Bedsteads, Valances for
Bedsteads, Ventilating
Bedsteads, Wardrobe
Bed-wrenches.
Camp-beds.
Canopies for bedsteads.
Camp-cots.
Clothes-clamps.
Couches.
Cradles.
Cribs.
Curled hair and its substitutes.
Hammocks.
Mattresses.
Mattresses, Materials for
Mattresses, Machines for making
Mosquito-netting, Frames for
Nursing couches.
Palliasses.
Pillows.
Sponge for stuffing, Preparing
Springs for beds.
Stuffing for mattresses, pillows, &c.
Stuffing for upholstering purposes.
Upholstering stuffing.
Valances.

# CLASS 6.

## BEE-HIVES.

Apiaries and Bee-culture.

---

Apiaries.
Bee-culture.
Bee-feeders.
Bee-fumigators.
Honey-boxes for bee-hives.
Moth-destroyers.
Moth-traps.
Swarm-indicators.

# CLASS 7.

## BEER AND WINE.

Beer, Wine, Cider, and Vinegar.

[*Except Mills and Presses, Class* 100.]

Ageing liquors.
Anti-fermentation processes and applications.
Beer-coolers. (*See Class* 62.)
Beer-cooling vats.
Beer-preserving vessels.
Brewery apparatus.
Brewing.
Cider-making, (*except Presses, Class* 100.)
Clarifying wines, &c.
Cooling-wort.
Fermentation-checks.
Fermented liquors, Cooling
Fermented liquors, Preserving
Hop-backs.
Hop-extracts.
Hops, Treatment of
Malting.
Malt liquors, Cooling
Malt liquors, Preserving
Mash-tubs.
Preserving fermented liquors.
Racking wines, &c.
Vinegar apparatus.
Vinegar processes.
Wine-ageing.
Wine-making.

# CLASS 8.

## BLEACHING, DYEING, ETC.

Bleaching, Dyeing, Sizing, and Printing Fabrics. Starch.

Aniline dyes.
Bucking-kiers.
Bleaching materials.
Bleaching processes.
Calico-printing, Composition for
Calico-printing, Processes for
Dextrine.
Dyeing apparatus.
Dyeing materials.
Dyeing processes.
Dyes.
Dye-vats.
Dye-wood extracts.
Glazed fabrics.
Paste, Compositions for
Paste-making.
Printing fabrics.
Resist printing.
Size, Preparation of
Sizing fabrics.
Starch, Manufacture of

# CLASS 9.

## BOATS.

Boats, Rafts, and Life-preservers.

Animals across rivers, Floating
Boat-bridges.
Boat-building.
Boat-detaching apparatus.
Boat-hoists.
Boat-hooks.
Boats.
Boats, Folding
Boats, Life
Boats, Metallic
Boats, Sectional
Boats, Surf
Buoys.
Buoy-safes.
Davits.
Detaching boats from davits.
Ferry-boats. (*See Class* 114.)
Floating safes.
Folding-boats.
Ice-boats.
Life-boats.
Life-buckets.
Life-preservers.
Life-rafts.
Logs, Booming
Logs, Rafting
Lowering boats.
Marine safes.
Metallic boats.
Oar-locks.
Oars.
Oars, Feathering
Outriggers.
Ponton-bridges.
Pontons.
Rafting-dogs.
Rafting logs.
Rafts.
Row-locks.
Sectional boats.
Surf-boats.
Suspending ships' boats.

# CLASS 10.

## BOLTS, NUTS, AND RIVETS.

Articles, Varieties, and Machines for Making. Taps, Screw-dies, and Lock-nuts.

---

Bolt-heading machines.
Bolt-making machines.
Bolts, Screw
Bolt-threading machines.
Dies for cutting screws.
Die-stocks.
Locking-nuts.
Nut-blanks, Making
Nut-locks.
Nut-making machines.
Nuts.
Nut-threading machines.
Riveting-machines.
Rivet making.
Rivets.
Screw-bolts.
Screw-nuts.
Screw-taps.
Stocks and dies.
Taps.
Washer-making machines, Metallic
Washers, Metallic

# CLASS 11.

## BOOK-BINDING.

Binding Books and Ruling Paper.

Account-books, Binding of
Albums, Binding of
Book-binders' presses.
Book-binding materials, processes, styles, tools, and machines.
Book-cover fastenings.
Book-covers.
Book-covers, Machines for making
Books in the round, Machines for cutting
Book-sewing machines.
Embossing-presses.
Indexing.
Plows, Book-binders'
Portfolios, Making
Ruling-machines.
Ruling-machines, Pens for
Sewing-presses.

# CLASS 12.

## BOOTS AND SHOES.

Boots, Shoes, their Parts, Varieties, and Materials, Tools and Machines for making.

[*For Manufacture of Shoemakers' Hardware, see Class 53, Section G.*]

Anklets.
Awls, Shoemakers'
Boot-channeling machines.
Boot-clamps.
Boot-crimps.
Boot-grooving machines.
Boot-hooks.
Boot-linings.
Boot-patterns.
Boots and shoes.
Boot-seam rubbers.
Boots, Metallic braces for
Boots, Metallic shanks for
Boots, Metallic soles for
Boot-soles.
Boot-stretchers.
Boots, Wooden soles for
Boot-trees.
Boot-uppers.
Boot-ventilators.
Burnishing machines, Boot-
Calks, Boot
Channeling-machines.
Counters.
Crimping-machines and tools.
Cutting-boards for leather.
Edge-planes for soles.
Floats, Boot
Gaiters, Leather
Galoshes.
Heel and toe caps, Materials for
Heel-caps
Heel-fastenings.
Heel-irons.
Heel-plates.
Heel-polishing machines.
Heel-protectors.
Heels.
Heel-shaves.
Heels, Machines for cutting
Heels, Machines for dressing
Heel-stiffeners.
Heel-trimming machines.
Ice-creepers.
Ice-sandals.
Insoles.
Jacks, Pegging
Lacing, Leather shoe
Lap-shavers.
Last-holders.
Lasting-machines.
Lasts.
Lasts, Toe-pieces for
Leather pegs for shoes.
Leggings.
Moccasins.
Nailing machines, Boot-
Overshoes.
Peg-cutters.
Pegging-jacks.
Pegging-machines.
Pegging soles.
Pegs, Leather
Pegs, Machines for arranging
Pegs, Machines for driving.
Pegs, Shoe
Pinchers, Lasting
Pinchers, Shoemakers'
Rand-machines.
Rands.
Rubber boots.
Rubber shoes.
Rubber soles.
Sandals.
Sewing-machines for boots and shoes. (*See Class* 112.)
Shank-cutters.
Shank-lasters.
Shank-markers.
Shanks, Boot
Shank-stiffeners.
Shoe-binding.
Shoe-clasps.
Shoe-distenders.
Shoe-fasteners.
Shoe-hammers.
Shoe-holders.
Shoe-hooks.
Shoe-horns.
Shoe-lacing.

Shoe-lacing, Machines for tagging
Shoe-lacing hooks.
Shoe-linings.
Shoemakers' benches.
Shoemakers' tools.
Shoe-patterns.
Shoes.
Shoe-soles.
Shoe-stays.
Shoe-straps.
Slippers.
Soles for boots and shoes.
Tags, Machines for cutting shoe
Tips, Shoe
Toe-caps.
Trees, Boot
Uppers, Machines for cutting
Ventilators for boots.
Waxed-ends, Machines for twisting
Waxing thread, Machines for
Welts.
Welts, Machines for forming
Welts, Machines for trimming

# CLASS 13.

## BRAKES AND GINS.

Treatment of Raw Flax, Hemp, and Cotton, Hair and Oakum Pickers, and Husk-Splitters.

---

Braking-machines.
Corn-husk splitters.
Cotton-gins.
Fabrics for re-use of fiber, Tearing up
Fibers of plants, Separating
Flax-brakes.
Flax rotting.
Flax-scutchers.
Gins, Cotton
Hackles.
Hackling corn-husks.
Hackling-machines.
Hair-pickers.
Hemp-brakes.
Hemp-hackling.
Hemp-rotting.
Husk-splitting.
Oakum-pickers.
Ripples for flax, broom-corn, &c.
Rotting flax and hemp.
Scutching-machines.
Separating fibers of plants for use in textile fabrics.

# CLASS 14.

## BRIDGES.

Bridges and Arches, their Piers and Abutments, Trusses and Girders for Bridges, Floors, and Roofs, Iron Trusses, Piers, and Columns.

---

Abutments for arches.
Anchors for suspension-cables.
Arches.
Bascule bridges.
Boat-bridges. (*See Class* 9.)
Bow-string bridges.
Bridges.
Cables, Suspension-bridge. (*See Class* 140.)
Centering for arches.
Columns, Iron
Draw-bridges.
Ferry-boats. (*See Class* 114.)
Floating bridges. (*See Class* 114.)
Flying bridges.
Girders for bridges, floors, roofs.
Girders of iron or wood.
Iron bridges.
Iron columns.
Iron girders.
Iron piers.
Iron trusses.
Piers, Bridge and abutment
Ponton bridges. (*See Class* 9.)
Saddles for suspension-bridges.
Stone bridges.
Suspension-bridges.
Suspension-cable anchors.
Suspension-bridge cables. (*See Class* 140.)
Suspension-cable saddles.
Suspension-cable wire-carriers.
Swing bridges.
Truss bridges.
Trusses for bridges, floors, and roofs.
Trusses of iron or wood.
Tubular bridges.
Viaducts.
Wire cables. (*See Class* 140.)
Wooden bridges.

# CLASS 15.

## BRUSHES AND BROOMS.

Brooms, Brushes, Mats, Mops, and Machines for making.

[*Excepting Machines for Wood-working, Class* 144.]

Boot-cleaners.
Bottle-brushes.
Bristle-assorting machines.
Bristle-bunching machines.
Bristle-cleaning machines.
Bristle-washing machines.
Broom-handles.
Broom-heads.
Brooms.
Brush backs.
Brushes, (*all kinds except Wire, Class* 140.)
Brush-handles.
Brush-heads.
Brush-making machines.
Coir brooms.
Corn brooms.
Door-mats.
Flesh-brushes.
Floor covering, Slatted
Hair-brooms.
Hair-brushes.
Hickory brooms.
Horse-brushes.
Husk brooms.
Knuckle-shields for scrubbing.
Mats, Car
Mats, Door
Mats for cars, Wooden
Mats, Slat
Mop-handles.
Mop-heads.
Mops.
Mop-wringers.
Nail-brushes.
Paint-brushes.
Rotary brushes.
Scrubbing-brushes.
Scrubbing-machines.
Scrubbing-pails.
Shoe-brushes.
Slat-mats.
Slatted floor-covering.
Splint brooms.
Tooth-brushes.

# CLASS 16.

## BUILDERS' HARDWARE.

Metallic Trimmings of Houses and Furniture.

[*Except Bells and Bell-hanging*.. *Class* 116.
*Locks and Latches* ..... *Class* 70.
*Nails* ................ *Class* 85.]

[*For Manufacture of Builders' Hardware, see Class* 53, *Section B.*]

---

Barn-door hangers.
Barn-door rollers.
Bell-hanging. (*See Class* 116.)
Bell-pulls.
Bells. (*See Class* 116.)
Bells, Cow. (*See Class* 54.)
Bells, Sleigh. (*See Class* 54.)
Blind-adjusters.
Blind-fasteners.
Blind-hinges.
Blind-operators.
Bolts for doors, shutters, and blinds.
Bolts, Chain
Bolts, Spring
Buffers for doors.
Cabin-door hooks.
Car-window fastenings.
Casters, Furniture
Chocks, Door
Coffin handles.
Coffin hardware.
Door-bolts.
Door-buttons.
Door-checks.
Door-hangers.
Door-hinges.
Door-holders.
Door-plates, Attaching
Door-pulls.
Door-securers.
Door-sheaves.
Door-springs.
Door-stops.
Drawer-pulls.
Furniture-casters.
Furniture-hooks.
Furniture-knobs.
Gate-hinges.
Hangings for doors, gates, and sash.
Hat and coat hooks.
Heads for nails and screws.
Hinges.
Hinges, Blind
Hinges, Door
Hinges, Gate
Hinges, Loose-joint
Hinges, Self-shutting
Hinges, Shutter
Hinges, Springs
Hooks, Coat and hat
Hooks, Door
Hooks, Furniture
Hooks, Shutter
Pulleys, Sash
Rollers, Barn-door
Roller-sheaves for doors.
Rollers, Sash
Rollers, Sliding-door
Sash-bolts.
Sash-centers.
Sashes, Fasteners for meeting rails of
Sash-fasteners.
Sash-hangings.
Sash-pulleys.
Sash-rollers.
Sash-springs.
Sash-stops.
Sash-weights.
Shutter-bars.
Shutter-fasteners.
Shutter-hinges.
Shutter-lifts.
Shutter-operators.
Sliding-door rollers.
Spring-bolts.
Spring-hinges.
Springs, Door
Stair-rod fastenings.
Stair-rods.

Stops, Door
Supporters, Window
Trunk-rollers.
Weights, Sash
Window-buttons.
Window-fasteners.
Window-operators.
Window-supporters.

# CLASS 17.

## BUTCHERING.

Catching, Slaughtering, and Skinning Animals, Dressing Carcasses, Cutting Meat, Sausage-Making.

---

Animal-catching.
Brine-tubs.
Butchering appliances.
Catching animals for slaughtering.
Cleaning carcasses.
Clutches, Animal
Gallows for suspending carcasses.
Gambrels.
Hair-cleaning.
Hog-elevators.
Hog-slaughtering apparatus.
Knives, Butcher. (*See Class* 30.)
Lard-squeezers. (*See Class* 100.)
Meat-cutters.
Meat-grinders.
Mincing-machines.
Packing-houses. (*See Class* 99.)
Sausage-cutters.
Sausage-stuffers.
Scalding-tubs and appliances.
Singeing apparatus.
Skinning appliances.
Slaughtering apparatus.
Steels. (*See Class* 51.)

# CLASS 18.

## CAOUTCHOUC.

Preparation and Manufacture of Caoutchouc, Gutta-Percha, and Vulcanite.

---

Caoutchouc manufactures.
Gutta-Percha manufactures.
Hard rubber.
India rubber.
Molds for rubber goods.
Rubber, Hard
Rubber, India
Rubber, Vulcanizing
Rubber, White
Vulcanite.
Vulcanizing furnaces.
Vulcanizing processes.
White rubber.

# CLASS 19.

## CARDING.

Preparation of Cotton and Wool for Spinning.

---

Alarms for stop-motions of carding-machines.
Batting-machines.
Burring-machines.
Cans for slivers.
Cards, Cleaning
Card-clothing.
Carding-machines.
Carding-machines, Alarms for
Carding-machines, Cylinders for
Carding-machines, Stop-motions for
Cards, (Wool and Cotton)
Card-strippers.
Coating or covering yarn.
Combing-machines.
Coiling or laying slivers.
Cotton-gins. (*See Class* 13.)
Cotton-cleaners.
Cotton-pickers, (Factory)
Doffing apparatus.
Feeding apparatus for carding-machines.
Flat-strippers.
Lapping-machines.
Pickers, Wool or Cotton, (Factory.)
Slivers, Cans for
Stop-motions for carding-machines.
Wadding.
Wadding-machines.
Wool-burring machines.
Wool-cleaning machines.
Wool-oiling machines.
Wool-washing machines. (*See Laundry, Class* 68.)
Yarn coating or covering.

# CLASS 20.

## CARPENTRY.

Wood-work of Houses.

[*Except Trusses and Girders, Class* 14.]

Awnings.
Beams of wood, Compound
Blind-operators. (*See Class* 16.)
Blinds.
Blinds, Metallic
Brackets for staging.
Carpenters' tools. (*See Class* 145.)
Carpenters' work of buildings.
Ceilings.
Centering of arches.
Door-hangers.
Door-operators. (*See Class* 16.)
Doors.
Doors, Sliding
Elevating scaffolds.
Extension ladders.
Fire-escapes.
Fire-ladders.
Fire-proof shutters.
Frames of wood, (*except Trusses and Girders, Class* 14.)
Floors.
Floors, Fire-proof
Floors, Wooden
Girders. (*See Class* 14.)
Gratings for windows.
Guards for windows.
Hatchways, Devices for operating
Hatchways, Safety-guards for
Houses.
Houses, Construction of
Houses, Portable
Ladders.
Ladders, Combination
Ladders, Extension
Ladders, Fire
Ladders, Garden
Ladders, Step
Lathing.
Lathing-machines.
Lathing, Metallic
Joiners' work of buildings.
Joints for timbers.
Joists.
Metallic blinds, doors, shutters, and gratings.
Partitions of buildings, Wooden
Portable houses.
Rinks.
Roofs. (*See Class* 108.)
Safety-guards for hatchways.
Safety-guards for windows.
Sash.
Sash-frames.
Scaffolding.
Scarfing and other timber-joints.
Screens for window.
Sheds and stables, Portable
Shutter-operators. (*See Class* 16.)
Shutters.
Shutters, Fire-proof
Siding of houses.
Skylights, Operating.
Sliding hills.
Skating-rinks.
Splicing timbers.
Stairs.
Step-ladders.
Store-houses.
Thresholds.
Trap doors.
Trusses for bridges, roofs, &c. (*See Class* 14.)
Weather-boarding.
Weather-strips.
Window-blinds.
Windows.
Window-screens.
Window-shutters.
Wood-work of buildings.

# CLASS 21.

## CARRIAGES AND WAGONS.

Construction of Wheeled Vehicles, including Sleighs, Trucks, Barrows, Litters, and Yokes.

[*Except Locomotives and Traction Engines*.. *Class* 123.
*Railway Cars*.................... *Class* 105.]

[*For Manufacture of Carriage and Wagon Hardware, see Class* 53, *Section C.*]

---

Ambulances.
Anti-friction axle-bearings.
Axle-arms.
Axle-boxes, Compositions for
Axle-boxes for carriages.
Axles, Lubricators for carriage
Axle-trees.
Barrows.
Biers.
Bodies of vehicles.
Boot-boxes of vehicles.
Bow-irons of vehicles.
Bows, Wagon
Boxes of wheel-hubs.
Brakes, Carriage
Brakes, Wagon
Calash-tops for carriages.
Carriage-couplings.
Carriage-curtain fastenings.
Carriage hardware.
Carriage-irons.
Carriages.
Carriages, Dress-guards for
Carriage-seat fastenings.
Carriage-seats.
Carriages, Fire-engine
Carriage-springs.
Carriage-step protectors.
Carriage-steps.
Carriages, Top-blocks for
Carriage-tops.
Carriage-trimmings.
Carriage-window frames.
Carts.
Chairs, Locomotive.
Children's carriages.
Children's hand-cars.
Couplings, Thill
Detaching horses from vehicles.
Double-trees.
Dumping-carts.
Dumping-wagons.
Equalizing-bars.
Fellies.
Fellies, Attaching
Fellies, Expanding
Felly-clamps.
Fifth-wheels.
Hand-barrows.
Hand-bearers.
Hand-trucks.
Hanging carriage-bodies.
Hay-loaders for wagons.
Hay-racks for wagons.
Hearses.
Hold-backs for vehicles.
Hose-carriages.
Hubs, Carriage
Ice-cars.
Ice-chairs and carriages.
Irons, Carriage
Irons, Carriage-top
Irons, Setting carriage
Linch-pins.
Litters.
Loading wagons, Machinery for
Lumber-wagons.
Neck-yokes.
Night-carts.
Omnibusses.
Ox-yokes, bows, and pins.
Perches, Wagon
Poles, Wagon
Racks, Hay-wagon
Rein-holders.
Running-gear of vehicles.
Safety-straps of vehicles.
Sailing-carriages.
Seats, Carriage
Seats, Spring-wagon

Self-loading carts.
Single-trees.
Skeins of axles.
Sled-brakes.
Sled-runners.
Sleds.
Sled-soles.
Sleighs.
Spokes.
Spoke-sockets.
Springs, Carriage
Stretchers.
Sulkies.
Sulky-seats.
Tail-boards for wagons.
Thill-couplings.
Thills.
Thill-tugs.
Tire-making. (*See Class* 78.)
Tires, Securing
Tires, Tightening
Tire-upsetting. (*See Class* 78.)
Tongues, Carriage and wagon
Tops, Carriage
Traces, Safety-attachment of
Trucks.
Umbrella-supporters on carriages.
Upsetting-machines. (*See Class* 78.)
Vehicles.
Velocipedes.
Wagon-bolsters.
Wagons.
Wagons, Dumping
Wagon-seats.
Wagons, Unloading-attachment for
Wagon-tongue supporters.
Wheelbarrows.
Wheels for vehicles.
Wheel-spokes.
Wheels to axles, Securing
Wheelwright's wood working machines. (*See Class* 144.)
Wheelwright's wood-working tools. (*See Class* 145.)
Whiffletrees.
Whip-sockets.
Yokes, Neck
Yokes, Ox

# CLASS 22.

## CASTING.

Appliances, Machines, Modes, and Tools of the Foundry.

Bell-founding.
Black-washing molds, Apparatus for
Brass-founding.
Britannia-ware, Casting processes for
Bullet-casting.
Casting metals, Appliances for
Casting metals, Modes of
Casting metal on metal.
Casting metal under pressure.
Casting soft metals.
Castings, Tumbling-boxes for
Chills.
Clamps for molds.
Cleaning castings.
Compression-casting.
Core-boxes.
Cores.
Cores, Ventilating
Crucibles for metal. (*See Class 25.*)
Facing compositions.
Facing molds, Machines for
Flask-clamps.
Flasks, Molders'
Foundry, Appliances of the
Ladles for molten metal.
Match-plates.
Molders' flasks.
Molders' tables.
Mold-facing.
Molding pipe, Machines for
Molds for casting metals
Pattern-lifters for molders.
Pewter-ware, Casting processes for
Projectiles, Casting
Shotting machines.
Shot making, Cast
Sprue-patterns.
Sprues.
Tables for molders.
Tumblers for small castings.
Tumbling-boxes.
Ventilating-cores.

# CLASS 23.

## CHEMICAL MISCELLANEOUS.

Chemical Apparatus, Processes, and Compositions, not otherwhere specifically assigned, as noted below.

Acids, Apparatus for making, (*except Stills, Class* 124.)
Acids, Processes for making
Air-proofing processes. (*See Class* 91.)
Anæsthetic refrigerators. (*See Class* 62.)
Animal and vegetable fiber, Separating
Antiseptics.
Atomizers. (*See Class* 62.)
Beer and Wine. (*See Class* 7.)
Beverages, (*except Aerated, Class* 1; *Fermented, Class* 7.)
Billiard Balls, Compositions for
Bitters, Tonic
Blasting compounds. (*See Class* 52.)
Bleaching. (*See Class* 8.)
Bread-raising compounds.
Brewing. (*See Class* 7.)
Calico printing. (*See Class* 8.)
Candles. (*See Class* 87.)
Caoutchouc. (*See Class* 18.)
Capsules, Medicinal
Carbureters. (*See Class* 48.)
Chemical apparatus.
Chemical furnaces.
Chemicals, Preparing
Corpse-embalming.
Cosmetics.
Dentifrices.
Deodorizers.
Depilatories.
Detergents.
Disinfecting apparatus, (*except Atomizers, Class* 62.)
Distillation. (*See Class* 123.)
Dolls, Composition for
Drugs.
Dyeing. (*See Class* 8.)
Embalming compositions.
Embalming processes.
Fats, Treatment of. (*See Class* 87.)
Fertilizers. (*See Class* 71.)
Fire annihilators. (*See Class* 1.)
Fire-proofing compounds. (*See Class* 91.)
Fire-proofing processes. (*See Class* 91.)
Fire-proofing wood, fabrics, &c.
Fish-oil, Treatment of. (*See Class* 87.)
Fluids, Writing
Food, Preservation of. (*See Class* 99.)
Foot-rot curatives.
Fuel. (*See Class* 44.)
Furnaces, Bone-black. (*See Class* 127.)
Furnaces, Chemical
Gas. (*See Class* 48.)
Glue. (*See Class* 87.)
Gunpowder. (*See Class* 52.)
Hair-dyes.
Hair-restoratives.
Horn, Artificial.
Ice. (*See Class* 62.)
India rubber. (*See Class* 18.)
Ink, Printing (*See Class* 91.)
Ink, Writing
Insect-destroying compounds.
Ivory, Artificial
Lamp wicks, Chemical treatment of
Lard, Treatment of. (*See Class* 87.)
Liniments.
Malt liquors. (*See Class* 7.)
Manures. (*See Class* 71.)
Matches. (*See Class* 52.)
Medical compounds.
Medicines.
Oils, fats, and glue. (*See Class* 87.)
Oils, Refining. (*See Class* 124.)
Ointments.
Paint. (*See Class* 91.)
Pastilles.
Perfuming apparatus.
Plasters, Adhesive
Poultices, Compositions for
Preserving food. (*See Class* 99.)
Preserving timber.

Pyroxyline.
Rectification of spirits. (*See Class* 124.)
Refining oils, &c. (*See Class* 124.)
Retorts, (*except Coal, Class* 48.)
Retort stands.
Salves.
Soap. (*See Class* 87.)
Stains, Composition for removing
Starch. (*See Class* 8.)
Stearine. (*See Class* 87.)
Stills. (*See Class* 124.)
Sugar. (*See Class* 127.)
Sulphuric acid, Recovering. *See Class* 124.)
Tallow. (*See Class* 87.)
Tanning. (*See Class* 129.)
Timber, Preservation of
Tobacco, Flavoring
Tooth-powder.
Vermin destroying compounds.
Vinegar. (*See Class* 7.)
Washing compounds.
Water-proofing compounds. (*See Class* 91.)
Wax. (*See Class* 87.)
Wines, Treatment of. (*See Class* 7.)
Wood, Artificial
Wood, Preservation of
Wool, Compositions for treating
Writing ink.
Yeast and its substitutes.

# CLASS 24.

## CLASPS AND BUCKLES.

Small Metallic Attachments; Buckles, Buttons, Clasps, Eyelets, Hooks, Links, Rings, Studs, &c., and Implements and Machines for Attaching and Closing.

[*For Manufacture of these Small Metallic Attachments, &c., see Class 53 in its several sections.*]

---

Band punches.
Belt-punches.
Brequet-chain hooks.
Buckles.
Button-hole punches.
Button-hooks.
Button keys or fasteners.
Buttons of any material, (*except Turning, Class* 142.)
Buttons to cards, Securing
Buttons to fabrics, Eyeleting
Buttons to fabrics, Riveting
Clasps.
Clasps for garments.
Clasps for skirts.
Clinch-rings.
Cock-eyes.
Collar fastenings, Shirt
Conductors' punches.
Cuff-buttons.
Eyelet-punches.
Eyeleting-machines.
Eyeleting-stamps.
Eyelet-presses.
Eyelets.
Garment-fasteners.
Glove-buttoners.
Glove-buttons.
Glove-fasteners.
Glove-fastenings.
Hook-buttons.
Hook-buttons, Implements for setting
Hooks.
Hooks, Mousing
Hooks, Snap
Hooks, Swivel
Key-rings.
Lacing-hooks.
Lap-rings.
Leather-punches.
Mousing-hooks.
Neck-tie clasps.
Open links.
Open rings.
Ornaments, Small metallic
Paper-fasteners.
Paper-punches.
Pinking-punches.
Punches, Leather, paper, and ticket
Rings.
Self-fastening buttons.
Shoe-buttoners.
Skirts, Clasps for
Sleeve-buttons.
Snap-hooks.
Snap-rings.
Split-rings.
Spring-punches.
Studs.
Swivel-hooks.
Ticket-punches.
Wad-punches.
Washer-punches.

# CLASS 25.

## CLAY.

Bricks, Clay-mills, Brick and Tile Machines and Presses, Ceramic Manufactures, Earthenware, Pottery, Crucibles, Glazes, Kilns.

---

Adobes.
Biscuit (clay) ovens. (*See Class* 34.)
Brick kilns. (*See Class* 34.)
Brick-machines.
Brick-molds.
Brick-presses.
Bricks.
Bricks, Fire
Cement-mixing machines.
Cement-pipe, Machines for making
Cement-pipes.
Cement-pipes, Coating
Ceramic-enamels.
Ceramic manufactures.
Chimney-pots.
China-ware.
Clay-mills.
Clay-pipes.
Clay-pulverizers.
Clay-screening machines.
Crucible molds.
Crucibles.
Draining-pipes.
Draining tiles.
Earthenware.
Enameling culinary vessels.
Enameling hardware.
Enameling-ovens. (*See Class* 34.)
Fire-back bricks.
Fire-bricks.
Fire-tiles.
Flower-pots, (Clay.)
Garden-pots.
Garden tiles.
Glazes for pottery, bricks, &c.
Jars, Earthen
Kaolin, Purifying
Kilns for bricks and ceramics. (*See Class* 34.)
Malaxators.
Mortar-machines.
Mortar-mixers.
Mills, Clay
Mills, Pug
Molds for bricks and ceramics.
Mosaic-tiles.
Ovens for biscuit, (Clay.) (*See Class* 34.)
Ovens for enameling pottery, &c. (*See Class* 34.)
Paving-tiles.
Pipes, Clay
Pipes, Compositions for making
Pipes, Draining
Porcelain materials and manufacture.
Pottery materials and manufacture.
Presses, Brick
Pug-mills.
Roofing-tiles.
Sarcophagi, Clay
Stone-ware.
Terra cotta.
Tile-machines.
Tiles, Drain
Tiles, Fire
Tiles, Floor
Tiles, Garden
Tiles, Mosaic
Tiles, Paving
Tiles, Roofing
Tools for molding clay.

# CLASS 26.

## CLOTH.

Dressing and Finishing of Woven Fabrics.

Beaming cloth.
Calendering cloth.
Carpet-rag cutter.
Cloth-cutting machines.
Cloth-dressing.
Cloth-drying.
Cloth-finishing.
Cloth-folding.
Cloth gassing.
Cloth-ironing.
Cloth-measuring.
Cloth-napping.
Cloth-pressing. (*See Class* 100.)
Cloth-shearing.
Cloth-sponging.
Cloth-steaming.
Cloth-stretching.
Cloth-teazeling.
Cloth-tentering.
Dressing cloth.
Evening cloth.
Finishing cloth.
Gig-mills.
Hair cloth, Machines for trimming
Ironing cloth.
Measuring cloth, Machines for
Milling cloth.
Mills, Gig
Napping-machines for cloth.
Packaging fabrics.
Pressing cloth.
Rag carpet, Cutting cloth for
Scouring cloth.
Shearing cloth.
Singeing cloth.
Sponging cloth.
Steaming cloth.
Stretching cloth.
Teazel grading machines.
Teazeling cloth.
Tenter-bars for cloth.
Whipping cloth.

# CLASS 27.

## COFFINS.

Antiseptics. (*See Class* 23.)
Burial-cases, Compositions for making
Burial-caskets.
Caskets, Burial
Coffin-handles. (*See Class* 16.)
Coffin-hardware. (*See Class* 16.)
Coffins.
Coffins, Compositions for covering
Coffins, Compositions for making
Coffin-screws.
Corpse-coolers. (*See Class* 62.)
Embalming-compositions. (*See Class* 23.)
Embalming-processes. (*See Class* 23.)
Sarcophagi clay. (*See Class* 25.)

# CLASS 28.

## CORDAGE.

Braid, Cord, Rope, Thread, Twine, and Yarn.

Assorting yarns and threads.
Balling cord and twine.
Balling yarns and threads.
Banding-machines.
Bindings for carpet, &c.
Bobbins.
Braid.
Braid-holders.
Braiding cords.
Braiding-machine carriers.
Braiding-machines.
Braiding tapes.
Braiding whips.
Bundling-presses. (*See Class* 100.)
Cordage.
Corded binding, Machine for making
Cords.
Dress cord.
Dressing yarn, warp, thread, &c.
Drying yarn.
Fishing-lines.
Folding or tying yarn, (*except Bundling Presses, Class* 100.)
Fringe.
Fringe-making machines.
Fringe-twisters.
Fuse-making machines.
Gassing yarns, threads, &c.
Gimp-machines.
Hawsers.
Laying rope.
Laying-top for cordage-machines.
Loop-banding machines.
Measuring yarns, threads, &c.
Ornamental yarns.
Packaging yarns, threads, etc., (*except Bundling Presses, Class* 100.)
Paper-twine machines.
Parceling rope.
Picking rope.
Polishing yarns, threads, &c.
Quilling yarn.
Quills, Yarn and silk
Quill thread fastener.
Reels for winding yarns and thread.
Rope coiling.
Rope-making machines.
Scraping and dressing cord.
Serving rope.
Sizing yarn, thread, &c.
Slashers.
Spooling-machines.
Spools.
Tassels.
Thread assorting, balling, dressing, gassing, measuring, packaging, polishing, reeling, sizing, spooling, tying, winding.
Twine-machines.
Tying yarn.
Untwisting rope, Machines for
Warp-dressing.
Warping-machines.
Welting cord.
Winding-bobbins.
Winding cord, cordage, fishing-lines, rope, thread, warp, yarn.
Winding-spools.
Wire, Braided thread-covering
Worming rope.
Yarn assorting, balling, dressing, drying, folding, gassing, measuring, packaging, polishing, reeling, sizing, spooling, tying, winding.
Yarn-hooks.

# CLASS 29.

## CRINOLINE AND CORSETS.

Hoop-skirts, Corsets.

---

Bosom-expanders.
Breast-pads.
Busks.
Bustles.
Corset-clasps.
Corset-fastenings.
Corsets.
Crinoline.
Hoops for skirts.
Hoops for skirts, Coating or covering.
Hoop-skirt formers.
Hoop-skirts.
Hoop-skirts, Machines for making
Panniers.
Skirts and corsets combined.
Skirts, Hoop
Skirts, Covered wire for hoop
Spangles to skirts, Attaching
Tapes for skirts.
Tournures.
Wire for hoops of skirts.

# CLASS 30.

## CUTLERY.

Knives, Scissors, and Shears.

[*Except Machine Shears, Class* 79.]

[*For Manufacture of Cutlery, see Class* 53, *Section A.*]

---

Bench-shears.
Bill-hooks.
Bistouries.
Cane-knives.
Can-openers.
Corn-knives.
Forks, Culinary
Forks, Flesh
Forks, Table
Fruit-knives.
Hay-knives.
Knives.
Knives, Amputating
Knives and forks combined.
Knives and spoons combined.
Knives, Attaching handles to
Knives, Bread
Knives, Budding
Knives, Butcher
Knives, Cane
Knives, Carving
Knives, Corn
Knives, Curriers'
Knives, Dirk
Knives for cutting corn from the cob.
Knives, Fruit
Knives, Grafting
Knives, Handles for
Knives, Hay
Knives, Hoof-paring
Knives, Machete
Knives, Meat
Knives, Oyster
Knives, Pocket
Knives, Pruning
Knives, Shoe
Lampwick scissors.
Lancets.
Machetes.
Pikes.
Pruning-knives.
Pruning-shears.
Razors.
Scalpels.
Scissors.
Shears.
Shears, Bench
Shears, Button-hole
Shears, Cattle-marking
Shears, Clipping
Shears, Edging
Shears, Garden
Shears, Grass-edging
Shears, Hair-clipping
Shears, Hand
Shears, Hedging
Shears, Lopping
Shears, Pruning
Shears, Sheep
Shears, Sheet-metal
Shears, Tailors'
Shears, Tinmen's
Sheep-shears.
Table-forks.
Table-knives.
Tinmen's Shears.
Twine-cutters.

# CLASS 31.

## DAIRY.

Machines and Appliances for Milking; Butter and Cheese Making.

---

Butter-molds.
Butter-stampers.
Butter-workers,
Cheese-hoops.
Cheese-making.
Cheese-presses. (*See Class* 100.)
Cheese-shelves.
Cheese-tables.
Cheese-turners.
Cheese-vats.
Churn-dashers.
Churns.
Cow-milkers.
Cream-heaters.
Cream-stands.
Cream-strainers.
Curd-cutters.
Dairies.
Dairy implements.
Milk-cans.
Milk-carriers.
Milk-closets.
Milk-coolers. (*See Class* 62.)
Milkers, Cow
Milk-houses.
Milking-stools.
Milk-pails.
Milk-pans.
Milk-safes.
Milk-shelves.
Milk-strainers.
Milk-testers, (*except Lactometers, Class* 73.)
Strainers, Milk

# CLASS 32.

## DENTAL.

Dental Instruments, Supplies, and Processes.

Artificial gums.
Artificial palates.
Artificial teeth.
Articulators for fitting dentures.
Casting dental plates.
Dental-compositions.
Dental-drills.
Dental-fillings.
Dental-forceps.
Dental-instruments.
Dental-mallets.
Dental-plates.
Dental-plates, Casting
Dentures.
Dies for dentists.
Drills, Dental
Forceps, Dental
Gold-leaf condensers.
Gums, Artificial
Keys, Dental
Mallets, Dental
Molds, Dental
Nitrous oxide, Apparatus for administering. (*See Class* 128.)
Palates, Artificial
Plates for artificial teeth.
Saliva-removers.
Springs for artificial teeth.
Swages, Dentists'
Teeth, Artificial
Teeth, Artificial, Compositions for
Vulcanizing-flasks for dentists' uses.

# CLASS 33.

## DRAFTING.

Drafting, Plotting, and Mathematical Instruments.

Bow pencils and pens.
Calipers.
Compasses, Drafting
Copying-instruments.
Dividers, Measuring
Drafting-instruments.
Drafting-scales.
Ellipsographs.
Geometrical constructors.
Graduated rules.
Map projectors.
Mathematical instruments.
Mathematical scales.
Pantographs.
Parallel rulers.
Pens, Ruling
Perspective apparatus.
Planning-instruments.
Plotting-instruments.
Protractors.
Roulettes.
Rulers, Parallel
Ruling-pens.
Scales, Drafting
Spirals, Instruments for drawing.
Trammels.

# CLASS 34.

## DRYERS AND KILNS.

Apparatus and Machines for Drying.

[*Except Clothes-dryers*.. *Class* 68.
*Paper-dryers*... *Class* 92.]

---

Bagasse-dryers.
Barrel-dryers.
Barrel-fumigators.
Barrel-heaters.
Biscuit (clay) ovens.
Brick-kilns.
Carbonizing wood, Furnaces and kilns for
Card-board dryers.
Charcoal-kilns.
Cigar-dryers.
Cloth-dryers. (*See Class* 26.)
Coal-dryers.
Coke-ovens.
Dessicating grain, food, farina, &c.
Drying apparatus, (*except Clothes*, *Class* 68; *Paper*, *Class* 92.)
Drying-houses.
Drying-kilns.
Enameling-ovens.
Feather-renovators.
Fertilizer-dryers.
Fish-curing and drying.
Fish-flakes.
Flour-dryers.
Fruit-dryers.
Fuel-dryers.
Glue-dryers.
Grain-dryers.
Hop-kilns.
Kilns, Brick
Kilns, Charcoal
Kilns for drying grain, peat, &c.
Kilns, Hop
Kilns, Lime
Kilns, Malt
Kilns, Pottery
Lime-kilns.
Lumber-dryers.
Malt-kilns.
Meat-dryers.
Oasts.
Offal-dryers.
Ovens, Coke
Ovens for biscuit, (Clay)
Ovens for enameling.
Ovens for pottery.
Peat-dryers.
Smoke-houses.
Smoking meat and fish.
Tobacco-dryers.
Wool-dryers.

# CLASS 35.

## EDUCATIONAL.

Devices for Teaching Reading, Writing, Arithmetic, Astronomy, Geography, Penmanship, &c.

---

Alphabet-blocks.
Astronomical charts.
Astronomical lanterns.
Blackboard-rubbers.
Blackboards.
Blackboards, Compositions for
Blind, Writing apparatus for the
Chart-holders.
Charts.
Copy-books.
Dissected maps.
Educational devices.
Frames, Numeral
Globes, Astronomical
Globes, Geographical
Map-holders.
Maps.
Maps, Dissected
Maps, Outline
Numeral-frames.
Object-teaching, Devices for
Orreries.
Penmanship, Appliances for teaching
Planetariums.
Planispheres.
School charts and maps.
Slate-cleaners.
Slate-frames.
Slate-pencils.
Slates.
Slates, Artificial
Tellurians.
Writing apparatus for the blind.

# CLASS 36.

## ELECTRICITY.

Batteries, Instruments; Applications of Electricity, Galvanism, and Magnetism; Electric Telegraphs.

---

Alarms, Electro-magnetic
Alarms, Telegraphic
Anti-incrustators, Galvanic
Anti-incrustators, Magnetic
Armatures.
Arresters.
Astatic needles.
Astronomical instruments, Electro-magnetic
Baths, Electro-voltaic
Batteries, Voltaic
Batteries, Thermo-electric
Buoys, Electrical
Cables, Submarine telegraph
Cables, Telegraph
Calendars, Electrical
Car-brakes, Electrical
Car-starters, Electrical
Chronographs, Electro-magnetic
Chronoscopes, Electro-magnetic
Circuit-breakers.
Circuit-closers.
Clocks, Electrical
Commutators.
Compasses, Mariners'.
Compasses, Surveyors'. (*See Class* 88.)
Condensers.
Conductors, Electro-telegraphic
Connectors for Voltaic batteries.
Current-regulators.
Electrical apparatus.
Electrical balances.
Electrical machines.
Electric fuses.
Electric lights.
Electrizing blood, Instruments for
Electrodes.
Electro-magnetic apparatus.
Electro-magnetic engines.
Electro-magnetic machines.
Electro-magnetic meteorological instruments.
Electro-magnets.
Electrometers.
Electronomes.
Electrophorus.
Electro-plating, Baths and batteries for
Electrotyping, Baths and batteries for
Engraving, Electric apparatus for
Escapements, Electrical
Fire-alarm telegraphs.
Fire-alarm telegraphs, Signal-boxes for
Galvanic appliances.
Galvanic attachments.
Galvanometers.
Galvanoscopes.
Gas-engines, Electrical apparatus for
Heaters, Electrical
Helices, Electrical
Hones, Magnetic
Hydro-electric machines.
Induction apparatus.
Insulated wire.
Insulators.
Iron filings, Separators for
Key-boards, Telegraphic
Keys, Telegraphic
Lamps, Electric
Lights, Electric
Lighting gas by electricity.
Lightning-arresters.
Lightning-protectors.
Lightning-rods.
Line-wires.
Line-wires, Covering
Line-wires, Insulating
Line-wires, Protecting
Line-wires, Supporting
Line-wires, Underground
Logs, Electrical
Looms, Electrical
Magnetic apparatus.
Magnetic needles.
Magneto-electric machines.
Magnets.
Magnets, Receiving
Magnets, Sounding

Manipulators
Mariners' compasses.
Medical-electric apparatus.
Meteorometers, Telegraphic
Meters, Electrical
Musical instruments, Electric attachments to
Pendulums, Electrical
Pens, Galvanic
Railway-signals, Electrical
Railway-switches, Electrical
Railway-telegraphs, Electrical
Reels, Telegraph
Registers, Telegraph
Relays.
Repeaters.
Resistance-boxes.
Resistance-coils.
Rheometers.
Rheotomes.
Rheotropes.
Rheostats.
Ruling-apparatus, Electrical
Semaphores, Electrical
Signal-boxes for electric telegraphs.
Signals, Electrical
Sorting apparatus, Electrical
Sounders.
Storm-signals, Electrical
Surveyors' compasses. (*See Class* 88.)
Switch-boards.
Switches, Electrical
Telegraphic codes.
Telegraphic signals.
Telegraph instruments.
Telegraphs, Electric
Telegraphs, Electro-chemical
Telegraphs, Electro-magnetic
Telegraphs, Magneto-electric
Telegraphs, Printing.
Telegraphs, Submarine
Telephones.
Time-balls, Electrical
Voltaic apparatus.
Voltaic baths.
Voltaic batteries.
Voltaic instruments.
Voltameters.
Watch-clocks, Electrical
Weighing apparatus, Electrical
Whaling apparatus, Electrical
Whistles, Electrical

# CLASS 37.

## EXCAVATING.

Earth Excavating, Grading, and Boring.

Artesian wells.
Augers for earth-boring.
Bog cutters.
Boring, Earth
Curbs for wells.
Ditching machines.
Dredging-machines.
Driven-wells.
Dumping-buckets.
Earth-boring.
Earth-removing machines.
Earth-works .
Excavators.
Grading.
Grading-scrapers.
Inclined planes.
Post hole borers.
Railroad earthworks.
Railroad excavating.
Road-making.
Road-scrapers.
Sand-pumps.
Scrapers, Earth
Scrapers, Grading
Submarine excavators.
Trenches.
Well-boring, (*except Rock, Class* 125.)
Well-digging.
Wells, Artesian
Wells, Driven
Well-tubing.

# CLASS 38.

## FELTING AND HATS.

Materials, Apparatus, Machines, and Processes for Felting Wool and Hair; including the manufacture of Hats, Caps, and other Head-coverings.

Bats, Forming
Blocks for hats and bonnets.
Blowing fur.
Bonnet caps.
Bonnet clasps.
Bonnet frames.
Bonnet fronts.
Bonnet-ironing machines.
Bonnet-presses.
Bonnets.
Bonnet-shaping machines.
Caps.
Cloth, Felt
Conformators.
Cork hats.
Curling hat-brims.
Cutting flock.
Cutting fur.
Dies for pressing hats.
Embossing hats.
Felt.
Felted fabrics.
Felt hats.
Felting cloth.
Felting hats.
Felting, Materials for
Felting yarn.
Flock-grinding.
Flocking fabrics.
Flocking hats.
Forming bats.
Fulling.
Fulling mills.
Fur-blowing.
Fur-cutting.
Fur from pelts, Cutting
Fur hats.
Fur-picking.
Fur, Treating
Grinding flock.
Hair from fur, Separating
Hair from skins, Pulling
Hair from skins, Separating
Hardening-machines for hats.
Hat-bodies.
Hat-bodies, Forming
Hat-bodies,-Hardening
Hat-bodies, Napping
Hat-blocking machines.
Hat-blocks.
Hat-brushing machines.
Hat-conformators.
Hat-finishing.
Hat-frames.
Hat-ironing machines.
Hat-lining.
Hat-measures.
Hat-napping.
Hat-planking.
Hat-pressing.
Hat-protectors.
Hats.
Hat-shaping machines.
Hat-shrinking machines.
Hat-sizing machines.
Hat-stiffening compositions.
Hat-stiffening machines.
Hat-stretching machines.
Hat supporters.
Hatters' kettles.
Hat tips.
Hat ventilators.
Head coverings.
Head-dresses.
Hoods, Ladies'
Horse-hair hats.
Jacks for hats, Rounding
Leather hats.
Napping-machines, Hat
Nets, Ladies' head-
Nubias.
Palm-leaf hats.
Paper hats. (*See Class* 93.)
Picking fur.
Planking hats.
Pouncing hats.
Size-markers for hats.
Spring-brim hats.
Sweat-leathers for hats.
Ventilators, Hat

# CLASS 39.

## FENCES.

Fences, Gates, Posts, Post-hole Diggers.

---

Fence-posts.
Fences.
Fences, Flood
Fences, Portable
Fences, Wire
Flood-fences.
Gate-posts.
Gates.
Hop-pole extractors.
Picket-fences.
Post-drivers.
Post-hole diggers and borers. (*See Class* 37.)
Post-pullers.
Wire-fences.
Wire-stretchers for fences.

# CLASS 40.

## FILES.

Files, Rasps, and Machines for making.

---

File blanks.
File chisels.
File cutters.
File-cutting machines.
File hammers.
Files.
Files, Machines for making
Files, Redressing
Files, Sharpening
Filing-machines. (*See Class* 51.)
Rasps.
Rasps, Machines for making

# CLASS 41.

## FINE ARTS.

Sculpture, Carving, Painting, Engraving, Lithography, and allied arts; Stenciling, Ornamentation, Decoration, Die-sinking, Artists' Appliances.

---

Aquariums.
Artificial flowers.
Artists' appliances (*except Brushes, Class* 15.)
Bank-notes.
Bronzing-machines.
Buhl work.
Canvas-stretchers.
Card-mounts.
Casts, Instruments for obtaining
Casts, Processes for obtaining
Copying figures.
Crayons.
Decorative gilding.
Decorative painting.
Die-sinking.
Draftsman's tools and appliances. (*See Class* 33.)
Drawing boards.
Easels.
Enameling on metal.
Enchasing.
Engravers' clutches.
Engravers' tools.
Engravers' vises.
Engraving, Copper-plate
Engraving-machines.
Engraving pantographs.
Engraving processes.
Engraving, Steel-plate
Engraving, Wood
Etching.
Flowers, Artificial
Flowers, Preserving
Frames, Picture
Gilding, Leaf
Graining-tools and appliances.
Gravers.
Herbariums.
Inlaying.
Japanning. (*See Class* 91.)
Likenesses, Mounting
Lithographic printing.
Lithography.
Marquetry.
Medals.
Monuments.
Mosaic.
Ornamentation.
Painting, Decorative
Painting on glass.
Paper-hanging appliances.
Parquetry.
Photographic card-mounts.
Picture-cases.
Picture-frames.
Picture-mounting.
Plate-engraving.
Postage-stamps.
Posting bills and placards.
Revenue-stamps.
Staining glass, wood, etc.
Stamps, Postage
Stamps, Revenue
Stencil-plates.
Stereoscopic picture-holders.
Stretchers for pictures.
Theatrical scenery and effects.
Tombstones.
Transfer, Methods of pictorial
Vitreography.
Wall-covering, Ornamental
Wall-paper hanging.
Wood-engraving.
Zincography.

# CLASS 42.

## FIRE-ARMS.

[*Excepting Ordnance and Projectiles of all kinds.*]

[*For Manufacture of Arms, see Class* 53, *Section A.*]

---

Air-guns. (*See Class* 98.)
Armor, Personal
Ball-screws for fire-arms.
Barrels to stocks, Attaching
Bayonet attachments.
Bayonets.
Breech-loading small-arms.
Cane-guns.
Cartridge-loaders.
Cartridge-shell retractors.
Coupling pistols to gun-stocks.
Fire-arms.
Fire-arms, Converting
Fire-arms, Safety-guards for
Guards for locks of fire-arms.
Guards for triggers of fire-arms.
Gun-covers.
Gun-hammers.
Gun-locks.
Gun-scrapers.
Gun-stocks.
Gun-wipers.
Gun-worms.
Hammers, Gun
Hammers, Safety-guards for gun
Locks for fire-arms.
Magazine small-arms.
Needle-guns.
Nipple-guards for fire-arms.
Nipples of fire-arms.
Pistols.
Pistol-swords.
Rammers for small arms.
Ramrods.
Repeating fire-arms.
Revolving fire-arms.
Rifles.
Rifling small-arms.
Saber-bayonets.
Safety-guards for gun-hammers.
Scrapers for fire-arms.
Self-loading fire-arms.
Sights for small-arms.
Small-arms.
Spade-bayonets.
Stocks for fire-arms.
Worms for fire-arms.

# CLASS 43.

## FISHING.

Dredges, Oyster
Eel traps.
Fish-decoys.
Fish-hooks.
Fishing.
Fishing-floats.
Fishing-line reels.
Fishing-lines. (*See Class* 28.)
Fishing-line sinkers.
Fishing-line swivels.
Fishing-nets.
Fishing-rods.
Fishing-seines.
Fish-spears.
Fish-traps.
Fishways.
Harpoons.
Nets, Fishing
Oyster culture.
Oyster-dredging
Oyster-rakes.
Oyster-tongs.
Pisciculture.
Rakes, Oyster
Reels, Fishing-line
Tongs, Oyster
Traps, Fish
Trolling devices.

# CLASS 44.

## FUEL.

Artificial Fuel; Treatment of Peat and Coal; Kindling Composts.

Coal, Purification of
Composts for fuel.
Fire-kindling compounds.
Fuel, Artificial
Fuel assorter.
Kindling materials.
Kindling packages.
Peat-machines.
Peat for fuel, Treatment of
Peat-presses. (*See Class* 100.)

# CLASS 45.

## FURNITURE.

Household Furniture, Upholstery, Window-shades, Mirrors.

[*Except Beds and Bedding* ....... *Class* 5.
*Kitchen Utensils* .. ..... *Class* 65.
*Stoves and Stove-fittings* .. *Class* 126.]

[*For Machines and Tools for Making Furniture, see Wood-working, Classes* 142–145; *for Manufacture of Furniture Hardware, see Class* 53, *Section D.*]

---

Baby-jumpers.
Basin clamps.
Blacking-boxes.
Blacking-box holders.
Book-cases.
Book-shelves.
Boot-jacks.
Boot-racks.
Bracket-shelves.
Broom-hangers.
Brush-racks.
Bureaus.
Cabinets.
Camp-kits.
Camp-stools.
Carpet-cleaners.
Carpet-fasteners.
Carpet-stretchers.
Chair-bottoms.
Chairs.
Chairs and seats, Head-rests for
Chairs, Barbers'
Chairs, Camp
Chairs, Dentists'
Chairs, Enema
Chairs, Folding
Chairs, Invalid
Chairs, Nursery
Chairs, Obstetrical
Chairs, Operating
Chairs, Pew
Cigar-racks.
Cloak-stands.
Closets, Portable
Closets, Revolving
Counters, Store and shop
Curtain-fixtures.
Curtains.
Cuspadores.
Desk-covers.
Desks.
Drapery-fasteners.
Drawers for closets.
Drawers for furniture.
Drawing-tables.
Dust-pans.
Dust receptacles.
Fastenings, Upholstery
Fenders, Furniture
Fire-screens.
Fly-traps.
Foot-stools.
Frames, Hanging picture
Furniture, Feet for legs of
Furniture pads.
Furniture-springs.
Hand-mirrors.
Hanging looking-glasses.
Hanging picture-frames.
Hanging-shelves.
Hat and coat racks.
Head-rests.
Insect-traps, Household
Lambrequins.
Looking-glasses.
Looking-glasses, Hanging
Market-stalls.
Mirrors.
Mirrors for windows.
Music-stands.
Nursery-gates.
Oil-cloth cutters.
Ottomans.
Packing furniture.
Pew fittings.
Piano-movers.
Piano-stools.
Picture-hangers.

Pier-glasses.
Portfolio-stands.
Racks, Boot, hat, and brush
Railway-car spittoons.
Reading-desks.
Sample-frames.
Seats, Garden
Seats, School
School furniture.
Screens, Fire
Settees.
Shelf-brackets.
Shelves, Folding
Show-cases.
Show-stands.
Shelving for stores, etc.
Soap-holders.
Sofas.
Spittoon-envelopes.
Spittoons, Railway-car
Spittoons.
Springs for sofas, chairs, &c.
Stair-covers.
Stair-pads.
Stair-plates.
Stair-rods.
Stands, Hat and cloak
Stands, Towel
Stands, Umbrella
Stools.
Stools, Camp
Stools, Folding
Stools, Foot
Stools, Piano
Store shelving.
Table-covers.
Table-leaf supporters.
Tables.
Tables, Card
Tables, Drawing
Tables, Extension
Tables, Ironing
Tables, Operating
Tables, Self-waiting
Tables, Writing
Toilet-mirrors.
Towel-racks.
Towel-stands.
Traps, Household insect
Twine-holders.
Upholstery.
Wall-protectors.
Wall-shields.
Wardrobes.
Window-blinds, Movable
Window-mirrors.
Window shades.
Window-washers.
Writing-tables.

# CLASS 46.

## GAMES AND TOYS.

Games, Gymnastic and Exercising Apparatus, Traps, and Nets.

---

Alphabet-blocks.
Animal-traps.
Archery.
Automata.
Base-balls.
Batons.
Bats for balls.
Bells, Dumb
Billiard-balls.
Billiard chalk-cups.
Billiard-cue tips.
Billiard-cues.
Billiard-markers.
Billiard-table chalk-holders.
Billiard-table pockets.
Billiard-tables.
Cards, Playing
Cards, Reading
Checker-boards.
Chess.
Children's toys.
Clubs, Exercising.
Croquet.
Croquet-mallets, etc.
Dancing figures.
Dice.
Dice-throwers.
Dolls.
Dumb-bells.
Exercising devices.
Fishing. (*See Class* 43.)
Game boards.
Game-bags.
Games.
Gymnastic apparatus.
Hobby-horses.
Mallets, Croquet
Masks.
Nets, Bird
Philosophical toys.
Playing-cards.
Police batons.
Pulleys, Calisthenic
Puzzles.
Reading-cards.
Rocking-horses.
Skate-fasteners.
Skates.
Skates, Parlor
Skates, Roller
Stage properties.
Stilts.
Swimming, Appliances for
Swings.
Targets.
Tops, Spinning
Toy-blocks.
Toys.
Traps, Animal
Traps, Bird

# CLASS 47.

## GARDEN AND ORCHARD.

Tools (not machines) for Digging, Cultivating, and Preparing the Soil, Planting, Transplanting, Weeding, Protecting, Potting, and Forcing Plants; Orchard-culture, Destroying Insects, Gathering Fruit, Sorghum-strippers, and Maple-sap gathering.

[*Except Pruning Knives, Shears, etc.; see* CUTLERY, *Class* 30.]

[*For Manufacture of Husbandry Tools, see Class* 81, *Section B.*]

---

Bird-houses.
Bouquet-holders.
Boxes for plants.
Budding knives. (*See Class* 80.)
Cane-mills. (*See Class* 83.)
Cranberry-gatherers.
Curculio-traps.
Diggers, Hand
Diggers, Thistle
Digging forks.
Ditching spades and scoops.
Earth-worm traps.
Edging-irons.
Flower-boxes.
Flower-pots.
Flower-stands.
Forcing-pits.
Forks, Digging
Forks, Hand hay-
Forks, Manure
Frames for plants.
Fruit-assorters.
Fruit-frames.
Fruit-gatherers.
Fruit-ripeners.
Fruit-separators.
Fumigators for insects.
Gardeners' basket-supports.
Garden-hoes.
Garden implements, (*except Shears, Class* 30.)
Garden-forks.
Grafting-knives. (*See Class* 30.)
Grafting-machines.
Grape-trellises.
Grape-vine protectors.
Grass-burners.
Green-houses.
Grubbing-hoes.
Hanging-baskets.
Hay-forks.
Hay-rakes, Hand
Hedge-benders.
Hedge-fasteners.
Hedge-planters.
Hedge-setters.
Hoes.
Hop-frames.
Hop-strippers.
Hot-bed frames.
Hot-houses.
Insect-destroyers.
Insect-fumigators.
Insect-lamps.
Insect-traps.
Layering.
Line reel, Gardeners'
Lopping-shears. (*See Class* 30.)
Maple-sugar buckets.
Maple-sugar spiles.
Manure-forks.
Manure-hooks.
Mattocks.
Osier-cutting.
Picks.
Pitchforks.
Plant-boxes.
Plant-frames.
Planting-tools.
Potato-forks.
Potato-hooks.
Potato-planters.
Potato-scoops.
Powder-blowers.
Propagating-boxes.
Propagating by cuttings.
Protectors, Plant and shrub
Protectors, Tree
Pruning knives and shears (*See Class* 30.)

Pruning-saws. (*See Class* 110.)
Rakes, Garden
Rakes, Hand hay
Sap-bucket hooks.
Sap-conducters.
Sap-spiles.
Sap-troughs.
Scrapers, Tree
Scuffle-hoes.
Shears for lopping and pruning. (*See Class* 30.)
Shovels.
Sorghum-strippers.
Spades.
Sprouting-beds for plants.
Spuds.
Teazel-trimmers.
Tobacco-curing.
Transplanting-machines.
Transplanting-tools.
Traps, Curculio
Traps, Earth-worm
Tree-boxes.
Tree-diggers.
Tree-dwarfers.
Tree-planter.
Tree-protectors.
Tree-training.
Trellises.
Trellises, Grape
Trellises, Hop
Trowels, Garden
Turf-cutters.
Turnip-toppers.
Vegetable assorters.
Vine locks.
Vine-supports.
Vine trellises.
Weeding-implements.
Weed-pullers.

# CLASS 48.

## GAS.

Manufacture of Gas. Carburetors.

Air-carbureting machines.
Carburetors for air or gas.
Gas-carburetors.
Gases, Methods and processes for obtaining
Gas-making apparatus.
Gas manufacture.
Gas-purifiers.
Gas-retorts.
Hydraulic mains.
Nitrous-oxide, Apparatus for making
Oxygen, Processes and apparatus for obtaining
Ozone, Apparatus for obtaining
Purifiers, Gas
Retorts for gas-making.

# CLASS 49.

## GLASS.

Compositions, Tools, and Appliances for Glass Manufacture, Modes of Manufacturing Articles of Glass.

[*For Manufacture of Glass-makers' Tools, see Class* 81, *Section B.*]

[*Excepting Glass Cutting, Grinding, and Polishing, Class* 51.]

---

Annealing glass.
Artificial gems.
Beads, Glass
Blowing-apparatus for glass-making. (*See Class* 98.)
Bottle-holders for glass-making.
Bottle-making, Glass
Button-making, Glass
Carboys.
Casting glass.
Chairs for glass-makers.
Crown glass.
Cutting glass. (*See Class* 51.)
Cutting jewels. (*See Class* 51.)
Cylinder glass.
Diamonds, Glaziers'
Embossing glass.
Enamels, Vitreous
Filigree glass.
Fluxing sand for glass.
Frit.
Frit-furnaces.
Frosting glass. (*See Class* 51.)
Furnaces, Glass
Gems, Artificial
Glass-blowing.
Glass-cementing.
Glass, Compositions for
Glass-cutting. (*See Class* 51.)
Glass furnaces.
Glass-house pots.
Glass houses.
Glass molding.
Glass pots.
Glass pressing.
Glaziers' diamonds.
Glaziers' diamonds, Setting
Grinding glass. (*See Class* 51.)
Incrustation of glass.
Jewels, Cutting (*See Class* 51.)
Kilns for glass.
Ladling glass.
Lapidary work. (*See Class* 51.)
Leers.
Lenses, Grinding (*See Class* 51.)
Lenses, Polishing (*See Class* 51.)
Melting glass.
Molding glass.
Molds for glass.
Molds, Composition for glass-
Ornamenting glass.
Packing glass.
Plate glass.
Platening glass.
Polishing glass. (*See Class* 51.)
Pontils.
Pots, Glass
Pressing glass.
Puttying machines for sash.
Retorts for glass.
Rods for gathering glass.
Rolling glass.
Roughening glass. (*See Class* 51.)
Scalloping glass.
Scouring glass. (*See Class* 51.)
Setting glaziers' diamonds.
Sheet glass.
Smelting glass.
Spinning glass.
Tables for glass-manufacture.
Tempering glass.
Tools for glass-manufacture.
Vitreous enamels.
Window glass.

# CLASS 50.

## GOVERNORS.

Governors for Steam-engine and other Machinery.

[*Excepting Brakes for Machinery, Class* 74.]

---

Cut-offs, Governor
Governor cut-off motions.
Governors.
Steam-governors.

# CLASS 51.

## GRINDING AND POLISHING.

Modes, Apparatus, Tools, Processes, and Appliances for glass, metal stone, and wood.

[*Excepting Clay and Mortar Machines*.. *Class* 25.
*Fruit Grinding* ............ *Class* 100.
*Grinding Mills* ............ *Class* 83.
*Ore and Stone Mills* ....... *Class* 90.]

---

Buffers.
Buffing-machines.
Burnishers.
Burnishing-machines.
Card-grinding.
Clamps for saw-filing, gumming, and sharpening.
Cutting glass.
Cutting jewels.
Emery-paper.
Emery-wheels.
Emery-wheels, Compositions for
Filing-machines.
Glass-cutting instruments.
Glass-cutting processes.
Glass-grinding.
Glass-polishing.
Glass-roughening.
Grinding glass.
Grinding-machines.
Grinding materials.
Grindstones.
Grindstones, Artificial
Grindstones, Hanging
Hanging grindstones.
Harvesters, Grinding-apparatus for
Hones.
Jewels, Cutting
Knife-sharpeners.
Lapidary tools.
Lapidary wheels.
Laps.
Lenses, Grinding
Lenses, Polishing
Marble-grinding.
Marble-grinding, Compositions for
Marble-polishing.
Oil stones.
Polishing-compounds.
Polishing glass.
Polishing-machines.
Polishing marble.
Polishing-materials.
Polishing metals.
Polishing-slate.
Polishing-stone.
Razor-strops.
Razor-strops, Compositions for
Reaper-knife sharpeners.
Rifles for whetting.
Roughening glass.
Sand-paper.
Saw-filing.
Saw-grinding.
Scissors-sharpeners.
Scouring glass.
Scouring stone.
Scythe-rifles.
Scythe-stones.
Skate-sharpeners.
Steels.
Stone-polishing.
Stove polish.
Strops.
Tool-grinding.
Tool-holders for grindstones.
Whetstones.

# CLASS 52.

## GUNPOWDER.

Explosive Compounds, Compositions, and Manufactures, Pyrotechnics, Compositions for Matches.

---

Ballistic pendulums. (*See Class* 73.)
Blasting compounds.
Cartridges, Compositions for
Cigars, Compositions for lighting
Corning gunpowder.
Detonating compositions.
Drying gunpowder.
Dynamite.
Eprouvettes. (*See Class* 73.)
Fulminates.
Glazing gunpowder.
Graining gunpowder.
Gun-cotton, Manufacture of
Gunpowder, Compositions for
Gunpowder, Corning
Gunpowder, Drying
Gunpowder, Glazing
Gunpowder, Graining
Gunpowder-mills. (*See Class* 83.)
Gunpowder-polishing.
Gunpowder-presses. (*See Class* 100.)
Incendiary compounds.
Matches, Compositions for
Matches, Machines for making and dipping (*See Class* 144.)
Nitro-glycerine.
Powder-kegs.
Pyrotechnic compounds.
Pyrotechnics.
Rockets.

# CLASS 53.

## HARDWARE MANUFACTURE.

Manufacture of Hardware Fittings and Trimmings, Arms, Cutlery, and Jewelry.

*Under the following heads:—*

A.—Arms and cutlery.

B.—Builders' hardware.

C.—Carriage hardware.

D.—Furniture hardware.

E.—Personal wear.

F.—Saddle and harness hardware.

G.—Shoemakers' hardware.

H.—Spinning and weaving hardware.

I.—Stationery and miscellaneous hardware.

J.—Trunk and carpet-bag hardware.

K.—Water and gas fittings.

[*For Manufacture of Railway-track and Car-irons and Fittings, see Railways* 4, *Class* 107; *see also* METAL-WORKING, *page* 106.]

---

### A.—*Arms and Cutlery.*

Bayonets, Manufacture of
Bullets, Manufacture of
Cartridge-shells, Manufacture of
Clipping-tools, Manufacture of
Cutlery, Manufacture of
Dies for making articles of cutlery.
Forks, Manufacture of
Gun-locks, Manufacture of
Knives, Manufacture of
Percussion-caps, Manufacture of
Sabers, Manufacture of
Scissors, Manufacture of
Shears, Manufacture of
Shot, Manufacture of
Shot-sorting machines.
Swords, Manufacture of

### B.—*Builders' Hardware.*

Alarm-bells, Manufacture of
Bells, (household,) Manufacture of
Blind-fittings, Manufacture of
Bolts, (household,) Manufacture of
Buttons, (door,) Manufacture of
Clock-bells, Manufacture of
Cow-bells, Manufacture of
Door-bells, Manufacture of
Door-checks, Manufacture of
Door-springs, Manufacture of
Door-strips, Manufacture of
Elbow-levers for pulls, Manufacture of
Hinges, Manufacture of
Keys, (door,) Manufacture of
Knobs for doors and bell-pulls, Manufacture of
Latches, Manufacture of
Locks, Manufacture of
Lock-trimmings, Manufacture of
Pulls, (bell,) Manufacture of
Sash-holders, Manufacture of
Sash-pulleys, Manufacture of
Sash-weights, Manufacture of
Sheaves, Manufacture of
Shutter-fasteners, Manufacture of

Shutter-workers, Manufacture of
Sleigh-bells, Manufacture of
Table-bells, Manufacture of
Weather-strips, (metallic,) Manufacture of
Window-sash, (metallic,) Manufacture of

C.—*Carriage and Wagon Hardware.*

Axle-boxes, (carriage,) manufacture of
Braces for carriages, Manufacture of
Carriage-bow joints, Manufacture of
Carriage-door handles, Manufacture of
Carriage-springs, Manufacture of
Carriage-steps, Manufacture of
Carriage-top braces, Manufacture of
Felly-plates, Manufacture of
Fifth-wheels, Manufacture of
Hub-bands, Manufacture of
King-bolts, Manufacture of
Perch-plates, Manufacture of
Shaft-couplings, Manufacture of
Spring-heads, Manufacture of
Thill-couplings, Manufacture of
Wheels, (metallic,) Manufacture of

D.—*Furniture Hardware.*

Cabinet-hooks, Manufacture of
Casters, Manufacture of
Clothes-hooks, Manufacture of
Hat-hooks, Manufacture of
Coffin-handles, Manufacture of
Coffin hardware, Manufacture of
Stair-rods, Manufacture of

E.—*Personal Wear.*

Beads, (metallic,) Manufacture of
Brooches, Manufacture of
Buckles, Manufacture of
Button rings, Manufacture of
Buttons, (metallic,) Manufacture of
Chains, (fob and watch,) Manufacture of
Clasps, Manufacture of
Clothes pins, Manufacture of
Collar-fastenings, Manufacture of
Combs, (metallic,) Manufacture of
Cuff buttons, Manufacture of
Eyelets, Manufacture of
Finger-rings, Manufacture of
Glove-buttons, Manufacture of
Hooks and eyes, Manufacture of
Jewelry, Manufacture of
Key-rings, Manufacture of
Lockets, Manufacture of
Ornaments, (small metallic,) Manufacture of
Shirt and collar fastenings, Manufacture of
Shoe-buttons, Manufacture of
Split-rings, Manufacture of
Watch and fob chains, Manufacture of
Watches, Manufacture of

F.—*Saddle and Harness Hardware.*

Buckles, (harness,) Manufacture of
Collar shells, Manufacture of
Hames, Manufacture of
Rivets, (harness,) Manufacture of
Saddle-trees, Manufacture of
Snap-hooks, Manufacture of
Stirrups, Manufacture of
Swivel-hooks, Manufacture of
Terrets, Manufacture of

### G.—*Shoemakers' Hardware.*

Boot-shanks, Manufacture of
Clogs, (metallic,) Manufacture of
Screw-nails, Manufacture of
Shoe-tips, Manufacture of

### H.—*Spinning and Weaving Hardware.*

Cop-tubes, Manufacture of
Eyes for harness, Manufacture of
Flyer-guides, Manufacture of
Knitting-burrs, Manufacture of
Ring-travelers, Manufacture of
Shuttle-frames, Manufacture of
Shuttle-tips, Manufacture of
Speeder-flyers, Manufacture of
Spindle heads for flyers, Manufacture of

### I.—*Stationery and Miscellaneous Hardware.*

Clock-pillars, Manufacture of
Fish-hooks, Manufacture of
Ferrules, Manufacture of
Gold pens, Manufacture of
Paper-clasps, Manufacture of
Pen-holders, Manufacture of
Pens, Manufacture of
Stationery, (metallic articles of,) Manufacture of
Steel Pens, Manufacture of
Umbrella-trimmings, Manufacture of

# CLASS 54.

## HARNESS.

Tools, Machines, and Special Hardware, Saddles, Harness, Muzzles, Pokes, Gags, and Whips.

[*Except Hooks, Snap-hooks, and Rings, Class* 24.]

[*For Manufacture of Saddlery and Harness Hardware, see Class* 53, *Section F.*]

[*For Manufacture of Leather-working Tools, see Class* 81, *Section B.*]

---

Back-bands.
Bells, Cow
Bells, Sleigh
Bell straps.
Bits, Bridle
Bitting-rigging.
Blind-bridles.
Blind-folding devices for animals.
Blinds for horses.
Bonnets for horses.
Boots for horses.
Breaking horses, Apparatus for
Breast bails.
Breast-slides.
Breast-straps.
Breast-strap slides.
Breech-straps.
Bridles.
Bridle-winkers.
Bull-rings.
Cart-saddles.
Cattle-gags.
Cattle-leaders.
Check-hooks.
Check-reins.
Clamps for sewing harness.
Clamps, Stitching
Collar-blocks.
Collars, Horse
Collars, Machines for stuffing
Collar straps, Horse
Cribbing, Devices to prevent
Crupper-loops.
Cruppers.
Dog-collars.
Dog-muzzles.
Draw-gage cutters.
Dumb jockeys.
Fetters for animals.
Fetters.
Fly-nets.
Fly-nets, Machines for making leather
Foot-pads for horses.
Gage-knives.
Gag-runners.
Gags for animals.
Gig-trees.
Halters.
Hame-clips.
Hame-fasteners.
Hames.
Harness.
Harness-chains.
Harness, Detaching
Harness-hooks.
Harness-makers' tools.
Harness-mountings.
Harness-oiling machines.
Harness pads.
Harness-reins, Machines for rounding
Harness-swivels.
Harness-wardrobes.
Head-protectors for horses.
Hog-rings.
Hopples.
Horn-knobs, Cattle
Horse-blankets.
Horse-protectors.
Horse-taming.
Horse-training apparatus.
Horse-yokes.
Housings for harness.
Interfering, Devices to prevent
Knee-caps.
Knee guards for horses.
Lassoes.

Martingales.
Muzzle-rings.
Muzzles.
Nose-rings.
Nostril-pinchers.
Pack-saddles.
Pad-crimps.
Pads.
Pad-trees.
Pokes for animals.
Rein-pulls.
Reins, Safety
Reins, Slides for
Rosettes for horses.
Rounding machines, Rein-
Shackles.
Saddle-blankets.
Saddle-cloths.
Saddle-girths.
Saddle-pad machines.
Saddles.
Saddles, Cart
Saddles, Harness
Saddles, Pack
Saddles, Pad
Saddles, Riding
Saddles, Side
Saddle-trees.
Saddle trees, Side-
Shaft tugs.
Shields for horses.
Sleigh-bell attachments.
Sleigh-bells.
Snap-hooks. (*See Class* 24.)
Snout-rings.
Spurs.
Stirrups.
Stirrups, Safety
Stitching horses.
Stuffers, Collar
Sun-shades for horses.
Surcingles.
Tail-supporters.
Tethers.
Throwing horses, Apparatus for
Trace-couplings.
Traces.
Trees, Saddle
Tug-carriers.
Whip-hangers.
Whip-racks.
Whips.
Whip-stocks.
Yokes for breechy animals.

# CLASS 55.

## HARROWS.

Machines for Rolling and Pulverizing the Soil, and Preparing the Surface.

[*For Manufacture of Husbandry Hardware, see Class* 81, *Section B.*]

---

Brush-pullers.
Clod-crushers.
Corn-stalk choppers.
Cotton-stalk choppers.
Grubbing-harrows.
Grubbing-machines.
Harrows.
Harrows, Rotary
Harrow-teeth.
Roller and harrow, Combined
Rollers, Flanged
Rollers, Land
Rotary harrows.
Scarifiers.
Scufflers.
Stalk-choppers.
Stalk-pullers.
Stone-gatherers.

# CLASS 56.

## HARVESTERS.

Machines and Implements (*except Hand Hay Rakes and Forks*) for Gathering and Securing Crops.

[*For the Manufacture of Iron Parts of Harvesters, see Class* 81, *Section B.*]

Bands for binding grain.
Barley-forks.
Bean-harvesters.
Binders for harvesters.
Cane-harvesters.
Clover-headers.
Clover-seed harvesters.
Combined reapers and mowers.
Corn-cutting machines.
Corn-harvesters.
Corn-shocking machines.
Corn-shock tyers.
Cotton-pickers.
Cotton toppers.
Cradles, Grain
Cut-offs for harvesters.
Cutters, Harvester
Dropping-platforms for harvesters.
Dumping-reels for harvesters.
Fingers, Harvester
Flax-pullers.
Forks, Horse hay
Forks, Grain
Gaveling-attachments for harvesters.
Gaveling-forks.
Gearing, Harvester
Grain-binders.
Grain-cradles.
Grain-forks.
Grain-harvesters.
Grain-rakes.
Grass-harvesters.
Guard-fingers for harvesters.
Harvesters, Bean
Harvesters, Binders for
Harvesters, Cane
Harvesters, Clover-seed
Harvesters, Corn
Harvesters, Cutters for
Harvesters, Dropping platforms for
Harvesters, Dumping-reels for
Harvesters, Fingers for
Harvesters, Gaveling-attachments for
Harvesters, Gearing for
Harvesters, Grain
Harvesters, Grass
Harvesters, Guard-fingers for
Harvesters, Hemp
Harvesters, Pitmans for (*See Class* 74.)
Harvesters, Rakes for
Harvesters, Reels for
Harvesters, Track-clearers for
Hay-caps.
Hay-forks, Horse
Hay rakers and cockers.
Hay rakers and loaders.
Hay-rakes, Horse
Hay-spreaders.
Hay-tedders.
Heading-machines, Grain
Hedge-trimming machines
Hemp-harvesters.
Hemp pullers.
Lawn-mowers.
Manure-drags.
Manure-loading rakes.
Mowing-machines.
Pitmans for harvesters. (*See Class* 74.)
Rakes for harvesters.
Rakes, Horse hay
Reaper-knife sharpeners. (*See Class* 51.)
Reaping-hooks.
Reaping-machines.
Reels for harvesters.
Seats for harvesters.
Scythe-fastenings.
Scythe-nibs.
Scythes.
Scythe-snaths.

Shock tyers.
Sickles.
Straw scatterers.
Sugar-cane cutters.
Sward-cutters.
Tedders, Hay
Timothy-headers.
Track clearers for harvesters.

# CLASS 57.

## HOISTING.

Blocks, Pulley.
Brakes for hoisting machinery. (*See Class* 74.)
Buckets, Dumping
Buckets, Hoisting
Bush-extractors.
Capstans.
Capstans for mole-plows.
Car-jacks.
Carriage-jacks.
Clamps for raising buildings.
Cranes.
Derricks.
Derricks, Floating
Dumb waiters.
Dumping tubs and buckets.
Elevators.
Elevators for bricks.
Elevators for grain.
Elevators for houses.
Elevators for mortar.
Elevators for warehouses.
Floating derricks.
Grain-conveyors.
Grain-elevators.
Hay-elevators.
Hay-loading derricks.
Hay-unloaders.
Hod-unloaders.
Hoisting apparatus.
Hoisting-buckets.
Ice-elevators.
Ice-runs.
Jacks, Carriage
Jacks, Lifting
Jacks, Railway-car
Jacks, Screw
Jacks, Wagon
Lifting apparatus.
Lifting-jacks.
Lifting-screws.
Loading machinery for wagons, vessels, and cars.
Logs, Implements for loading
Logs, Implements for piling
Logs, Implements for rolling
Lumber, Implements for loading and unloading
Lumber, Implements for piling
Masting-sheers.
Pulley-blocks.
Rail jacks.
Sheers, Masting
Ships' capstans.
Skids.
Skips.
Speeders for capstans.
Stump-extractors.
Tackle and blocks.
Tackle-blocks.
Tender-loaders.
Track-raisers, Railway
Unloading-machines.
Vessels, Loading and unloading, Machinery for
Winches.
Windlasses.

# CLASS 58.

## HOROLOGY.

Clocks, Watches, Time-keepers, and Registers, and cases for the same.

[*For Watchmakers' Tools, see Class* 81, *Section A.*]

---

Chronographs.
Chronometers.
Chronoscopes.
Clepsydras.
Clock-cases.
Clock-keys.
Clocks.
Dials.
Dials, Time-piece.
Escapements.
Hands for clocks and watches.
Horological-instruments.
Hour-glasses.
Illuminated clocks.
Indicators, Automatic time
Keys, Clock
Keys, Watch
Metronomes.
Pendulums.
Registers, Time
Springs, Watch and Clock
Stem-winders.
Time-indicators, Automatic
Time-registers.
Time-reporters.
Watch-cases.
Watch-dials.
Watches.
Watch-keys.
Watchman's time-registers.
Watch-motions.
Watch-springs.

# CLASS 59.

## HORSE SHOES.

Shoes and Machines for making.

[*For Blacksmiths' Forges and Machines, see* METAL-WORKING 3, *Class* 78; *for Blacksmiths' Tools, see Class* 81, *Section A.*]

---

Horseshoe-bars.
Horseshoe bars, Machines for bending
Horseshoe bars, Machines for shaping
Horseshoe blanks, Making
Horseshoe-machines.
Horseshoes.
Horseshoes, Calks for
Horseshoes, Pads for
Oxshoes.

# CLASS 60.

## HOSE AND BELTING.

Hose and Belting, Compositions and Machinery for making, Couplings, and Fasteners.

[*For Manufacture of Leather-working Tools, see Class* 81, *Section B.*]

Band and cord fasteners.
Band punches. (*See Class* 24.)
Bands for pulleys.
Belt-clasps.
Belt-couplings.
Belt-cutting machines.
Belt-fasteners.
Belting, Compositions for
Belting-Compound
Belting, Modes of making
Belting, Rubber
Belt lacers.
Belt-lacing, Machines for cutting
Belt-punches. (*See Class* 24.)
Belts.
Belt-shifters. (*See Class* 64.)
Belt-splicing.
Belt-tighteners. (*See Class* 64.)
Compositions for belting.
Compositions for hose.
Compositions for elastic tubing.
Cord and belt fasteners.
Couplings, Band
Couplings, Belt
Couplings, Cord
Couplings, Hose
Cuttingmachines, Belt-
Cutting machines, Belt-lacing
Elastic tubing.
Fasteners for bands, belts, and cords.
Flexible gas tubing.
Hose.
Hose-bridges.
Hose, Compositions for
Hose-couplings.
Hose, Leather
Hose, Modes of making
Hose-protectors.
Hose, Rubber
Materials for bands, belts, and hose.
Rubber hose.
Spanners for hose-couplings.
Tubing, Coating of elastic
Tubing, Compositions for elastic
Tubing, Elastic
Tubing, Lining of elastic

# CLASS 61.

## HYDRAULIC ENGINEERING.

Aqueducts, Canels, Dikes, Harbors, Breakwaters, Docks, Quays Sub-aqueous Explorations and Works, Piles, Improvement of Rivers.

---

Aqueducts.
Armor, Submarine
Booms for logs.
Breakwaters.
Cables, Laying submarine
Caissons for piers, &c.
Camels.
Canal hoists.
Canal lining.
Canal locks.
Canal-lock gates.
Canal propulsion, apart from the boat.
Canals.
Chutes for carrying logs.
Chutes for coal.
Chutes for water-courses.
Coal-chutes.
Coffer-dams.
Cores for dikes.
Cradles for building vessels.
Cradles for launching vessels.
Cradles for raising vessels.
Current fenders.
Dams.
Dikes.
Diving apparatus.
Diving-bells.
Docks, Dry
Docks, Tidal
Docks, Wet
Draining cellars, Automatic devices for
Drain-tiles, Machines for, and modes of laying
Dredging-machines. (*See Class* 37.)
Dry-docks.
Embankments, Making and repairing
Ferry-boat platforms.
Fish-ways.
Flood-gates.
Flumes for logs.
Flushing apparatus.
Gates, Canal-lock
Gates, Flood
Gates, Sluice
Graving docks.
Harbor defenses.
Harbor obstructions, Removing
Harbors.
Harbors, Obstructing
Hydraulic mining.
Inclined planes for canals.
Iron tunnels.
Irrigating apparatus.
Landing platforms.
Launching vessels.
Laying submarine cables.
Levees.
Lining canals.
Locks, Canal
Lock valves, Canal
Marine railways.
Marshes, Machines for filling
Marshes, Reclaiming
Mill-dams.
Mud-machines.
Obstructing rivers and harbors.
Piers, Landing
Pile-cutters.
Pile-drivers.
Piles.
Piles, Protection of
Piles, Screw
Quays.
Raising sunken vessels.
Reservoirs, Canal
Rivers bars, Removing
Rivers, Improving channels of
Rivers, Protection against inundations of
Screw-piles.
Shields for tunneling.
Slips, Marine
Sluice-gates.
Snags, Removing
Staging for boats.
Staging for loading and unloading vessels.
Staiths.

Stopping crevasses.
Sub-aqueous foundations.
Sub-aqueous railways.
Sub-aqueous tunneling.
Sub-fluvian tunnels.
Submarine armor.
Submarine cables, Laying
Submarine excavating.
Submarine helmets.
Submarine operators.
Tidal docks.
Tidal valves.
Tide basins.
Tide-gates.
Tide powers.
Trainways for ferry-boats.
Tunnels, Iron
Tunnels, Sub-aqueous
Valves for draining tidal lands.
Wave-powers.
Weirs.
Well-curbs.
Wet-docks.
Wharfs.
Wharfs, Floating (See Class 114.)
Wire ferriage.

# CLASS 62.

## ICE.

Refrigerative Processes and Machinery, Gathering, Manufacturing, Storing, and Applications of Ice.

---

Air-cooling apparatus.
Air-cooling processes.
Anæsthetic refrigerators.
Atomizers.
Beer-coolers.
Boxes for transportation, Refrigerating
Cars for transportation, Refrigerating
Cooling apparatus.
Cooling mixtures.
Cooling processes.
Corpse-coolers.
Filter and cooler combined.
Freezing-liquids.
Freezing-mixtures.
Ice-cream freezers.
Ice-crushers.
Ice-cutters.
Ice-elevators. (*See Class* 57.)
Ice-houses.
Ice-levelers.
Ice-making machines.
Ice manufacture.
Ice-picks.
Ice-plows.
Ice-safes.
Ice-shavers.
Ice-smoothers.
Milk-coolers.
Railway cars, Refrigerating
Refrigerating-cars.
Refrigerating-cellars.
Refrigerators.
Water-coolers.

# CLASS 63.

## JEWELRY.

Bracelets, Brooches, Dress-pins, Rings, and Ornaments.

[*For Manufacture of Jewelry, see Class* 53, *Section E.*]

---

Bracelets.
Breast-pins.
Diaper-pins.
Dress-pins.
Ear drops and rings.
Finger-rings.
Fob-chains.
Gems, Setting
Hooks for watch-chains. (*See Class* 24.)
Jewelers' tools and appliances. (*See Class* 81, *A.*)
Lapidaries' work and appliances. (*See Class* 51.)
Locket-cases.
Lockets.
Pins, Breast
Pins, Diaper
Pins, Dress
Pins, Scarf
Pins, Shawl
Precious stones, Setting
Scarf-pins.
Setting gems.
Shawl-pins.
Watch-chains.

# CLASS 64.

## JOURNALS AND BEARINGS.

Journals, Bearings, Shafting, Couplings, and Lubricators.

Anti-friction compounds.
Anti-friction metals.
Anti-friction rollers.
Band pulleys.
Bands. (*See Class* 60.)
Band wheels.
Bearings for machinery.
Belt couplings. (*See Class* 60.)
Belts. (*See Class* 60.)
Belt-shifters.
Belt-tighteners.
Boxes, Journal
Clutches for shafting.
Clutches, Friction
Coupling, Shaft
Drums, Belt
Friction-clutches.
Friction-couplings.
Grease cups
Hangers for shafting.
Journal-bearings.
Journal-bearings, Compositions for
Journal-boxes.
Journal-box lining.
Journals.
Lagging for drums.
Lubricants.
Lubricating compounds.
Lubricators.
Lubricators for spindles.
Lubricators, Mechanical
Pulleys, Belt
Pulleys, Fast and loose
Oil-cups.
Oilers.
Shaft-couplings.
Shaft-hangers.
Shafting.
Spindles, Lubricating
Universal joints.

# CLASS 65.

## KITCHEN UTENSILS.

Machines and Appliances for Preparing Bread, Meat, Fruit, and Vegetables for Cooking or for the Table; Table-ware, (*except Cutlery,*) Baskets, Boxes, and Crates for Packing and Transporting Provisions, (*not including Stoves and their appliances, Class* 126.)

---

Almond-peelers.
Apple-corers.
Apple-parers.
Apple-slicers.
Barrel-covers.
Barrel-handles.
Baskets.
Baskets for fruit, fish, grain, and provisions.
Batter-machines.
Bean-shellers.
Beer cups.
Beer-pitchers.
Bins, Meal
Biscuit-boards.
Biscuit-cutters.
Bottle-holders.
Boxes, Egg-packing
Boxes, Fish
Boxes, Food
Boxes for provisions.
Boxes for vegetables.
Boxes, Fruit
Boxes, Lunch
Boxes, Paper. (*See Class* 93.)
Boxes, Pepper
Boxes, Spice
Bread and pastry boards.
Bread-cutters.
Bread-machines.
Buckets.
Buckets, Folding
Buckets, Dinner
Butter-buckets.
Butter-dishes.
Butter-tongs.
Cake-cutters.
Cake-turners.
Camp-kits.
Canisters.
Can-openers.
Canteens.
Caster-bottles.
Caster-stands.
Cheese-covers.
Cheese-cutters.
Cheese-grinders.
Cherry-stoners.
Chimney-cleaners, Lamp
Coffee-cups.
Confectionery-molds.
Cork-pullers.
Cork-screws.
Corn-ball molds.
Corn-graters.
Cracker-crushers.
Cracker-machines.
Crates for fruit, fish, vegetables, &c.
Cruet-stands.
Crumb-removers.
Culinary ladles.
Cups.
Dinner-buckets.
Dinner-pails.
Dippers.
Dish-cleaners.
Dishcloth-holders.
Dish-covers.
Dish-drainers.
Dishes.
Dish-fasteners.
Dish-holders.
Dish-stands.
Dish-trays.
Dish-washers.
Dough-kneaders.
Dough-machines.
Dough-raisers.
Dough-rollers.
Dredging-boxes, Flour
Drinking-cups.
Egg-beaters.
Egg-cups.
Egg-pans.
Egg-testers.
Epergnes.
Fish-scalers.

Flour-chests.
Flour-dredgers.
Flour-scoops.
Fruit-baskets.
Fruit-boxes.
Fruit-crates.
Fruit-mashers.
Fruit-panniers.
Fruit-peelers.
Fruit-strainers.
Glass-cleaners.
Graters, Corn
Graters, Spice
Hampers.
Haversacks.
Kitchen-safes.
Kneading-machines.
Knife-cleaners.
Knife-racks.
Knife-rests.
Lamp-chimney cleaners.
Lemon-squeezers.
Liquor-pourers.
Lunch-boxes.
Meal-bins.
Meal-chests.
Meat-cutters. (*For Mincers, see Class* 17.)
Meat-holders.
Meat-safes.
Meat-spits.
Meat-tenderers.
Mess-chests.
Mess-kits.
Napkin-holders.
Napkins.
Nut-crackers.
Nutmeg-graters.
Nut-pickers.
Oyster-openers.
Pans, Kitchen and pantry.
Pea-shellers.
Peach-parers.
Peach-stoners.
Peeling almonds.
Pepper-boxes.
Pie-crimpers.
Pie-trimmers.
Pitchers, Beer
Pitchers, Ice
Pitchers, Molasses
Plates.
Potato-mashers.
Potato-peelers.
Powder-mixers.
Provision-safes.
Raisin-seeders.
Rolling-pins.
Root-peelers.
Safes, Meat
Sauce-pans.
Scoops, Flour
Sifters of all kinds. (*See Class* 140.)
Sinks, Kitchen. (*See Class* 4.)
Skimmers, Culinary
Slop-buckets.
Slop-jars.
Spice-boxes.
Stove-pipe shelves.
Table-cars.
Table-mats.
Table-trays.
Trays, Dish
Trays, Food
Trays, Fruit
Trays, Tea
Tumbler-washers.
Turnip-peelers.
Urn-stands.
Vegetable-peelers.
Vegetable-washers.

# CLASS 66.

## KNITTING AND NETTING.

Knitting, Netting, Lace, and Hosiery.

[*For Manufacture of Parts of Knitting-machines, see Class* 53, *Section H.*

---

Bobbinet-machines.
Darning-machines.
Dressing and finishing hosiery.
Fishing-nets, Making
Gimp.
Hosiery.
Knitted goods.
Knitting-machines.
Knitting-machines, Burrs for
Knitting-machines, Needles for.
Knitting-machines, Setting up devices for
Knitting-machines, Sinkers for
Knitting-machines, Stop-motions for
Knitting-machines, Take-up devices for
Knitting-machines, Tension devices for
Knitting-machines, Transferring devices for
Lace-machines.
Looped fabrics.
Needles for knitting-machines.
Netting-machines.
Sinkers for knitting-machines.
Stocking and sock machines.
Stop-motions for knitting-machines.
Tatting-machines.
Twist-lace machines.

# CLASS 67.

## LAMPS AND GAS-FITTINGS.

Lamps, Lanterns, Lights, Gas-fittings, and Lighting devices.

---

Alarms, Gas
Boxes, Match
Burners.
Calcium lights.
Candle-holders.
Candlesticks.
Catoptric lights. (*See Class* 88.)
Chandeliers.
Chimney-cleaners, Lamp. (*See Class* 65.)
Chimneys, Lamp
Cigar-lighters.
Clocks, Illuminated. (*See Class* 58.)
Cressets.
Dioptric lights. (*See Class* 88.)
Drop-lights.
Electric lights. (*See Class* 36.)
Extinguishers.
Flexible gas-tubing. (*See Class* 60.)
Gasaliers.
Gas-brackets.
Gas-burners.
Gas-fittings.
Gas-lighters.
Gas-lighting by electricity. (*See Class* 36.)
Gas-meters. (*See Class* 98.)
Gas-regulators.
Gas-tubing, Flexible (*See Class* 60.)
Hangers for lamps.
Head-lights, Locomotive
Illuminated clocks. (*See Class* 58.)
Illuminating-reflectors.
Illuminated signs. (*See Class* 120.)
Lamp-chimney adjusters.
Lamp-chimney cleaners. (*See Class* 65.)
Lamp-chimneys.
Lamp-cones.
Lamp-elevators.
Lamp-posts.
Lamp-reflectors.
Lamp-shade holders.
Lamp-shades.
Lamp-stands.
Lamp-wick adjusters.
Lamp-wick trimmers.
Lamp-wicks.
Lamps.
Lamps, Miners'
Lamps, Vapor
Lanterns.
Lighting devices.
Lighting devices, Electric. (*See Class* 36.)
Locomotive head-lights.
Lusters.
Magnesium-lights.
Match-boxes.
Match-safes.
Mica lamp-chimneys.
Oxyhydrogen lights.
Reflectors for lamps.
Reflectors for windows.
Safes, Match
Shades, Lamp
Signs, Illuminated (*See Class* 120.)
Smoke-bells.
Snuffers.
Tubing for gas, Flexible (*See Class* 60.)
Vapor-burners.
Vapor-lamps.
Wicks, Compositions for treating
Wick-trimmers.
Window-illuminators.
Window-reflectors.

# CLASS 68.

## LAUNDRY.

Washing, Wringing, and Drying Machines, and the Appliances of the Laundry.

---

Clamps for wringers.
Clothes-driers.
Clothes-frames.
Clothes-horses.
Clothes-line reels.
Clothes-lines.
Clothes-pins.
Clothes sprinklers.
Clothes-sticks.
Clothes-tongs.
Clothes-washers.
Clothes-wringers.
Fluting-irons.
Ironing apparatus.
Ironing-tables.
Italian irons.
Laundries.
Line-fasteners.
Line-reels.
Mangles.
Reels for clothes-lines.
Rubber rolls for wringers.
Sad-irons.
Sad-iron holders.
Smoothing-irons.
Sprinklers for clothes.
Wash-benches.
Wash-boards.
Wash-boilers.
Washing compounds. (*See Class* 23.)
Washing-machines.
Wool-washing machines.
Wringers.
Wringers, Clamps for
Wringers, Rolls for

# CLASS 69.

## LEATHER.

Machines for operating upon leather irrespective of its specific application or use.

[*For Applications of Leather, see—*
*Boots and Shoes* .. .... *Class* 12.
*Harness and Saddlery*.. *Class* 54.
*Hose and Belting* ...... *Class* 60.
*Trunks and Satchels* ... *Class* 133.]

---

Boarding-machines, Leather (*See Class* 129.)
Corrugating machines, Leather-
Creasing machines, Leather-
Cutting machines, Leather-
Embossing machines, Leather-
Glueing machines, Leather-
Hammering machines, Leather-
Leather carpet.
Leather-cutting machines.
Leather-dressing machines. (*See Class* 129.)
Leather-embossing machines.
Leather-glueing machines.
Leather-hammering machines.
Leathering tacks.
Leather-piercing machines.
Leather-pressing machines.
Leather-punching machines.
Leather-quilting machines. (*See Class* 112.)
Leather rings, Tools for cutting
Leather-rolling machines.
Leather-rounding machines.
Leather-scalloping machines.
Leather-shaving machines.
Leather-skiving machines.
Leather-sorting machines.
Leather-splitting machines.
Leather straps.
Leather-stretching machines.
Leather-stripping machines.
Leather tubs, Machines for making
Leather washers and wads, Tools for cutting
Leather whips, Machines for making
Pressing machines, Leather-
Punching machines, Leather-
Rings, Tools for cutting leather
Rolling machines, Leather-
Skiving machines, Leather-
Sorting machines, Leather-
Splitting machines, Leather-
Stretching machines, Leather-
Tubs, Machines for making leather
Wads, Machines for making leather
Washers, Machines for making leather
Whips, Machines for making leather

# CLASS 70.

## LOCKS AND LATCHES.

Locks, Latches, their Trimmings and Accessories.

[*For Manufacture of Locks, see Class* 53, *Section B.*]

---

Alarm-locks.
Bag-locks.
Bolt-locks.
Car-door latches.
Car-door locks.
Car-seat arm-locks.
Chest-locks.
Cupboard-catches.
Door-fasteners.
Door-knobs.
Door-locks.
Drawer-locks.
Gate-latches.
Hand-cuffs, Locks for
Hasp-locks.
Keepers for door-locks.
Key-fasteners.
Key-hole guards.
Key-hole guides.
Keys.
Knob-fasteners.
Knob-latches.
Knobs, Door
Knobs to spindles, Attaching
Latches.
Lock-bolt keepers.
Lock-bolts.
Locks.
Locks, Bag
Locks, Car-door
Locks, Car-seat arm
Locks, Chest
Locks, Door
Locks, Drawer
Locks, Hasp
Locks, Pad
Locks, Permutation
Locks, Piano
Locks, Safe
Locks, Sash
Locks, Seal
Locks, Shutter
Locks, Time
Locks, Trunk
Locks, Valise
Locks, Window
Lock-trimmings.
Night-latches.
Pad-locks.
Permutation-locks.
Piano-locks.
Register-locks.
Roses for door-knobs.
Safe-locks.
Safety-guards to locks.
Scutcheons for key-holes.
Seal-locks.
Shanks for door-knobs.
Spindles for door-knobs.
Time-locks.
Trunk-locks.
Valise-locks.
Window-locks.

# CLASS 71.

## MANURES.

Composition and Preparation of Fertilizers. Treatment of Guano, Sewage, Offal, Poudrette, Marl, Composts, etc.

Composts, Fertilizing
Deodorizing excrements.
Deodorizing offal.
Deodorizing vaults.
Fertilizers, Manufacture of
Fertilizer mixers.
Fertilizers, Packaging
Fertilizers, Treating
Guano, Treatment of
Manure, Artificial
Manures, Composition, processes, machinery.
Marl, Treatment of
Night-soil, Treatment of
Offal, Deodorizing
Offal for fertilizers, Treatment of
Phosphates for fertilizers, Treatment of
Privies, Cleaning
Poudrette, Treatment of
Sewage for fertilizers, Treatment of
Sewer cleaners.

# CLASS 72.

## MASONRY.

Masonry, Brick-work, Structures of Concrete, Plastering, Iron Houses.

---

Arches. (*See Class* 14.)
Basins for sewers.
Béton, Structures of
Brick-work of buildings.
Cellar-floors.
Centering for arches. (*See Class* 20.)
Cistern covers.
Cisterns.
Cistern tops.
Columns, (*except Iron, Class* 14.)
Concrete, Structures of
Coping.
Cornice, Metallic
Culverts.
Domes.
Fire-proof building.
Fire-proof floors.
Fire-proof walls.
Fire-proof walls, Portable
Floats, Plasterers'
Floors of stone or concrete.
Fortifications, (*except Earthworks, Class* 37.)
Foundations of stone or concrete.
Garbage-sinks.
Harbor defenses.
Hawks.
Hods.
House-fronts, Iron
House-fronts, Stone-facing
Houses, Iron, (*except Trusses, Girders, Piers, and Columns, Class* 14.)
Iron facing for houses.
Iron fronts for houses.
Iron houses.
Iron houses, Portable.
Jails.
Light-houses.
Man-holes for sewers.
Masonry.
Molds for cornice.
Mortar, Applications of
Pisé-building, Implements and modes for
Plastering-machines.
Plastering-tools.
Portable houses, (metallic.)
Privies.
Sewers.
Stone facing for houses.
Stone fronts for houses.
Stone and slate tanks, Construction of
Stone-work of buildings.
Tanks for storage, Excavated
Trowels, Bricklayers'
Trowels, Plasterers'
Trusses, Wood or iron. (*See Class* 14.)
Vaults.
Walls, (other than wooden.)

# CLASS 73.

## MEASURING-INSTRUMENTS.

Instruments of Precision, including Counting-machines and Registers, Meteorological Instruments, Measures and Measurers of Number, Distance, Weight, Capacity, Speed, Temperature, Moisture, &c.

[*Excepting—*
*Drafting and Mathematical*.. *Class* 33.
*Electrical* ................. *Class* 36.
*Horological* ................ *Class* 58.
*Optical and Surveying* ...... *Class* 88.]

Adding-machines.
Alcoholmeters.
Alkalimeters.
Anemometers.
Anemoscopes.
Arithmometers.
Balances.
Ballistic pendulums.
Barometers.
Bilgewater-gages.
Bottles, Dropping
Bottles, Graduated medicine
Burettes.
Calculating-machines.
Coin-assorters.
Coin-balances.
Coin-counting machines.
Coin detecting, Counterfeit
Computing-tables.
Counting-machines.
Drop-meters.
Dropping bottles.
Dropping-tubes.
Dynamometers.
Eprouvettes.
Eudiometers.
Gages, Bilgewater
Gages, Barrel-filling
Gages, Measuring
Gages, Registering devices of pressure
Gages, Rain
Gages, Sliding
Gages, Registering devices of steam
Gages, Registering devices of steam-boiler
Gages, Registering devices of vacuum
Gages, Registering devices of water
Gaging-instruments.
Game-counters.
Gas-meters, Registering devices of
Goniometers.
Graduated glasses.
Graduated measures.
Grain-tallies.
Grain-weighers.
Hydrometers.
Hydrostatic balances.
Hygrometers.
Hypsometers.
Inclinometers.
Indicators, Automatic station
Indicators in cars, Automatic street
Indicators, Leak
Indicators, Lee-way
Indicators, Registering devices of steam-pressure
Indicators, Weather
Lactometers.
Lee-way indicators.
Letter-balances.
Logs, Ships'
Lumber-tallies.
Measures, Board
Measures, Cloth
Measures, Dry
Measures, Liquid
Measures, Lumber
Measures of capacity.
Measures, Tailors'
Measures, Tape

Measures, Tire
Measuring devices of faucets.
Measuring devices of funnels.
Measuring devices of pumps.
Measuring-sticks.
Medicine glasses, Graduated
Meteorological instruments.
Meters.
Meters, Current
Meters, Drop
Meters, fluid, Registering devices of
Meters, gas, Registering devices of
Meters, liquid, Registering devices of
Meters, spirit, Registering devices of
Meters, Registering devices of water
Odometers.
Oleometers.
Pedometers.
Perambulators.
Photometers.
Pipettes.
Planimeters.
Plumbs.
Pressure-indicators, Registering devices of steam
Pyrometers.
Reckoning-instruments.
Recording-instruments.
Registers.
Registers, Billiard
Registers, Counting
Registers, Grain
Registers, Measuring
Registers, Passenger
Registers, Printing-press
Registers, Steam-engine
Registers, Street-car
Registers, Weighing
Registering devices of fluid-meters.
Registering devices of air, steam, and water gages.
Rules, Measuring
Saccharometers.
Salinometers.
Scales, Sliding
Scales, Tailors'
Scales, Weighing
Ships' logs.
Sliding-gages.
Sounding, Instruments for marine
Specific-gravity apparatus.
Speed-indicators.
Spirometers.
Squares.
Steam-engine indicators.
Steam-gages, Registering devices of
Steam-pressure indicators, Registering devices of
Sphygmographs.
Stethometers.
Sympiesometers.
Tables, Arithmetical
Tables, Computing
Tallies for grain, lumber, &c.
Thermometers.
Thermostats.
Urinometers.
Vanes.
Velocimeters.
Verniers.
Votes, Instruments for recording
Water-gages, Registering devices of
Water-gages for steam-boilers, Registering devices of
Water-meters, Registering devices of
Weather-guides.
Weighing-apparatus.
Weighing-buckets.
Weighing-scales.

# CLASS 74.

## MECHANICAL POWERS.

Horse-powers, Arrangements of Gearing, Brakes for Machinery, Cranks, Pitmans, Treadles; Modes of Converting, Multiplying, and Transmitting Motion.

[*For Excavating* .......... *see Class* 37.
*Hoisting* .............. *see Class* 57.
*Journals and Shafting*.. *see Class* 64.
*Mills* ................. *see Class* 83.
*Presses* ............... *see Class* 100.]

---

Animal powers.
Bearings for machinery. (*See Class* 64.)
Brakes for horse-powers.
Brakes for machinery.
Buildings, Moving
Cam-rods.
Cams, Variable
Changing speed, Apparatus for
Churn-powers.
Claw-bars.
Clutches, Rope
Cog-gearing.
Connecting-rods.
Converting motion, Apparatus for
Crank-connections.
Crank-motions.
Crank, Overcoming the dead-points of
Crank-pins.
Cranks.
Crank, Substitutes for the
Crow-bars.
Differential levers.
Dog-powers.
Eccentrics.
Fly-wheels.
Gearing, Arrangements of
Gibs for pitmans, etc.
Harvester-gearing. (*See Class* 56.)
Hoisting-machines. (*See Class* 57.)
Hooks for cam-rods.
Horse-powers.
Horse-powers, Endless chains for
Horse-powers, Links for
Horse-powers, Treads for
Intermittent-gearing.
Journals. (*See Class* 64.)
Lever-powers.
Levers, Differential
Lubricators. (*See Class* 64.)
Mechanical movements.
Mill-gearing.
Mills. (*See Class* 83.)
Motion, Converting
Motion, Multiplying
Motion, Transmitting
Motors, Mechanical
Pawls and ratchets.
Pendulum-powers.
Pitman connections.
Pitmans.
Pitmans for harvesters.
Pitmans for sewing-machines.
Presses. (*See Class* 100.)
Reversing motion, Apparatus for
Rope-clutches.
Shafting. (*See Class* 64.)
Starting engines off their centers.
Tide-powers.
Traction-wheels.
Transmitting motion, Apparatus for
Treadle-machines.
Treadles.
Wheels, Fly

# CLASS 75.

## METALLURGY.

Furnaces, Operations, Processes, and Machinery for the Reduction and Manufacture of Metals.

---

Alloys.
Amalgamators. (*See Class* 90.)
Annealing metals.
Assaying.
Blast furnaces.
Carbonizing furnaces.
Case-hardening compounds.
Case-hardening processes.
Cementation furnaces.
Coloring metals.
Converting iron into steel.
Copper furnaces.
Cupola furnaces.
Decarbonizing iron.
Desulphurizing coal.
Desulphurizing-furnaces.
Desulphurizing ores.
Enameling furnaces for metals.
Files, Tempering
Furnace-lining.
Furnaces, Boiling
Furnaces, Blast
Furnaces, Catalan
Furnaces, Copper
Furnaces, Cupola
Furnaces, Desulphurizing
Furnaces for reducing metals.
Furnaces for roasting ores.
Furnaces for treatment of iron.
Furnaces, Hot-blast
Furnaces, Iron
Furnaces, Lead
Furnaces, Puddling
Furnaces, Smelting
Furnaces, Zinc
Iron, Manufacture of
Malleable iron castings, Furnaces for
Manufacture of aluminium.
Manufacture of cobalt.
Manufacture of copper.
Manufacture of iron.
Manufacture of lead.
Manufacture of magnesium.
Manufacture of nickel.
Manufacture of potassium.
Manufacture of sodium.
Manufacture of steel.
Manufacture of tin.
Manufacture of zinc.
Metallurgic furnaces.
Metallurgic processes.
Metals, Furnaces for reducing
Metals, Manufacture of
Metals, Purification of
Ores, Calcining
Ores, Desulphurizing
Ores, Furnaces for reducing
Ores, Oxidizing
Ores, Roasting
Ores, Smelting
Puddling.
Puddling furnaces.
Puddling processes.
Reducing furnaces, Metal
Reducing processes, Metal
Refining furnaces.
Roasting ores.
Saws, Tempering
Sheet iron, Furnaces for
Smelting furnaces.
Steel, Compositions in manufacturing
Steel, Machinery in manufacturing
Steel, Processes in manufacturing
Tempering steel.
Tempering tools.
Welding compounds.

## METAL-WORKING.

*Under the following heads:—*

Metal-working, 1.—Bending and Straightening .......Class 76.
Metal-working, 2.—Boring and Drilling ..............Class 77.
Metal-working, 3.—Forging, Swaging, and Riveting...Class 78.
Metal-working, 4.—Punching, Cutting, and Shearing..Class 79.
Metal-working, 5.—Rolling..........................Class 80.
Metal-working, 6.—Tools ...........................Class 81.
Metal-working, 7.—Turning, Planing, and Milling.....Class 82.

*See also—*

Bolts, Nuts, and Rivets .....Class 10.
Casting ....................Class 22.
Files and Rasps ............Class 40.
Grinding and Polishing .....Class 51.
Hardware Manufacture .....Class 53.

[*Thus subdivided:* A.—Arms and Cutlery.
B.—Builders' Hardware.
C.—Carriage Hardware.
D.—Furniture Hardware.
E.—Personal Wear.
F.—Saddle and Harness Hardware.
G.—Shoemakers' Hardware.
H.—Spinning and Weaving Hardware.
I.—Stationery and Miscellaneous Hardware.
J.—Trunk and Carpet-bag Hardware.
K.—Water and Gas Fittings.]

Horse-shoes ...............Class 59.
Locks and Latches .........Class 70.
Nails and Spikes ...........Class 85.
Needles and Pins ..........Class 86.
Plating .................. Class 96.
Railway-track and Car-iron..Class 107.
Safes .....................Class 109.
Saws ......................Class 110.
Sheet-metal ...............Class 113.
Tubing and Wire-making ...Class 134.
Wire-working .............Class 140.
Wood Screws .............Class 141.

# CLASS 76.

## METAL-WORKING, 1.

### Bending and Straightening.

Modes, Apparatus, Machines, Tools, and Appliances, (except as otherwise specifically classified,) including—

Angle-iron, Bending or straightening
Armor-plates, Bending or straightening
Bars, Bending or straightening
Bending bars and plates.
Bending rails.
Corrugating plates.
Plates, Bending metallic
Plates, Corrugating metallic
Plates, Straightening metallic
Rails, Bending or straightening
Rods, Bending or straightening
Shafts, Bending or straightening
Straightening bars and plates.
Straightening gun-barrels.
Straightening rails.
Straightening rods and shafts.
Twisting bars, rods, and fagots.

# CLASS 77.

## METAL-WORKING, 2.

### Boring and Drilling.

Modes, Apparatus, Machines, Tools, and Appliances, (excepting thos otherwise specifically classified,) including—

Annular boring-bits for metal.
Boring cannon.
Boring cylinders.
Boring-machines, Metal
Bow-drills.
Countersink-tools for metal.
Cylinders, Boring
Drifts.
Drill-bits.
Drill-gages.
Drilling-jigs.
Drilling machines, Metal
Drilling metals.
Drill-rests, Metal
Drills, Gages for
Drills, Ratchet
Drill-stocks.
Drills, Twist
Drills, Watchmakers'
Expanding drills for metal.
Ratchet-drills.
Reamers for metal.
Spiral stock-drills.
Twist-drills.

# CLASS 78.

## METAL-WORKING, 3.

Forging, Swaging, and Riveting.

Modes, Apparatus, Machines, Tools, and Appliances, (except as otherwise specially classified,) including—

Anchors, Manufacture of
Anvils, Manufacture of
Axles, Manufacture of carriage
Blacksmiths' forges.
Blacksmiths' tool-stands.
Boiler making, Steam
Chain-making.
Dies for swaging.
Drop-hammers.
Drop-presses.
Forges, Portable
Forging.
Furnaces for blanks.
Furnaces for skelps.
Furnaces for tires.
Gold-beating.
Gun-carriages, Making of iron
Hammers, Drop
Hammers, Hydraulic
Hammers, Pneumatic
Hammers, Power
Hammers, Steam
Hammers, Water
Olivers.
Ordnance-manufacturing.
Pneumatic hammers.
Portable forges.
Power-hammers.
Pressing metals, scrap, &c.
Riveting.
Riveting-machines.
Skelp and blank furnaces.
Stamping-mills.
Staving metals.
Steam-hammers.
Steam-hammers, Engines of (*See Class* 121.)
Tire coolers.
Swaging metals.
Tires, Bending and upsetting
Tires, Making
Upsetting-machines, Tire
Upsetting metals.
Vise-making, Machinists'
Welding apparatus.

# CLASS 79.

## METAL-WORKING, 4.

Punching, Cutting, and Shearing.

Modes, Apparatus, Machines, Tools, and Appliances, (except as otherwise specifically classified,) including—

---

Blanks from bars and plates, Cutting
Cutting blanks from bars or plates.
Cutting planchets from bars or plates.
Gumming-saws.
Planchets from bars and plates, Cutting
Punching-machines for bar or plate.
Saw-gummers.
Saw-making.
Saw-sets.
Saw-swages.
Shearing-machines for bar or plate.
Tube-sheets, Punching

# CLASS 80.

## METAL-WORKING, 5.

### Rolling.

Machines, Modes, Tools, Apparatus, and Appliances, (except as otherwise specifically classified,) including—

Angle-iron, Rolling
Armor-plates, Rolling
Bars and rods, Rolling
Beams, Rolling
Bundling scrap.
Casting metal rules between rollers.
Casting sheet metal between rollers.
Chemical processes in sheet metal making.
Die-rolling.
Fagots and piles for rolling.
Girder-bar, Rolling
Hoop-iron, Rolling
Horseshoe-bar, Rolling
Lighting-rods, Rolling
Piles and fagots for rolling
Puddlers' balls, Rolling and squeezing
Railroad-railpieces, Piling
Railroad-rails, Capping
Railroad-rails, Repairing
Railroad rails, Re-rolling
Railroad-rails, Rolling
Rolling metals.
"Russian" sheet-iron, Imitations of
Scrap, Bundling and rolling
Sheet-iron, Chemical processes in manufacture of
Sheet-iron, Rolling
Sheet-iron, Scraping and cleansing
Sheet-lead, Rolling
Sheet-metal making.
Sheet-metal planishing.
Spike-bar rolling
Squeezers for puddlers' balls.

# CLASS 81.

## METAL-WORKING, 6.

Tools.

Under the following heads:

A.—Metal-working tools.

B.—Manufacture of metallic tools for purposes other than metal-working.

---

A.—*Metal-working tools for hand use, (not elsewhere specially classified,) and their analogues, including the manufacture of the same.*

Blacksmiths' tools.
Boiler-makers' tools.
Bolt-cutters.
Clinchers, Blacksmiths'
Cutters, Bolt
Cutters, Pipe
Cutters, Rod
Farriers' tools.
Foot-rests for horses.
Gas-fitters' tools.
Hand-hammers.
Hand-vises.
Hoof-spreaders.
Jewelers' tools.
Jeweling watches, Tools for
Locksmiths' tools.
Machinists' tools.
Metal-working tools.
Monkey-wrenches.
Pipe-cutters.
Pipe-tongs.
Pipe-wrenches.
Plumbers' tools.
Ratchet-wrenches.
Rod-cutters.
Sheet-metal workers' tools.
Shoeing-tools, Blacksmiths'
Squares, Making
Tinners' tools.
Tongs, Blacksmiths'
Tongs, Pipe
Tube-expanders.
Vises, Machinists'
Watchmakers' tools.
Wrenches, Monkey
Wrenches, Pipe

B.—*Tool-Manufacture.—The making of Metallic Tools, or parts of Tools, and Implements, for Working in the Earth, in Wood, Stone, Clay, Mortar, Leather, and Glass; the making of iron parts of Harvesters and other Agricultural Machinery.*

Adzes, Manufacture of
Agricultural tools of metal, Manufacture of
Augers, Manufacture of
Axes, Manufacture of
Bricklayers' tools, Manufacture of
Cabinet-makers' tools of metal, Manufacture of
Carpenters' tools of metal, Manufacture of
Chisels, Manufacture of
Clay-working tools, Manufacture of
Coopers' tools, Manufacture of
Cutlery, Manufacture of (*See Class 53, Section A.*)
Edge-tools, (not cutlery,) Manufacture of
Farm implements of metal, Manufacture of
Gardeners' implements of metal, Manufacture of
Glass-working tools, Manufacture of
Joiners' tools, Manufacture of
Leather-working tools, Manufacture of

Lumbermen's tools, Manufacture of
Miners' tools, Manufacture of
Shipwrights' tools, Manufacture of
Stone-working tools, Manufacture of
Track-layers' tools, Manufacture of
Trowels, Manufacture of
Tuyeres, Manufacture of
Wheelwrights' tools of metal, Manufacture of
Wood-workers' tools of metal, Manufacture of

# CLASS 82.

## METAL-WORKING, 7.

Turning, Planing, Slotting, and Milling.

(Except as otherwise specifically stated,) including—

Burr-cutters.
Centering-plates.
Centering-tools.
Chucks for metal-lathes.
Cutter-stocks for metal-planers.
Cutting-tools for lathes and planers.
Dogs for metal-lathes and planers.
Engine-lathes.
Gear-cutting machines.
Groove-cutting machines.
Key-seat cutting-machines.
Lathes for milling.
Lathes for turning.
Lathes for screw-cutting.
Lathes for watchmakers.
Mandrels.
Milling-machines.
Planers.
Planing-machines, Metal
Rests for metal-lathes.
Rifling-machines.
Rose-engines.
Screw-cutting lathes.
Shaping-machines, Metal
Slotting-machines, Metal
Tool-holders for lathes.
Tool-rests for lathes.
Tools for metal-lathes, Turning
Turning cannon.
Turning balls, bullets, and shells.
Turning gun-barrels.
Turning-lathes, Metal
Type-cutting machines.
Watchmakers' lathes.

# CLASS 83.

## MILLS.

Grinding-Mills for Bark, Cane, Grain, Paint, Sugar; Flour-bolts, Smut and Hulling Mills.

[*Excepting Clay and Mortar Mills* ...... *Class* 25.
*Fruit Mills and Presses* ...... *Class* 100.
*Ore-mills and Stone-crushers* .. *Class* 90.]

---

Alarms, Grist
Bag-fasteners.
Bag-holders.
Balance-rynds.
Bark-mills.
Bolting meal, &c.
Bolts, Flour
Bran-dusters.
Burrs, Metallic
Cane-mills.
Cider-mills. (*See Class* 100.)
Coffee-cleaning machines.
Coffee-mills.
Coffee-polishers.
Corn-mills.
Drug-mills.
Exhaust for millstones.
Feed regulators for mills.
Flour-bolts.
Flour-coolers.
Flour-mills.
Flour-packers. (*See Class* 100.)
Grain-bagging.
Grain-bruisers.
Grain cleaners for mills.
Grain-conveyors. (*See Class* 57.)
Grain-dampeners.
Grain-distributors for mills.
Grain-elevators. (*See Class* 57.)
Grain-feeders.
Grain for grinding, Preparing
Grain-mills.
Grain-spouts for mills.
Grain tollers.
Grinding plates for mills.
Grist alarms.
Gunpowder-mills.
Hominy-machines.
Hopper-boys.
Hopper-shoes for mills.
Hulling-machines.
Huskers, Corn
Leveling millstones.
Millers' alarms.
Mill gearing. (*See Class* 74.)
Millstone-bridges.
Millstone-curbs.
Millstone-dress.
Millstone-eyes.
Millstones.
Millstones, Balancing
Millstones, Dressing
Millstones, Elevating
Millstones, Hanging
Millstones, Leveling
Millstones, Regulating
Millstones, Supporting
Millstones, Tramming
Millstones, Ventilating
Paint-mills.
Paint-mullers.
Pea-nut cleaners.
Picks, Millstone (*See Class* 125.)
Powder-mills.
Pug-mills. (*See Class* 25.)
Regulators for millstones.
Rice-cleaning machines.
Rice-hullers.
Rice-polishers.
Sack-fasteners.
Scouring-machines.
Shellers, Corn
Smut-machines.
Smut-mills.
Sorghum-mills.
Spice-mills.
Spindles, Mill
Stone-breakers. (*See Class* 90.)
Sugar cane mills.
Sugar-mills.
Tolling devices for grain-mills.
Tramming millstones.
Tram-staffs for mills.
Ventilating millstones.

# CLASS 84.

## MUSIC.

Instruments, Attachments, and Accessories.

Accordeons.
Banjos.
Cabinet-organs.
Chiroplasts.
Clarinets.
Cornets.
Cymbals.
Dampers for musical instruments.
Drums.
Dulcimers.
Eolians.
Fifes.
Flageolets.
Flutes.
Guitars.
Harmoniums.
Harps.
Key-boards for musical instruments.
Keys for musical instruments.
Knee-swells for organs.
Leaf-turners for music-books.
Melodeons.
Music.
Musical instruments.
Musical notation.
Music-boxes.
Organ-actions.
Organ-bellows.
Organ-couplers.
Organ key-boards.
Organ-pipes.
Organs.
Organ sound-boards.
Organ-stops.
Organs, Tremolo attachments to
Organ-swells.
Organ-tuning.
Organ wind-chests.
Piano-forte actions.
Piano-forte attachments.
Piano-forte cases.
Piano-forte frames.
Piano-forte keys.
Piano-fortes.
Piano-fortes combined with reed instruments.
Piano-forte sounding-boards.
Piano-fortes, Scales for
Piano-fortes, Stringing
Piano-fortes, Swell-attachment to
Piano-fortes, Tuning
Reed instruments.
Stringed instruments.
Strings for musical instruments.
Sounding-boards for musical instruments.
Swells for musical instruments.
Tremolo-attachments for musical instruments.
Trumpets.
Tuning instruments.
Violin-attachments.
Violin-bows.
Violins.
Violin-screws.
Violin sound-boards.
Violoncellos.
Wind-chests for organs.
Wind instruments.

# CLASS 85.

## NAILS.

Nails, Spikes, Tacks, and Staples, and Machines for making.

---

Blind-staples.
Capping nails and screws.
Cut-nails.
Dies for making spikes and nails.
Horseshoe-nails.
Horseshoe-nails, Machines for making
Leathering tacks.
Nail-plate feeders.
Nails.
Nails, Cut
Nail-distributing machines.
Nails, Machines for making
Nails, Wrought
Picture-knobs.
Picture-nails.
Railroad-spikes.
Sheathing nails.
Shoe-nails.
Spikes.
Spikes, Machines for making
Spikes, Self-clenching
Staples.
Staples, Machines for making
Tacks.
Tacks, Leathering
Tacks, Machines for making
Wrought-nails.

# CLASS 86.

## NEEDLES AND PINS.

The Manufacture and Preparation for market.

---

Crochet-needles, Making
Hair-pins, Making
Knitting-burs, Making
Knitting-needles, Making
Needles, Machines for making
Needles, Machines for papering
Pins, Machines for making
Pins, Machines for papering
Sewing-machines needles, Making

# CLASS 87.

## OIL, FAT, AND GLUE,

Including Glycerine, Paraffine, Soap, Wax, and Candles.

---

Ash-leaches.
Blanching oils, fats, wax, &c.
Candle machinery.
Candle-making.
Candle-molds.
Candle-molds, Compositions for
Candles.
Candles, Compositions for
Dubbing.
Fats, Treatment of
Fish-oil, Treatment of
Gelatine.
Glue-cutting machines.
Glue dryers. (*See Class* 34.)
Glue-making.
Glue pots. (*See Class* 126.)
Glycerine.
Grease from slush, Separating
Horn, Artificial
Isinglass.
Lard-coolers.
Lard-oil presses. (*See Class* 100.)
Lard-oil rendering-tanks.
Lard-presses. (*See Class* 100.)
Lard-renderers.
Leaches, Ash
Lye-extracts.
Offal, Treating
Oil extracting, Chemical modes of
Oil-filters.
Oil-presses. (*See Class* 100.)
Oils and fats from fiber and meal, Removing
Oils, Refining. (*See Class* 124.)
Oil-stills. (*See Class* 124.)
Oleometers. (*See Class* 73.)
Pans for manufacturing glue.
Paraffine.
Presses for oils, stearine, fats, &c. (*See Class* 100.)
Rendering lard and tallow.
Soap.
Soap-cutting machines.
Soap-making machines.
Soap-making processes.
Starch. (*See Class* 8.)
Stearine.
Tallow, Treatment of
Washing compositions.
Wax-cutting machines.
Wax, Treatment of
Wool, Removing suint and grease from

# CLASS 88.

## OPTICS.

Astronomical, Optical, and Surveying Instruments.

---

Achromatic glasses.
Altimeters.
Altitude instruments.
Astronomical instruments.
Azimuth instruments.
Cameras.
Cameras, Photographic
Cameras, Solar
Catoptric lights.
Chains, Surveyors'
Circumferentors.
Collimators.
Compasses, Surveyors'
Delineators of grades and routes.
Departure and distance instruments.
Dioptric lights.
Eye-glasses.
Heliometers.
Heliostats.
Heliotropes.
Horizons, Artificial
Lanterns, Magic
Latitude, Instruments for determining
Lenses.
Levels.
Level-staffs.
Longitude, Instruments for determining
Lorgnettes.
Magic-lanterns.
Magnifying-glasses.
Meridian-finders.
Micrometers.
Microscopes.
Octants.
Odometers. (*See Class* 73.)
Opera-glasses.
Optical instruments.
Pedometers. (*See Class* 73.)
Perambulators. (*See Class* 73.)
Photographic cameras.
Plane-tables.
Polariscopes.
Polarizing-apparatus.
Prisms.
Profiling-apparatus.
Quadrants.
Reflecting-circles.
Sextants.
Solar cameras.
Spectacles.
Stereoscopes.
Surveying instruments.
Surveyors' chains.
Surveyors' marks.
Telescopes.
Theodolites.
Transit instruments.
Tripods, Surveyors'

# CLASS 89.

## ORDNANCE.

[*Except Projectiles, Class* 102.]

Artillery.
Artillery-carriages.
Breech-loading cannon.
Bushings for gun-vents.
Cannons.
Cannon-sights.
Carriages for artillery.
Centrifugal guns.
Chassis for guns.
Counterpoise gun-carriages.
Disabling ordnance, Instruments for
Gas-checks for breech-loading ordnance.
Gun-carriages.
Gun-carriages, Counterpoise
Gun-chassis.
Gun-platforms.
Locks for ordnance.
Machine-guns.
Mantelets.
Mitrailleuses.
Mortars.
Mounting ordnance.
Operating ordnance.
Ordnance.
Ordnance, Attaching trunnions to
Ordnance, Instruments for disabling
Ordnance, Mounting
Ordnance, Operating
Quoins for ordnance.
Rammers for ordnance.
Revolving breech-cannon.
Rifled ordnance.
Sights for ordnance.
Sponges for ordnance.
Submarine ordnance.
Tompions.
Traverses of guns.
Valves for submarine ordnance.
Vents of ordnance.
Vent-stoppers.

# CLASS 90.

## ORE.

Apparatus, Machines, and Processes for crushing and grinding Ore, Stone, Coal, or Bone; for separating Ores of precious metals, mechanically or by amalgamation.

---

Amalgamators.
Arrastras.
Bone mills.
Coal-breakers.
Coal-washers.
Disintegrating ores.
Gold-washers.
Jiggers.
Marble-crushing machines.
Ore-crushers.
Ore-disintegrators.
Ore-dressers.
Ore-mills.
Ore-separators.
Ore-stamps.
Ore-washers.
Plaster-mills.
Quartz-mills.
Sand-washer.
Stamping-mills.
Stone-breakers.
Stone-crushers.
Washers, Gold
Washing-tables.

# CLASS 91.

## PAINT.

Paints, Pigments, Varnishes, Lacquers, and Staining Compositions, and their Manufacture; Painted and Enameled Fabrics; Lining Barrels; Leaf-gilding. Air, Fire, and Water Proofing Fabrics and Processes.

---

Air-proofing processes.
Artificial leather.
Barrel-lining.
Barrels impervious to liquids, Rendering
Barrels, Pitching
Brushes. (*See Class* 15.)
Coating and lining barrels.
Enameling of cloth.
Enameling of leather.
Fire-proofing compounds.
Fire-proofing processes.
Floor-cloth.
Gas-tubing.
Gilding, Leaf
Gum resins, Treatment of
Imitation leather.
Ink, Printers'
Japanning cloth.
Japanning leather.
Japanning wood.
Lacquers.
Lampblack.
Leather, Artificial
Leather-cloth.
Leather, Enameling
Leather, Japanning
Metallic surfaces, Compositions for coating
Oil-cloth.
Paint-brushes. (*See Class* 15.)
Paint-burners. (*See Class* 126.)
Painted fabrics.
Painters' tools, (*except Brushes, Class* 15.)
Painting-machines.
Paint-mills. (*See Class* 83.)
Paint-pot hangers.
Paints.
Paints, Apparatus for compounding
Paints, Apparatus for mixing
Pigments.
Pigments, Manufacture of
Priming for paint.
Sand-blowers. (*See Class* 98.)
Staining.
Varnishes.
Varnishing-machines.
Water-proofing compounds.
Water-proofing processes.
White-lead machinery.
White-lead processes.
Zinc, white, processes and machinery.

# CLASS 92.

## PAPER-MAKING.

Manufacture of Paper, Card-board, &c., and of Articles direct from the pulp.

---

Beater-engines.
Bed-plates of Rag-engines.
Binders' boards.
Boxes direct from pulp, Making paper
Card-board.
Enameling paper.
Flocking paper.
Glazing paper.
Mill-board.
Paper.
Paper, Calendering
Paper, Drying
Paper, Enameling
Paper, Flocking
Paper-making machines.
Paper, Materials for
Paper, Processes of making
Paper-pulp.
Paper-pulping machines.
Paper-sizing.
Paper water-proofing.
Papier-maché.
Paste-board.
Pulping-machines.
Pulp, Reducing materials to
Rag-engines.
Roofing-paper.
Safety-guards in paper.
Straw-boards.
Straw for paper, Treating
Tobacco-paper.
Vegetable parchment.
Wall-paper. (*See Class* 101.)
Water-marks in paper.
Water-proofing paper.
Wood for paper, Treating

# CLASS 93.

## PAPER MANUFACTURES.

Articles of Paper, Appliances, Machines, Processes, and Tools for manufacturing articles of Paper.

---

Apparel, Paper
Bag machines, Paper
Book-folders.
Box machines, Paper
Collars, Machines for boxing paper
Collars, Machines for cutting out paper
Collars, Machines for embossing paper
Collars, Machines for folding paper
Collars, Machines for patching button-holes of paper
Collars, Machines for planishing paper
Collars, Machines for punching paper
Collars, Machines for Shaping paper
Collars, Packing-envelopes for paper
Collars, Paper
Cuffs, Paper
Cutters, Paper
Envelope-making machines.
Envelopes.
Envelopes and letter sheets combined.
Feeders, Paper
Floor-cloth, Paper
Label-machines, Paper
Lace-paper machines.
Newspaper-folders.
Paper-apparel.
Paper-bag machines.
Paper-bags.
Paper-box machines.
Paper-boxes.
Paper-collars.
Paper-cuffs.
Paper-cutters.
Paper-cutting machines.
Paper-envelope machines.
Paper-feeders.
Paper floor-cloth.
Paper-folders.
Paper-labels, Manufacture of
Paper, Lace
Paper-manufactures.
Paper-perforating machines.
Paper-tags, Manufacture of
Paper-twine.
Tag-machines, Paper
Wall-paper trimmers.

# CLASS 94.

## PAVING.

Materials, Compositions and Varieties, Streets and Sidewalks, Making, Repairing, and Sweeping.

---

Asphalt and wood pavements combined.
Asphalt pavements.
Cement floors.
Cement paving.
Coal-hole covers.
Coal-hole guards.
Coal-traps in pavements.
Concrete pavement.
Concrete pavement, Devices for laying
Curbs, Pavement
Floors, Cement
Garbage-boxes in pavement.
Gutter-clearers.
Gutters, Pavement
Inlets to sewers.
Iron and wood pavements combined.
Iron pavements.
Laying pavements.
Pavements, Asphalt
Pavements, Composition
Pavements, Concrete
Pavements, Iron
Pavements, Stone
Pavements, Wooden
Paviors' tools, machines, and processes.
Paving compositions.
Paving-tiles. (*See Class* 25.)
Rammers, Paviors'
Road-paving.
Road-rollers.
Roughening pavements.
Sidewalks.
Stone, Artificial. (*See Class* 125.)
Stone, Manufacture of paving. (*See Class* 125.)
Street-levelers.
Street-paving.
Street rammers.
Street-scrapers.
Street-sweepers.
Tiles. (*See Class* 25.)
Traps for gutters.
Traps in pavements, Coal
Vault-covers.
Vault-lights.
Wood and asphalt pavements combined.
Wood and iron pavements combined.
Wood and stone pavements combined.
Wooden pavements.

# CLASS 95.

## PHOTOGRAPHY.

Instruments and Processes.

---

Albumenizing paper.
Cameras, Photographic. (*See Class* 88.)
Cameras, Solar. (*See Class* 88.)
Camera-stands.
Developing, Photographic
Head-rests, Photographic
Heliotypes.
Photogalvanography.
Photographic apparatus.
Photographic cameras. (*See Class* 88.)
Photographic paper.
Photographic plate-holders.
Photographic printing.
Photographic prints, Coloring
Photographic processes.
Photographic toning.
Photography.
Photolithography.
Photo-sculpture.
Picture-holders, Photographic
Printing-frames, Photographic
Solar cameras. (*See Class* 88.)
Stereoscopes. (*See Class* 88.)

# CLASS 96.

## PLATING.

Metals on Metals.

Buttons, Gilding
Coating metallic surfaces.
Electro-plating.
Electrotyping.
Galvanizing metals.
Gilding.
Glass-gilding.
Glass-platinizing.
Glass-silvering.
Lead-pipe lining by electro-plating process.
Metals with metals, Coating
Moire metallique.
Platinizing.
Silvering.
Silver-plating.
Tinning.

# CLASS 97.

## PLOWS.

Machines for Breaking, Digging, Trenching, and Paring the Soil, Cultivating Crops, and Digging Roots.

[*Manufacture of Agricultural Tools, etc., of Metal, see Class* 81, *Section B.*]

---

Bog-cutters.
Corn-coverers.
Cotton-choppers.
Cotton-scrapers.
Cotton-toppers.
Cultivator-teeth.
Cultivators.
Cultivators, Frames for
Cultivators, Wheel
Cutters, Sod
Diggers, Peanut
Diggers, Potato
Diggers, Root
Digging-machines.
Ditching-plows.
Double shovel-plows.
Draining-plows.
Furrowing-machines.
Gang-plows.
Garden-walks, Machines for scraping
Hand-plows.
Hoeing-machines.
Hoes, Horse
Horse-hoes.
Land paring machines.
Mold-boards for plows.
Mold-boards, Materials for
Mole-plows.
Plow-beams.
Plow-cleaners.
Plow clevises.
Plow-colters.
Plow-holders.
Plow-irons.
Plow-landsides.
Plow mold-boards.
Plow-points.
Plow-regulators.
Plows, Ditching
Plows, Double mold-board
Plows, Double-shovel
Plows, Draining
Plows, Drain-tile
Plows, Expanding
Plows, Gang
Plows, Garden
Plows, Hand
Plowshares
Plows, Hill-side
Plows, Marking
Plows, Mole
Plow-soles.
Plows, Paring
Plows, Prairie
Plows, Rotary
Plows, Seeding
Plows, Single-shovel
Plow-standards.
Plows, Steam
Plows, Subsoil
Plows, Sulky
Plows, Tile-laying
Plows, Wheel
Plow-wheel colters.
Potato-digging machines.
Root-diggers.
Rotary spaders.
Scrapers, Cane
Scrapers, Cotton
Shovel-plows.
Single-shovel plows.
Sleds for furrowing ground.
Sod-cutters.
Spading-machines.
Steam-cultivators.
Steam-plows.
Steam spading-machines.
Subsoil plows.
Sweet potato cultivators.
Tile-laying machines.
Turf-cutting machines.
Vine cutter, Potato-

# CLASS 98.

## PNEUMATICS.

Mechanical Applications of Air and other Elastic Fluids; Ventilation.

[*Except Valves, Class* 136.]

[*For Manufacture of Pipe and Gas-fittings, see Class* 53, *Section K.*]

Aërial cars.
Aërostats.
Aëro steam-engines. (*See Class* 121.)
Air and gas meters.
Air-blast.
Air chambers.
Air-compressing machines and appliances.
Air-cushions.
Air-engines.
Air-filters.
Air-guns.
Air-heating traps for air-engines.
Air-meters.
Air-pumps.
Ammonia engines.
Atmospheric motors.
Balloons.
Bellows.
Bellow-nozzles.
Blast-machines.
Blast-regulators.
Blowers.
Blowers for furnaces.
Blowers for organs.
Blowers, Steam
Blowing apparatus for glass-makers.
Blow-pipes.
Blow-pipes for glass-workers.
Caloric engines.
Carbonic acid gas-engines, (motors.)
Car-brakes, Pneumatic
Compressed air-engines.
Dust from cars, Excluding
Engines, Air
Engines, Air-compressing
Engines, Ammonia
Engines, Blast
Engines, Caloric
Engines, Carbonic acid
Engines, Compressed air
Engines, Gas
Engines, Gunpowder
Engines, Vapor, (except steam.)
Exhaust-fans.
Fan-blowers.
Flying-machines.
Foot-bellows.
Forge-bellows.
Gas-compressing machines.
Gas-engines.
Gas-meters.
Gas-pumps.
Gunpowder-engines.
Hot-air engines.
Hot-blast engines.
Meters, Air
Meters, Gas
Meters, Registering devices of (*See Class* 73.)
Piston-blowers.
Pneumatic dispatch-tubes. (*See Class* 104.)
Pneumatic railways. (*See Class* 104.)
Railways, Pneumatic. (*See Class* 104.)
Rock-drills, Air-engines for
Rotary-blowers.
Sand blowers.
Tuyeres.
Valves. (*See Class* 136.)
Vapor-engines, (*except steam.*)
Ventilation by heated air.
Ventilation by hot water.
Ventilation by steam.
Ventilation of cars.
Ventilation of halls.
Ventilation of houses.
Ventilation of mines.
Ventilation of ships.
Ventilators.
Ventilators, Plenum
Ventilators, Vacuum
Wind-mills.
Wind wheels.

# CLASS 99.

## PRESERVING FOOD.

Preservative Processes and Hermetical Packages.

[*Excepting Bottling* .............. *Class* 1.
*Drying and Smoking* .. *Class* 34.
*Refrigerating* ......... *Class* 62.]

---

Acids, Air-tight packages for
Air-tight cans, jars, and packages.
Alkalies, Air-tight packages for
Bottle-stoppers. (*See Class* 1.)
Cans, Air-tight.
Cans for preserving provisions.
Can tops, caps, and stoppers.
Chemical preparation of food.
Chemicals, Air-tight packages for
Condensing milk.
Curing meat and fish, (*except Smoking and Drying, Class* 34.)
Drying fruit. (*See Class* 34.)
Eggs, Preserving
Fish, Cans for packing
Fish-carrying boxes.
Fish-flakes. (*See Class* 34.)
Food, Cans, jars, and air-tight packages for
Food, Chemical preparation of
Food, Hermetically sealed packages for
Fruit-cans.
Fruit-cars.
Fruit-jars.
Fruit cellars and houses.
Fruit-chambers.
Fruits, Chemical preparation of
Hermetically sealed packages for provisions.
Ice-houses and refrigerators. (*See Class* 62.)
Jars for fruit, meat, &c.
Jars, Self-sealing
Meat cans and jars.
Meat-curing.
Meat-packing.
Meats, Compositions for preserving
Meats, Processes for preserving
Milk, Air-tight packages for
Milk, Condensing
Oil-cans, Air-tight
Oil-tanks, Sealed
Oyster-cans.
Packing houses, Meat
Paint-cans, Air-tight
Powder-cans, Air-tight
Preserving-cans.
Preserving-cars.
Preserving-houses.
Preserving processes for food.
Provision cans, jars, &c, Air-tight
Provision-cars.
Refrigerating processes. (*See Class* 62.)
Refrigerators. (*See Class* 62.)
Sealing cans and jars.
Smoke-houses. (*See Class* 34.)
Smoking meat and fish. (*See Class* 34.)
Vegetables, Preserving

# CLASS 100.

## PRESSES.

Except Printing and Hydraulic Presses.

Automatic presses.
Bale-ties and hoops.
Baling hay.
Baling-presses.
Beater-presses.
Blubber-presses.
Box-packing presses.
Broom-corn presses.
Bundling-presses for yarns and threads.
Chain-testers.
Cheese-presses.
Cider-mills.
Cider-presses.
Cloth-presses.
Compressors for re-baling.
Cotton-bale ties.
Cotton-presses.
Dry-goods packers.
Filtering-presses.
Fleece-tables and tyers.
Flour-packers.
Fruit-packers.
Fruit-presses.
Grape-presses.
Gunpowder-presses.
Hayp-resses.
Hoop-locks.
Hoop-strainers.
Hop-presses.
Hot-presses for fabrics and paper.
Hydraulic presses. (*See Class* 103.)
Lard-oil presses.
Lard-presses.
Mats for oil presses.
Meat-presses.
Oil-cake trimmers.
Oil-mills.
Oil-presses.
Packers for flour, fruit, sugar, salt, etc.
Packing-bands.
Peat-presses.
Portable presses.
Press and strainer combined.
Printing-presses. (*See Class* 101.)
Re-baling presses.
Roping bales.
Salt-packers.
Screw-presses.
Spring-testing presses.
Steam-finishing presses for cloth.
Steam plates for cloth-presses.
Stearine-presses.
Strainer-presses.
Sugar-packers.
Sugar-presses.
Ties for bales.
Tobacco-presses.
Wheel-presses, (*not Hydraulic, Class* 103.)
Wine-presses.
Wool-packers.
Wool-presses.
Wool-tyers.
Yarn-presses.

# CLASS 101.

## PRINTING.

Appliances and Supplies of the Printing Office, Type, Stereotype, and Electrotype.

---

Addressing-machines.
Address-stamps.
Black-leading machines for electrotyping.
Calico-printing machines.
Caoutchouc-printing.
Color-printing.
Composing-machines.
Composing-sticks.
Compositors' appliances.
Compositors' cases.
Copper-plate presses.
Copper-plate printing.
Curved printing-surfaces, Making
Damping machines, Paper-
Distributing-machines.
Electrotyping, Producing the matrix for
Fabrics, Printing
Floor-cloth printing.
Flyers for printing-presses.
Furniture, Printers'
Gage-pins for tympan-sheets.
Galley-rests.
Galleys.
Glass, Printing
Hand-stamps.
Hat-printing machines.
Inking apparatus.
Inking-rollers, Compositions for
Lapping, Printers'
Lead-cutters. (*See Class* 113, *Section A.*)
Lithographic presses.
Mechanical typographers.
Mitering printers' rules. (*See Class* 113, *Section A.*)
Newspaper-folders. (*See Class* 93.)
Numbering-machines.
Oil cloth printing.
Paging-machines.
Paper-dampers.
Porcelain, Printing
Post-office letter-stamping apparatus.
Presses.
Presses, Canceling
Presses, Copper-plate
Presses, Copying
Presses, Embossing
Presses for obtaining relief-printing surfaces.
Presses, Hand printing
Presses, Lithographic
Presses, Printing
Presses, Seal
Printers' blankets.
Printers' blankets, Washing
Printers' boxes.
Printers' cases.
Printers' furniture.
Printers' lapping.
Printers' quoins.
Printers' rollers.
Printers' rules.
Printers' spaces, Cutting and mitering. (*See Class* 113, *Section A.*)
Printing cloth.
Printing floor-cloth.
Printing oil-cloth.
Printing-presses.
Printing-processes.
Printing-surfaces.
Printing surfaces, Presses for obtaining relief-
Printing wall-paper
Printing-wheels.
Quoins, Printers'
Register-points for printing-presses.
Revenue-stamp cancelers.
Rollers, Printers'
Rubber type.
Rule-cutters. (*See Class* 113, *Section A.*)
Rules, Printing
Seals.
Skirts, Printing
Spaces for printers.
Spool-printing.
Stamp-cancelers.

Stamps, Address
Stamps, Branding
Stamps, Hand
Stamps, Post-office
Stamps, Seal
Stencil-printing machines.
Stereotype molds, Machines for producing
Stereotyping.
Stereotyping, Compositions for
Stereotyping, Processes for
Tags, Printing
Turtles.
Type.
Type, Casting
Type, Composing
Type, Distributing
Type, Flexible
Type-finishing.
Type-founding.
Type-scouring.
Type-setting.
Type-writing machines.
Typography.
Wall-paper printing.

# CLASS 102.

## PROJECTILES.

Balls, Bombs, Bullets, Cartridges, Fuses, Rockets, Shell, Shot, and Torpedoes; Blasting.

---

Ammunition cases.
Balls for cartridges.
Balls for fire-arms.
Balls for ordnance.
Blasting-cartridges.
Blasting-compounds. (*See Class* 52.)
Blasting-fuses.
Blasting-plugs.
Blasting rock.
Bomb-lances.
Bullet-pressing machines. (*See Class* 53 *A.*)
Bullets.
Bullets, Lubricating
Canister-shot.
Cap-boxes, Percussion
Carcasses.
Cartridge-filling machines.
Cartridge-holders
Cartridge-linings.
Cartridge-pouches.
Cartridge-priming machines.
Cartridges.
Cartridges, Blasting.
Cartridges, Compositions for. (*See Class* 52.)
Cartridges for fire-arms.
Cartridges for ordnance.
Cartridge-shells, Making. (*See Class* 53 *A.*)
Cartridges, Metallic
Cartridges, Priming
Cartridge-wrappers.
Compositions for coating projectiles.
Compositions for packing projectiles.
Concussion-fuses.
Explosive balls.
Explosive projectiles.
Flasks, Powder
Friction-primers.
Fuses, Blasting
Fuses, Concussion
Fuses, Firing
Fuses, Safety
Fuses, Tape
Fuses, Time
Grape-shot.
Grenades.
Gun-wads.
Hand-grenades.
Harpoons, Gun
Incendiary shells.
Lances, Bomb
Lubricators for projectiles.
Oil-wells, Blasting
Oil-wells, Torpedoes for
Ordnance-projectiles.
Packing projectiles.
Percussion-cap-filling machines.
Percussion-cap holders.
Percussion-cap-lining machines.
Percussion-caps.
Percussion-cap shells. (*See Class* 53 *A.*)
Percussion-cartridges.
Percussion-shells.
Powder-flask chargers.
Powder-flasks.
Primers.
Primers, Friction
Projectiles, Casting. (*See Class* 22.)
Projectiles, Compositions for coating and packing
Projectiles for fire-arms.
Projectiles for ordnance.
Rock-blasting.
Rocket-harpoons.
Rockets, War
Sabots for projectiles.
Safety-fuses.
Shell-fuses.
Shells.
Shells, Incendiary
Shot.
Shot-cartridges.
Shot-chargers.

Shot, Making. (*Pressed*, *Class* 53 *A.*)
Shot, Making. (*Cast*, *Class* 22.)
Shot-pouches.
Spherical-case shot.
Tape-fuses.
Tape-primers.
Time-fuses.
Torpedoes.
Torpedoes for oil-wells.
Torpedoes, Submarine
Wad punches. (*See Class* 24.)

# CLASS 103.

## PUMPS.

Water-elevators, Hydraulic Engines, Jacks, Meters, Presses and Rams; Water and Spirit Meters, Injectors, and Ejectors.

[*For Manufacture of Valves and Stop-cocks, see Class* 136; *for Manufacture of Water-pipe Fittings, see Class* 53, *Section K.*]

---

Beer-fountains.
Beer pumps.
Cellars, Draining.
Cistern-pumps.
Draining cellars.
Drawing liquids.
Ejectors, Liquid
Elevators, Water
Fire-engines, Hydraulic
Fountain-levers.
Fountains, Beer
Gas-meters. (*See Class* 98.)
Hydraulic buffers.
Hydraulic elevators.
Hydraulic engines.
Hydraulic jacks.
Hydraulic meters.
Hydraulic motors.
Hydraulic presses.
Hydraulic rams.
Injectors for steam-boilers.
Jacks, Hydraulic
Liquid-meters.
Meters, Liquid
Meters, Registering devices for. (*See Class* 73.)
Meters, Spirit
Meters, Water
Presses, Hydraulic
Pump-brakes.
Pump-motions.
Pump piston-rods.
Pump-pistons.
Pumps.
Pumps, Air. (*See Class* 98.)
Pumps, Cattle
Pumps, Centrifugal
Pumps, Chain
Pumps, Circular brake
Pumps, Double-acting
Pumps, Feed
Pumps, Force
Pumps, Hydraulic-press
Pumps, Lift
Pumps, Oil
Pumps, Oscillating
Pumps, Packing for
Pumps, Rotary
Pumps, Ships'
Pumps, Submerged
Pumps, Water
Pumps, Wind-propelled
Rams, Water
Registering devices for meters. (*See Class* 73.)
Spirit-meters.
Steam fire-engines. (*For Engines, see Class* 121.)
Steam-pumps. (*For Engines, see Class* 121.)
Stop-cocks. (*See Class* 136.)
Syringes.
Valves. (*See Class* 136.)
Water-elevators.
Water-rams.

# CLASS 104.

## RAILWAYS.—1. THE "WAY."

Elevated, Pneumatic, Portable, and Street Railways; Crossings, Frogs, Gates, Guards, Joints, Rails, Splices, Switches, &c.

[*For Manufacture of Railway Tracks, Irons, etc., see Class* 107 ; *Rolling Railway Rails, see Class* 80.]

---

Bumping-posts.
Car-replacers.
Cattle-guards.
Chairs, Railway
Crossings, Railway
Elevated railways.
Fish-bars, Railway
Frogs, Railway
Gates, Railway
Guards, Railway
Joints, Railway
Platforms, Railway
Pneumatic dispatch-tubes.
Pneumatic railways.
Portable railways.
Rails, Railway
Railway bumping-posts.
Railway cattle-guards.
Railway-chairs.
Railway-crossings.
Railway fish-bars.
Railway-frogs.
Railway-gates.
Railway-joints.
Railway-rails.
Railways.
Railway safety-guards.
Railways, Elevated
Railway-sleepers.
Railway splice-pieces.
Railways, Pneumatic
Railways, Portable
Railways, Street
Railways, Track-laying
Railway-switches.
Railway-ties.
Railway-track layers.
Railway-track platforms.
Railway turn-outs.
Railway turn-tables.
Rope-ways.
Safety-guards for railway-tracks.
Sleepers, Railway
Splice-pieces, Railway
Street railways.
Switches, Railway
Ties, Railway
Track-laying railways.
Track-raisers, Railway
Turn-out, Railway
Turn-tables, Railway
Wire-ways.

# CLASS 105.

## RAILWAYS, 2.

Cars, their Varieties and Interior Fittings.

[*Manufacture of Railway-car Irons, see Class* 107.]

Adhesion-cars.
Baskets, Car
Berths, Car
Box-cars.
Car-couches.
Car-roofs.
Cars, Adhesion
Cars, Box
Cars, Cattle
Cars, Coal
Cars, Dummy
Cars, Dumping
Car-seats.
Cars, Freight
Cars, Gravel
Cars, Hand
Cars, Petroleum
Cars, Revolving
Cars, Safety
Cars, Sleeping
Cars, Stock
Cars, Street
Cars, Tank
Cars, Wrecking
Car-windows.
Cattle-cars.
Coal-cars.
Couches for railway-cars.
Dummy-cars.
Dumping cars.
Freight-cars.
Grain-cars.
Gravel cars.
Hand-cars.
Mail-bag receiving and delivering apparatus.
Petroleum-cars.
Platform-cars.
Revolving-cars.
Safety-cars.
Seats for railway-cars.
Sleeping-cars.
Stock cars.
Street-cars.
Tank-cars.
Windows for railway-cars.
Wrecking-cars.

# CLASS 106.

## RAILWAYS, 3.

### Car-mountings and Exterior Fittings.

[*Manufacture of Railway-car Irons, see Class* 107.]

Axle-boxes for railway-cars.
Axles, Car
Boxés, Railway-axle
Brakes, Car
Buffers, Car
Bumpers, Car
Cabs for railway-cars.
Car-axle boxes.
Car-axles.
Car-brakes.
Car-brakes, Automatic
Car-brake shoes.
Car-brakes, Pneumatic. (*See Class* 121.)
Car-brakes, Steam-actuated. (*See Class* 121.)
Car-buffers.
Car-bumper springs.
Car-cabs.
Car-couplings.
Car draw-bars.
Car draw-heads.
Car-pedestals.
Car-platforms.
Car-springs.
Car stake-holders.
Car-standards.
Car-starters.
Car-truck frames.
Car-trucks.
Car-trusses.
Car-wheels.
Couplings, Car
Cow-catchers.
Draw-bars for railway-cars.
Platform-bridges for railway-cars.
Platforms for railway-cars.
Plows, Snow
Pneumatic car-brakes. (*See Class* 121.)
Safety-guards for car-platforms.
Safety-guards on cars.
Snow-plows.
Springs, Car
Steam actuated car-brakes. (*See Class* 121.)
Track-clearers.
Trucks, Car
Wheels, Car

# CLASS 107.

## RAILWAY-IRONS. MANUFACTURE OF

Railway Track and Car-irons, Tires, Wheels, and Iron Fittings.

---

Axle-boxes, Making railway-car
Axles, Making railway-car
Axle, Making locomotive
Car-wheels, Making wrought
Chairs, Making railway
Crossings for railways, Making
Fish-bars for railways, Making
Frogs for railways, Making
Locomotive-axles, Making
Locomotive-frames, Making
Railway-car irons, Making
Railway-chairs, Making
Railway-crossing, Making
Railway-frogs, Making
Railway metallic mountings, Making
Railway-rails, Making
Railway splice-pieces, Making
Railway-wheels, Making
Spikes, Railroad
Spikes, Screw
Splices for railways, Making
Wrought-metal railway-wheels.

# CLASS 108.

## ROOFING.

Materials, Compositions, and Varieties.

(*Except Girders and Trusses, Class* 14.)

---

Brackets for eave-troughs.
Brackets for roofs.
Eave-troughs.
Fenders for roofs.
Girders for roofs. (*See Class* 14.)
Gravel-heaters. (*See Class* 126.)
Gutters for roofs.
Metallic roofing.
Paper for roofs, Machines for preparing
Roofing-compositions.
Roofing processes.
Roofing-tiles.
Roofing-tiles, Manufacture of. (*See Class* 25.)
Roofs, (*except trusses, Class* 14.)
Roofs, Compositions and means for preserving
Shingles, Compositions and means for preserving
Shingling brackets.
Skylights, Devices for operating
Slating.
Spouting for eaves.
Tiling for roofs.
Trusses for roofs. (*See Class* 14.)
Zinc roofing.

# CLASS 109.

## SAFES.

Burglar, Fire, and Water-proof Safes and Vaults.

Burglar-proof safes.
Coin safes.
Filling for safes.
Fire-proof safes.
Lining for safes.
Portable safes.
Safe-locks. (*See Class* 70.)
Safes.
Safes, Lining for
Steam fire-proof safes.
Vaults, Safe
Water-proof safes.

# CLASS 110.

## SAWS.

Saws, Saw-mills, Sawing-machines.

[*For Saw-filing and Grinding* ...... ..*see Class* 51.
*Saw-Gummers, Sets, and Swages*..*see Class* 79.
*Saw-making* ....................*see Class* 79.

---

Annular-saws.
Band-saws.
Cant-hooks.
Chair-back-sawing machines.
Circular-saws.
Clapboard-sawing machines.
Cross-cut saws.
Cylindrical saws.
Diamond-saws. (*See Class* 125.)
Dogs for saw-mills.
Drag-saws.
Feed-devices for saw-mills.
Felling trees, Machines for
Gages for sawing-machines.
Gate-saws.
Hand-saws.
Head-blocks for saw-mills.
Hoops, Machines for sawing
Ice-saws.
Insertable saw-teeth.
Meat-saws.
Metal-saws.
Muley-saws.
Oars, Machines for sawing
Pruning-saws.
Rossing-machines.
Saw-bucks.
Saw-dust carriers.
Saw-filing machines. (*See Class* 51.)
Saw-frames.
Saw-gages.
Saw-gates.
Saw-grinding machines. (*See Class* 51.)
Saw-gummers. (*See Class* 79.)
Saw-horses.
Sawing-machines for chair-backs.
Sawing-machines for clapboards.
Sawing-machines for combs.
Sawing-machines for hoops.
Sawing-machines for laths.
Sawing-machines for match-splints.
Sawing-machines for oars.
Sawing-machines for shingles.
Sawing-machines for splints.
Sawing-machines for staves.
Sawing-machines for veneers.
Saw jacks.
Saw-making. (*See Class* 79.)
Saw-mills.
Saw mills, Circular
Saw-mills, Dogs for
Saw-mills, Feed for
Saw-mills, Gates for
Saw-mills, Head-blocks for
Saws.
Saws, Amputation
Saws, Annular
Saws, Band
Saws, Bow
Saws, Circular
Saws, Cross-cut
Saws, Cylindrical
Saws, Diamond. (*See Class* 125.)
Saws, Drag
Saw-sets. (*See Class* 79.)
Saws, Gate
Saws, Hand
Saws, Hanging
Saws, Ice
Saws, Key-hole
Saws, Meat
Saws, Metal
Saws, Pruning
Saws, Scroll
Saws, Stone. (*See Class* 125.)
Saws, Straining
Saws, Tenon
Saws, Web
Saws, Whip
Saw-teeth.
Scroll-saws.
Spurs for saw-mills.
Stone-saws. (*See Class* 125.)
Teeth, Insertable saw
Tenon-saws.
Veneer-saws.
Web-saws.
Whip-saws.

# CLASS 111.

## SEEDERS AND PLANTERS.

Machines for Sowing and Planting Seeds and Distributing Fertilizers.

[*For Manufacture of Agricultural Tools of Metal, see Class* 81, *Section B.*]

---

Broad-cast sowers.
Cane Planters.
Cotton-seed for planting, Preparing
Dibblers, Hand
Dibbling-machines.
Drag-bars for drill-teeth.
Drill-rollers.
Drills, Grain
Drill-teeth.
Drop-drills.
Fertilizer-sowers.
Grain-drills.
Hand-drills for grain.
Hand-planters.
Land-markers.
Lime-spreaders.
Manure carts, Distributing
Manure-drills.
Manure drills, Liquid
Manure-screens.
Manure-sowers, Broad-cast
Manure-spreaders.
Manuring, Green
Markers for seeding-machines.
Planters, Corn
Planters, Cotton-seed
Planters, Hand
Planters, Seed
Planters, Tobacco-seed
Seed-drills.
Seed-droppers.
Seed for planting and sowing, Preparing
Seed-planters.
Sowers, Broad-cast
Sowers, Fertilizer
Sowers, Seed

# CLASS 112.

## SEWING-MACHINES.

Machines for Sewing, Stitching, Embroidery, Button-holing; Attachments and Appurtenances of Sewing-machines.

---

Binders for sewing-machines.
Bobbins for sewing-machines.
Bobbin-winders for sewing-machines.
Box-plaiting attachments for sewing-machines.
Braiders for sewing-machines.
Button-hole sewing-machines.
Clutch-pulleys for sewing-machines.
Corders for sewing-machines.
Creasers for sewing-machines.
Feed-devices for sewing-machines.
Fellers for sewing-machines.
Embroidery-attachments for sewing-machines.
Fringing attachments for sewing-machines.
Gages for sewing-machines.
Gatherers for sewing-machines.
Guides for sewing-machines.
Hemmers for sewing-machines.
Leather-quilting machines.
Markers for sewing-machines.
Needle-setters for sewing-machines.
Needle threaders for sewing-machines.
Needles for sewing-machines.
Plaiters for sewing-machines.
Presser-feet for sewing-machines.
Rufflers for sewing-machines.
Sewing-machines.
Sewing-machines for boots and shoes.
Sewing-machines for embroidery.
Sewing-machines for making button-holes.
Sewing-machines for sewing straw goods.
Shuttle-drivers for sewing-machines.
Shuttles for sewing-machines.
Stitches for sewing-machines.
Tape-trimming, Machines for making
Tension-devices for sewing-machines.
Thread-controllers for sewing-machines.
Thread-cutters for sewing-machines.
Thread-waxers for sewing-machines.
Tuckers for sewing-machines.
Wax-thread heaters for sewing-machines.

# CLASS 113.

## SHEET-METAL.

Modes, Machines, and Tools for the Manufacture of Sheet-metal Ware and other articles of Sheet-metal.

---

### A.—*Tools, Machines, and Processes.*

Bending sheet-metal.
Brazing.
Crimping sheet-metal.
Dies for sheet-metal ware.
Double-seaming machines for sheet-metal.
Embossing sheet-metal.
Folding sheet-metal.
Grooving sheet-metal.
Lead-cutters for printers.
Mitering printers' rules.
Ornaments for sheet-metal ware, Die
Printers' rules, Cutting and mitering
Punching sheet-metal.
Roof-seaming machines.
Rule-cutters for printers.
Screw forming, Sheet-metal-
Seaming-machines, Sheet-metal
Sheet-metal, Machines for working
Sheet-metal, Processes for working
Sheet-metal table-ware, Making
Sheet-metal, Tools for working
Sheet-metal ware, Making
Solder.
Soldering-furnaces.
Soldering-implements.
Soldering-processes.
Soldering-tools.
Spinning sheet-metal ware.
Stamps for sheet-metal working.
Swaging sheet-metal.
Tin-plate working.
Tin-scrap, Utilizing
Tools for working sheet-metal.
Wiring sheet-metal ware.

### B.—*Products.*

Barrels, Making sheet-metal
Boilers, Making sheet-metal domestic
Boxes, Making sheet-metal
Britannia-ware.
Buckets, Making sheet-metal
Builders' sheet-metal ware.
Cans, Making sheet-metal
Candlesticks, Making sheet-metal
Capsules for cans and bottles, Making
Cisterns, Making sheet-metal
Glaziers' points, Making
Gutters, Making sheet-metal
Gutter-tubes, Making sheet-metal
Heaters for stoves and ranges, Making sheet-metal
Kitchen furniture, Making sheet-metal
Screw-caps, Making sheet-metal
Screw-rings, Making sheet-metal
Sheet-metal barrels and casks, Making
Sheet-metal boilers and cisterns, Making
Sheet-metal boxes and cans, Making
Sheet-metal buckets and pans, Making
Sheet-metal capsules and rings, Making
Sheet-metal gutters and spouting, Making
Sheet-metal kitchen and table ware, Making
Sheet-metal screw caps and rings, Making
Spoons, Making
Spouting, Making
Stove-pipes, Making
Stove-furniture, Making sheet-metal
Table-furniture, Making sheet-metal
Tin cans and boxes, Making

# CLASS 114.

## SHIPS, 1.

Construction, Masting, Rigging, Fittings.

Anchor-balls.
Anchors.
Anchors, Drift
Anchor-shackles.
Anchor-trippers.
Architecture, Naval
Armor-plating.
Ballasting.
Batteries, Floating
Batteries, Submarine
Bending sails.
Berths, Floating
Berths, Ships'
Bilge-boards.
Bilge-water, Discharging
Binnacles.
Bitts, Riding and towing
Blocks, Bushings for
Blocks, Snatch
Blocks, Tackle
Boats, Canal
Boats, Ferry
Boats, Ice-breaking
Boats, Steam
Boats, Torpedo
Booms.
Bridges, Deck
Bridges, Paddle-box
Bulkheads.
Cabins, Floating
Cable-grippers.
Cable-hooks.
Cable-nippers.
Cables for vessels.
Cables, Links for
Cable-shackles.
Cables, Securing
Cable-stoppers.
Calking.
Calking-machines.
Camels.
Canal-boats.
Cargo-ports.
Center-boards.
Chimneys, Operating ships'
Clasps, Rigging
Cleats.
Construction of vessels.
Cringles.
Dead-eyes.
Dead-eyes for canal-boats.
Dead-lights.
Deck-bridges.
Deck-covers.
Deck-irons.
Decks.
Decks, Fire-proof
Decks, Water-proof
Docks, Floating
Docks, Graving
Draught of water, Ascertaining
Ferry-boat locks.
Ferry-boats.
Fids.
Fire-extinguishers on ships.
Floating-batteries.
Floating-berths.
Floating-bridges.
Floating-cabins.
Floating-wharves.
Floating-docks.
Furling sails.
Gaff-jaws.
Gaffs.
Galleys, Ships'
Grain-levelers in vessels.
Grapnels.
Grappling-irons.
Graving-docks.
Grommets.
Hanks for sails.
Hawse-pipes.
Hawser-clamps.
Ice-boats.
Ice-breakers.
Iron blocks, Nautical
Iron-clad vessels.
Iron frames, Ships'
Iron ribs, Ships'
Iron vessels.
Jaws for gaffs.
Keels.
Keelsons.
Knee-formers.

Knees.
Lazy-jacks for sails.
Leak-stoppers.
Lighters.
Lights, Deck
Light-ships.
Lights, Side
Lining for ships.
Locking apparatus for ferry-boats.
Logs, Ships'. (*See Class* 73.)
Mast-hoops.
Masts.
Monitors.
Monitor-turrets.
Mud-boats.
Naval architecture.
Paddle-box bridges.
Parrel-blocks.
Parrels.
Paying-seams.
Pilot-stands.
Pipes, Chain-locker
Port-hole closers.
Port-holes, Submarine
Propulsion of vessels. (*See Class* 115.)
Protecting bulwarks.
Protecting tiller-ropes.
Puttying-machines for vessel seams.
Railways, Marine
Rams, Naval
Reefing apparatus.
Reefing sails.
Ribs, Iron
Riding-bitts.
Rigging.
Rigging-clasps.
Rigging, Running
Rigging, Setting up
Rigging, Standing
Rigging-stoppers.
Rudders.
Rudders, Attachments for
Rudders, Hanging
Rudders, Protecting
Sail-clutches.
Sail-hoops.
Sails.
Sails, Furling
Sails, Making
Sails, Reefing
Scuppers.
Securing cables and hawsers.
Securing hatches.
Shaping ships' timbers.
Sheathing vessels.
Ships' armor.
Ships' berths.
Ships' knees.
Ships, Propulsion of
Shot-hole stoppers.
Snag-boats.
Snubbers.
Sounding apparatus. (*See Class* 73.)
Sparring steamboats.
Spars, Handling and hoisting
Steering apparatus.
Submarine vessels.
Supporting vessels.
Surge-relievers.
Tables, Ships' cabin
Tackle, Hoisting
Tackle, Safety
Thimble-machines for vessels.
Thimbles, Nautical
Tightening standing-rigging.
Torpedo vessels.
Tow-hooks.
Tow-lines.
Tracking vessels.
Train-ways for boats. (*See Class* 71.)
Travelers for masts and booms.
Treenails.
Trimming vessels.
Trusses for vessels.
Truss-hoops.
Vessels.
Vessels, Armor for
Vessels, Construction of
Vessels, Grain-levelers for
Vessels, Iron
Vessels, Iron-clad
Vessels, Launching
Vessels, Loading and unloading. (*See Class* 57.)
Vessels, Propulsion of. (*See Class* 115.)
Vessels, Steam
Vessels, Submarine
Vessels, Torpedo
Vessels, Trimming
Warping-bitts.
Warping-checks.
Wharf-boats.
Wharfs, Floating
Yards to masts, Attaching

# CLASS 115.

## SHIPS, 2.

Propulsion of Vessels.

Canal-boats, Propulsion of
Hydraulic-propellers.
Paddle-wheels.
Propellers.
Propellers, Screw
Propelling in shoal water.
Screw-propellers.
Ships, Propulsion of
Vessels, Propulsion of

# CLASS 116.

## SIGNALS.

Audible and Visual Signals and Alarms, except Electrical.

Bell-hanging and Telegraphs other than Electrical.

Alarms, Bilgewater
Alarms, Burglar
Alarms, Fire
Alarms, Fog
Alarms, Household fire
Alarms, Leak
Alarms, Nautical
Alarms, Shoal
Alarms, Temperature
Alarms, Till
Alarms, Window
Annunciators.
Beacons.
Beacons, Sonorific
Bell-hanging.
Bell-ringing.
Bells.
Bells, Clock
Bilgewater-alarms.
Bridge-signals.
Buglar-alarms.
Call-bells.
Clock-bells.
Crossing-signals, Railway
Door-alarms.
Door-bells.
Draw-bridge signals.
Fire-alarms, Household
Fog-alarms.
Fog-bells.
Gongs.
Hotel-annunciators.
Lanterns, Signal
Leak-signals.
Marine-signals.
Nautical-alarms.
Railroad-signals.
Shoal-alarms.
Signaling apparatus.
Signal-bells.
Signal-codes.
Signal-lanterns.
Signals, Fog
Signals, Marine
Signals, Pyrotechnic
Signals, Railroad
Signals, Semaphoric
Signal-towers.
Speaking-trumpets.
Speaking-tubes.
Switch-indicators.
Switch-signals.
Telegraphs, Acoustic
Telegraphs, Optical
Telegraphs, Pneumatic
Telegraphs, Semaphoric
Temperature-alarms.
Window-alarms.

# CLASS 117.

## SILK.

Special appliances of Silk Filature and Working.

Cleaning silk.
Clearing silk.
Cocooneries.
Cocoon-lodgments.
Cocoons, Unwinding
Doubling silk.
Gaging silk.
Glossing silk.
Hurdles for silk winders.
Pressing silk.
Reeling silk.
Rollers for delivering silk.
Silk filature.
Silk manufacture.
Silk-worm frames.
Sizing silk.
Sorting silk.
Spinning silk.
Stretching silk.
Throwing silk.
Twisting silk.
Winding silk.

# CLASS 118.

## SPINNING.

[*For Manufacture of Spinning Hardware, see Class 53, Section H.*]

Banding-spindles.
Bobbin-winders.
Bobbins.
Bolsters for spindles.
Cap-spinning machines.
Cop-tubes.
Cop-tubes of paper, Machines for making
Covering rollers.
Doubling-machines.
Drawing and twisting heads and tubes.
Drawing-machines.
Drawing-rollers.
Drawing-rollers, Cleaners for
Flyers.
Fly-frames.
Footsteps for spindles.
Jack-spinning.
Mottled yarns.
Mottled yarns, Machines for making.
Mules.
Pasting cops.
Presser-flyers.
Reels for spinning-machines.
Rings for spinning.
Ring-travelers.
Rollers, Covering
Roving-frames.
Slubbing-frames.
Spindles for spinning-machines.
Spindles, Banding
Speeders.
Spindle-steps.
Spinning-machinery.
Spinning-wheels.
Stop-motions for spinning-machines.
Swifts.
Throstles.
Wheel-heads.

# CLASS 119.

## STABLING.

Care of Domestic Animals; Shelters, Stalls, Preparation of Food, Feeding, Currying.

---

Agricultural furnaces and boilers. (*See Class* 126.)
Animals, Preparing food for
Cards, Currying
Cattle-feeders.
Cattle pens.
Cattle-stalls.
Cattle-ties.
Chicken-coops.
Chicken-houses.
Chicken-nests.
Chicken-raising apparatus.
Chicken-roosts
Corn-huskers and shellers. (*See Class* 130.)
Corn-mills. (*See Class* 83.)
Currycombs.
Currying-tools.
Cutters, Feed
Cutters, Root
Cutters, Straw, hay, &c.
Egg-assorters.
Egg-carriers.
Egg-hatching apparatus.
Egg-packers.
Feed-mixers.
Feed-racks.
Feed-troughs.
Fleece-tyers.
Food for animals, Preparing
Foot-lifters for horses.
Forage-rations.
Fowls, Coops, nests, and roosts for
Grooming apparatus.
Hay-racks for feeding.
Hen's-nests.
Hitching devices.
Hitching-posts.
Hog-pens.
Hog-troughs.
Horse-washers.
Incubators.
Mangers.
Nests for fowls.
Nicking-horses, Apparatus for
Pens, Cattle and hog
Poultry-coops.
Poultry-feeder, and houses.
Poultry-houses.
Poultry-nests.
Racks, Hay-feeding
Racks, Sheep
Ration-packages for animals.
Roosts for fowls.
Root-cutters.
Root-washers.
Sheep-holders.
Sheep-racks.
Sheep-shearing machines.
Sheep-shearing tables.
Sheep-shears. (*See Class* 30.)
Sheep-sheds.
Sheep-taggers.
Sheep-troughs.
Silk-worm culture.
Singeing tools.
Stable-cleaners.
Stables.
Stalls for horses and cattle.
Stanchions.
Stock-feeders.
Stocks for refractory animals.
Stocks for shoeing animals.
Straw-cutters.
Troughs, Feed
Vegetable-cutters.
Vegetable-slicers.
Vegetable-washers.
Ventilating hay and grain.

# CLASS 120.

## STATIONERY.

Writing Materials and Appliances; Desk and Office Supplies.

Address, Canceling, Numbering, and Seal Stamps.

Cards, Checks, Labels, Seals, Tickets, Advertising Appliances and Signs.

Mechanical Memoranda and Non-automatic Indicators.

[*For the Manufacture of Metallic Articles of Stationery, see Class* 53, *Section I.*]

---

Adhesive fastenings for paper, etc.
Advertising cards.
Advertising devices.
Advertising frames.
Animal labels.
Baggage-checks.
Ballot-boxes.
Bill-holders.
Binders, Temporary
Blotters, Book
Blotters.
Blotting-pads.
Book-binders, Temporary
Book-clamps.
Book-cover protectors.
Book-holders.
Book-holders, Portable
Book-markers.
Books, Copying
Books, Manifold-writing
Book-supports.
Bottles, Mucilage
Boxes, Ballot
Boxes, Conductor's check
Boxes, Fare
Boxes, Letter
Boxes, Mail-distributing
Boxes, Ticket
Bulletin-boards.
Bulletin-boards, Adjustable
Calendars.
Card-calendars.
Card-cases.
Card-holders.
Card-racks.
Cards.
Checks, Baggage
Checks, Conductors'
Conductors' check-boxes.
Clips, Paper
Copy-books.
Copy-holders.
Copying-books.
Currency-holders.
Desk paper-cutters.
Distributing boxes, Mail
Door-indicators.
Door-plates.
Envelope-openers.
Envelope-fasteners.
Erasers.
Fare-boxes.
Fasteners, Paper. (*See Class* 24.)
Filers, Paper
Frames, Advertising
Illuminated signs.
Indelible pencils.
Indicators, Door
Indicators, Station
Indicators, Street
Indicators, Time, (*non-automatic.*)
Inkstands.
Label-fasteners.
Label-holders.
Labels.
Labels, Adhesive
Letter-boxes.
Letter-clips.
Letter-files.
Letter-openers.
Letter-receiving boxes.
Mail-distributing boxes.
Mark-holders for bales.
Mechanical memoranda.
Memorandum-books.
Mucilage-bottles.
Mucilage-holders.
Newspaper-clamps.

Newspaper-covers.
Office time-indicators, (*non-automatic.*)
Paper-clamps.
Paper-clips.
Paper-fasteners. (*See Class* 24.)
Paper-files.
Pen-cases.
Pencil-cases.
Pencil-cleaners.
Pencil-mark erasers.
Pencils.
Pencils, Compositions for indelible
Pencil-sharpeners.
Pen-holders.
Pen makers, Quill
Pen-racks.
Pens.
Pens, Fountain
Pens, Rubber
Pen-trays.
Photographic albums.
Pigeon-holes, Desk
Plant-labels.
Pocket-book fastenings.
Pocket-books.
Porte-crayons.
Portemonnaies.
Portfolios.
Postal cards.
Postage-stamp holders.
Post-office boxes.
Quill-pen makers.
Railway-ticket holders.
Record-books.
Register-books.
Register books, Hotel
Ribbons for hand-stamps, Preparing
Roll-blotters.
Rubber erasers.
Rulers, Desk
Scrap-books.
Show-cards.
Signs.
Signs, Illuminating
Sponge cups.
Stamp-holders.
Stationery, Articles of
Station-indicators, (*non-automatic.*)
Street-indicators.
Tablets, Compositions for
Tablets, Writing
Tag-holders.
Tags.
Temporary binders.
Ticket-boxes.
Ticket-holders.
Ticket-racks.
Tickets, Railway
Time-indicators, (*non-automatic.*)
Time-tables.
Trunk-labels.
Visiting-cards.
Writing-cases.

# CLASS 121.

## STEAM, 1.

Engines, their Parts and Applications.

[*Except Valves, Class* 136.]

Aëro-steam engines.
Air brakes for car, Steam-actuated
Cataracts for steam-engines.
Cocks. (*See Class* 136.)
Cross-heads of steam-engines.
Cut-off apparatus.
Cut-off, Variable
Cylinders, Steam
Engines for operating rock-drills.
Engines for operating steam-hammers.
Engines for operating steam-pumps.
Engines, Marine steam
Engines, Oscillating
Engines, Reciprocating
Engines, Rotary
Engines, Stationary
Engines, Steam
Engines, Steam fire
Engines, Valve-gear for
Expansion joints for steam-engines.
Fire-engines, Steam
Gaskets for steam-joints.
Hammers, Steam-engines for
Lubricators for pistons.
Lubricators, Steam-operated
Oscillating steam-engines.
Packing for pistons.
Packing for steam-joints.
Piston-packing.
Piston-rod packing.
Piston-rods.
Pistons.
Pneumatic car-brakes.
Pumps, Steam
Reciprocating steam-engines.
Reversing gear for steam-engines.
Rock-drills, Steam-engines for
Rotary steam-engines.
Stationary steam-engines.
Steam-actuated car-brakes.
Steam chests of engines.
Steam-cylinder heads.
Steam-cylinders.
Steam-engines.
Steam-engines for car-brakes.
Steam-engines for hammers.
Steam-engines for hoisting.
Steam-engines for pumps.
Steam-engines, Lubricators for
Steam-engines, Valve-gear for
Steam fire-engines.
Steam hammers. (*See Class* 78.)
Steam-packing
Steam-piston packing.
Steam-pistons.
Steam-pumps.
Steam-traps.
Steam-valves. (*See Class* 136.)
Stop motions for steam-engines.
Stuffing-boxes.
Stuffing-box packing.
Traps, Steam
Trunnions for steam-engines.
Valve-gear for steam-engines.
Valves. (*See Class* 136.)

# CLASS 122.

## STEAM, 2.

Steam-Boilers, Super-heaters, Boiler-Setting, and Furnaces.

*Except Registering Devices of Water, Vacuum and Steam-pressure Gages.*]

---

Boiler covering.
Boiler-feeders.
Boiler-flue cleaners.
Boiler-flues.
Boiler-setting, Steam
Boilers, Furnace-doors for
Boilers of steam-heaters.
Boilers, Furnaces for
Boilers, Steam
Boilers, Setting steam
Boiler-tile arrangements.
Boiler-tubes, Devices for cleaning
Boiler-tubes, Ferrules for
Boiler-tubes, Tools for scaling
Bridges, Steam-furnace
Cleading for boilers.
Cleading for cylinders.
Collapse of boilers, Preventing
Condensers, Jet
Condensers of steam-engines.
Condensers, Packing-tubes for
Condensers, Surface
Condensers, Tubular
Covering for steam-boilers.
Doors for steam-boiler furnaces.
Draft-regulators for steam-boiler furnaces.
Feeders, Steam-boiler
Feed-water apparatus.
Feed-water heaters for steam-boilers.
Feed-water regulators.
Fire-bars.
Fire-extinguishers for boilers.
Fire-regulators for steam-boilers.
Floats for steam-boilers.
Flue-boilers.
Flue-cleaners.
Flue-scrapers.
Flues, Boiler
Foam-collectors for steam boilers.
Fuel, Apparatus for burning wet
Fuel-feeders for steam-boiler furnaces.
Furnace-doors for steam-boiler furnaces.
Furnace-grates for steam-boiler furnaces.
Furnaces, Steam-boiler
Fusible plugs for steam-boilers.
Gage-cocks. (*See Class* 136.)
Gages for steam-boilers, (*except registering devices.*)
Generators, Steam
Grate-bars for steam-boilers.
Hand-holes for boilers and tanks.
Heaters, Boilers of steam
High pressure alarms for steam-boilers.
Hydro-carbon furnaces for steam-boilers.
Incrustation in steam-boilers, Preventing
Incrustation in steam-boilers, Removing
Injectors for steam-boilers.
Low-water alarms for steam-boilers.
Low-water detectors for steam-boilers.
Low-water indicators for steam-boilers.
Manholes for boilers and tanks.
Marine steam-boilers.
Mud-collectors for steam-boilers.
Pipes, Non-conducting covering for steam
Regulators, Steam
Safety-plugs for steam-boilers.
Safety-tubes for steam-boilers.
Scaling boilers, Tools for
Sediment collectors for steam-boilers.
Skimmers for steam-boilers.
Smoke-pipes for boiler-furnaces.
Spark-arresters.
Spark-consumers.
Steam-blast for furnaces.
Steam-boiler-covering.
Steam-boiler furnaces.
Steam-boilers.
Steam-cocks. (*See Class* 136.)

Steam-generators.
Steam-heaters, Boilers of
Steam in furnaces, Use of
Steaming apparatus, Boilers of
Steam-pipes.
Steam-pipes, Non-conducting coving for
Steam-pressure gages.
Steam-pressure regulators.
Steam-pressure.
Steam-regulators
Steam-whistles.
Super-heaters for steam-boilers.
Tools for scaling boilers.
Tube-cleaners for steam-boilers.
Tube-plugs for steam-boilers.
Tubes for steam-boilers.
Tube-sheets.
Tubular boilers.
Vacuum-gages for steam-engines.
Water-gages for steam-boilers.
Water-heaters for steam-boilers.
Water-indicators for steam-boilers.
Water pumps, Feed
Water-purifiers for steam-boilers.
Whistles for steam-boilers.

# CLASS 123.

## STEAM, 3.

Locomotives, Traction Engines, Portable Engines, and their special parts.

Agricultural engines.
Ash-pans of locomotives.
Blast-nozzles for steam-engines.
Boilers, Locomotive
Cow catchers.
Dummies.
Engines, Agricultural
Engines, Dummy
Engines, Locomotive
Engines, Portable
Engines, Steam-plow
Engines, Traction
Escape-pipes for steam-engines.
Exhaust mechanism for steam-engines.
Exhaust-nozzles for steam-engines.
Exhaust-pipes for steam-engines.
Furnaces, Locomotive
Link-motion for steam-engines.
Locomotive-boilers.
Locomotive-engines.
Locomotive-furnaces.
Locomotive-pumps.
Locomotives.
Locomotives, Fire boxes for
Locomotives, Furnace-doors for
Locomotive smoke-stacks.
Locomotives, Reversing-gear for
Locomotive valve-gear.
Nozzles, Variable exhaust
Portable-engines.
Regulators, Exhaust
Tanks for tenders.
Tenders.
Traction-engines.
Variable exhaust for steam-engines.
Valve-motions for locomotives.

# CLASS 124.

## STILLS.

Distillation and Refining of Spirits; Oils and Acids; Burning-fluids.

Acids, Distillation of
Alcohol, Apparatus for making
Alcohol, Materials for making
Burning-fluids.
Condensers of stills.
Deodorizing benzole.
Distillation.
Fire tests for hydro-carbons.
Grain for distillation, Preparing
Hydro-carbons, Manufacture of
Oil-refining.
Oil-stills.
Petroleum, Apparatus for treating
Petroleum stills.
Rectifying liquors.
Refining liquors.
Refining oils.
Revenue guards.
Spirits, Distillation of
Spirits, Rectification of
Stills.
Sulphuric acid, Recovering
Tar, Extracting
Turpentine-stills.

# CLASS 125.

## STONE, LIME, AND CEMENT.

Mining; Quarrying; Boring Rock; Stone, Marble, and Slate Working; Lime, Mortar, Concrete, and Cement.

[*Excepting Mortar and Concrete Mills and Mixers..see Class* 25.
*Marble, Slate, and Stone Polishing .....see Class* 51.
*Kilns ..............................see Class* 34.

---

Artificial marble.
Artificial slate.
Artificial stone, (*except Grindstones, Class* 51.)
Béton.
Blasting. (*See Class* 102.)
Boring rock.
Burning lime.
Carving stone.
Cement, Fire-proof
Cement floors and paving. (*See Class* 94.)
Cement for lining pipes.
Cement, Indurating.
Cement-kilns.
Cement, Plastic.
Cements, Earthen
Channeling stone.
Coal-mining.
Coal-mining buckets..
Concretes.
Cutting slate and stone.
Diamond-drills.
Diamond-planers.
Diamond-saws.
Dressing slate and stone.
Drill-rod grabs.
Drills, Diamond
Drills, Expanding rock
Drills for wells.
Drills, Rock
Drills, Stone
Facing slate and stone.
Finishing stone.
Furrowing stone.
Grabs, Drill-rod
Grinding stone. (*See Class* 51.)
Grindstones. (*Sèe Class* 51.)
Hammers, Millstone
Hewing stone.
Hydraulic cement.
Lewis for lifting stone.
Lime-furnaces.
Lime-kilns. (*See Class* 24.)
Malaxators. (*See Class* 25.)
Marble, Artificial
Marble-polishing machines. (*See Class* 51.)
Marble-working machines.
Millstone bush-hammers.
Millstone-hammers.
Millstone-picks.
Millstones.
Millstone-scrapers.
Mining.
Mortar.
Mortar-machines. (*See Class* 25.)
Ornaments, Compositions for plaster
Paving. (*See Class* 94.)
Paving-stone, Preparation of
Picks, Millstone
Pipe, Cement
Planing stone.
Polishing slate and stone. (*See Class* 51.)
Quarrying.
Quarrying-machines.
Rock-drills.
Sand-cleaning machines.
Sawing stone.
Saws, Diamond
Saws, Stone
Scagliola.
Scouring stone. (*See Class* 51.)
Scrapers, Millstone
Slate, Artificial
Slate working machines.
Splitting slate.
Staining-compositions for marble and stone.
Stone and cement, Indurating

Stone, Artificial
Stone, Channeling
Stone, Compositions for porous
Stone-dressing.
Stone-drills.
Stone, Indurating
Stone-polishing. (*See Class* 51.)
Stone-saws.
Stone-staves for millers.
Stone-tools for working.
Stone-working machines.
Tunneling, Rock
Turning stone.

# CLASS 126.

## STOVES AND FURNACES.

Cooking and Heating Stoves, Furnaces and Ranges, Cooking Utensils, and Stove Appliances.

---

Agricultural boilers.
Agricultural furnaces.
Andirons.
Ash-pans.
Ash-pit covers.
Bake-pans.
Bakers' ovens.
Bakers' ovens, Portable
Base-burning fire-place heaters.
Base-burning stoves.
Bath-heaters.
Boilers, Agricultural
Boilers, Domestic
Boilers for heaters. (*See Class* 122.)
Broilers.
Car-heaters.
Carriage-heaters.
Casters, Stove-leg
Chafing-dishes.
Charcoal-furnaces for cooking.
Chimney-caps.
Chimney-cleaners.
Chimney-cowls.
Chimneys.
Coal-boxes.
Coal-buckets.
Coal-hods.
Coal-scuttles.
Coffee-percolators.
Coffee-pots.
Coffee-roasters.
Coffee-urns.
Cooking by gas.
Cooking by lamps.
Cooking by lime.
Cooking by steam.
Cooking-ranges.
Cooking-stoves.
Cooking-stoves, Base-burning
Cooking-stove shields.
Corn-poppers.
Cowls.
Cressets.
Crickets for stoves, Heating
Dampers for furnaces.
Dampers for stoves.
Digesters for bones, fats, etc.
Digesters for cooking.
Dish-heaters.
Domestic furnaces.
Draught-regulators.
Egg-boilers.
Fire-backs.
Fire-dogs.
Fire-fenders.
Fire-grates.
Fire-irons.
Fire-kindlers.
Fire-lighters.
Fire-place heaters, Base-burning
Fire-places.
Fire-plates for stoves.
Fire-pokers.
Fire-pots for stoves.
Fire-shovels.
Fire-tongs.
Flat-iron heaters.
Flue-cleaners.
Foot-stoves.
Foot-warmers.
Frying-pans.
Furnaces, Bath
Furnaces, Charcoal
Furnaces, Domestic
Furnaces, Heating
Furnaces, Hot-air
Furnaces, Portable
Gas-heaters.
Gas-stoves.
Glue pots.
Grates for stoves.
Grate-tippers.
Gravel-heaters.
Griddles.
Gridirons.
Handles for tea and coffee pots.
Heaters.
Heaters, Boilers of steam. (*See Class* 122.)
Heaters, Hot-water
Heaters, Railway-car
Heaters, Steam

Heaters, Street-car
Heating-furnaces.
Heating-irons.
Heating skates, boots, &c.
Heat-regulators.
Hot-air furnaces.
Hot-air furnace-doors.
Hot closets for stoves.
Hot-water heaters.
Illuminators, Stove
Kindling arrangements for stoves.
Kindling baskets.
Lard tanks for rendering.
Lid-lifters, Stove
Lifters for kettles, ovens, &c.
Lifters for stove-lids.
Linings for stoves.
Lunch-heaters.
Oils, Apparatus for boiling
Ovens, Bakers'
Ovens, Portable
Ovens, Reel
Ovens, Rotary
Ovens, Stove
Paint-burners.
Paste-heating pots.
Pea-nut roasters.
Petroleum-burners for stoves.
Pitch-heaters.
Pokers, Fire
Pot-covers.
Radiators.
Railway-car heaters.
Ranges.
Registers, Hot-air
Regulators, Draught
Regulators, Heat
Rendering tanks.
Roasters.
Roller heaters.
Sad-iron heaters.
Shovels and tongs.
Skillets.
Smoke-burners.
Smoke-jacks.
Spiders.
Spits for roasting meat.
Steam carving-tables.
Steam-cooking apparatus.
Steam-heaters.
Steam-heating tables.
Stove-aprons.
Stove-boilers.
Stove-cover holders.
Stove-covers.
Stove-door handles.
Stove-door linings.
Stove-doors.
Stove-drums.
Stove-fasteners.
Stove-flues.
Stove-illuminators.
Stove-irons.
Stove-leg casters.
Stove-legs.
Stove-lid holders.
Stove-lids.
Stove-linings.
Stove-pipe cleaners.
Stove-pipe drums.
Stove-pipe thimbles.
Stove-pipes.
Stove-plates.
Stoves.
Stoves, Base-burning
Stoves, Camp
Stoves, Coal
Stoves, Coal-oil
Stoves, Cooking
Stoves, Foot
Stoves, Gas
Stoves, Heating
Stoves, Lamp
Stoves, Soapstone
Stoves, Wood
Stoves, Utensils
Stoves, Window
Street-car heaters.
Tea-kettles.
Tea-pots.
Toasters.
Urns for coffee, &c.
Urn stands and heaters.
Vegetable-boilers.
Waffle-irons.
Water-backs for ranges and stoves.

# CLASS 127.

## SUGAR.

Evaporation, Refining, and Chemical Processes; Salt Furnaces and Pans; Bone-black; Candy.

---

Animal-black.
Beet-root sugar apparatus.
Black, Animal
Black, Bone
Black, Ivory
Bone-black filters.
Bone-black furnaces.
Bone-black, Processes for reviving
Bone-black, Purifying
Candy-cutters.
Candy-making machines.
Cane-juice, Treatment of
Centrifugal filters.
Confectionery.
Confection-pans.
Defecators for juices and sirups.
Evaporators for saccharine juices.
Evaporators for salines.
Evaporators for vegetable extracts.
Furnaces, Bone-black
Furnaces, Salt
Furnaces, Sugar
Glucose, Manufacture of
Grape-sugar, Manufacture of
Grain, Manufacture of sugar from
Lozenge-making machines.
Maple-sugar, Manufacture of
Mills, Cane. (*See Class* 83.)
Mills, Sorghum. (*See Class* 83.)
Mills, Sugar. (*See Class* 83.)
Molasses, Treatment of
Pans for cooking in vacuo.
Pans for making vegetable extracts.
Pans, Sugar
Pans, Vacuum
Salt-furnaces.
Sugar-furnaces.
Sirup, Treatment of
Skimmers for sugar-pans.
Sorghum-evaporators.
Sorghum manufacture.
Sugar-breakers.
Sugar-crushers.
Sugar-filters.
Sugar from grain, Obtaining
Sugar-furnaces.
Sugar-making processes.
Sugar-mills. (*See Class* 83.)
Sugar-mixing machine.
Sugar-packers. (*See Class* 83.)
Sugar-pans.
Sugar-pan skimmers.
Sugar-strainers.
Sugar-refining.
Vacuum-pans.

# CLASS 128.

## SURGERY.

Including Pharmaceutical Apparatus, Appliances for the Compounding and Administration of Medicine.

[*Except Artificial Limbs and other Prosthetic parts* .. *Class* 3.
*Dental Appliances* .............................. *Class* 32.
*Surgical Cutlery* ............................... *Class* 30.]

---

Acoustic instruments.
Acupuncturators.
Anæsthesia, Instruments for producing, (*except Refrigerative, Class* 62.)
Articulators.
Atmospheric baths.
Bandages.
Bandages, Catamenial
Bandages, Suspensory
Bed-pans.
Blister-spreaders.
Blood, Instruments for infusing
Blood, Instruments for propelling the
Bone-elevators.
Breast-cups.
Breast-pumps.
Calculi, Instruments for extracting
Canes, Physicians'
Capsules, Machines for making
Castrating-clamps.
Catamenial sacks.
Catarrhal-syringes.
Catheters.
Cauterizing-instruments.
Chest-protectors.
Chiropodic-instruments.
Compresses.
Corns, Instruments for removing
Cupping-instruments.
Cups, Medicinal
Depurators.
Dermopathic-instruments.
Douche-instruments.
Drinking-tubes for invalids.
Drug-crushers.
Dockholders for horses.
Dyspepsia, Instruments for treating
Ear, Apparatus for treating the
Ear-explorers.
Ear, Instruments for treating the
Ex-sections, Apparatus for
Eye, Apparatus for treating the
Eye-cups.
Eye, Instruments for treating the
Eye-protectors.
Eye-shades.
Filters for medicinal preparations.
Fissure-needles.
Fomentations, Modes of applying
Foot-stoppers for horses.
Forceps, Artery
Forceps, Calculi
Forceps, Obstetrical
Forceps, Polypus
Globule-machines.
Hemorrhage, Instruments for arresting
Hernia, Instruments, for treating
Hypodermic syringes.
Inhalers.
Injectors, Medical
Irrigators for nose, vagina, etc.
Leeches, Mechanical
Leucorrhea, Instruments for treating
Life, Instruments for stimulating
Ligatures.
Ligatures, Instruments for adjusting
Lithotriptic-instruments.
Lithotritors.
Mechanical leeches.
Medicators.
Medicinal powders, Instruments for putting up
Medicine-cases.
Medicines, Instruments for administering
Medicines, Instruments for compounding
Medicine-spoons.

Menstruation, Instruments for arresting
Milking-shields.
Mortars, Pharmaceutists'
Movement-cure apparatus.
Nasal-douches.
Nasal-tubes.
Needles, Surgeons'
Nipples, Artificial
Nipple-shields.
Nitrous-oxide gas, Instruments for administering
Nursery-bottles.
Obstetric-apparatus.
Ophthalmoscopes.
Otoscopes.
Pads for trusses.
Percolators for medicinal solutions.
Pessaries.
Pharmaceutical apparatus.
Piles, Instruments for treating
Pills, Machines for making
Plasters, Machines for spreading
Porte-caustic.
Powders, Appliances for putting up
Powder injectors.
Probangs.
Probes.
Prolapsus, Instruments for treating
Pulmometers.
Pumps, Stomach
Respirators.
Respiring apparatus.
Rubbing apparatus for medical treatment.
Scarificators.
Sounds.
Speculums, Anal
Speculums, Vaginal
Spermatorrhea, Instruments for treating
Spoons, Medicinal
Stethoscopes.
Stomach-pumps.
Strabismus-goggles.
Supporters, Abdominal
Supporters, Menstrual
Supporters, Obstetrical
Supporters, Pile
Supporters, Uterine
Supporters, Vaginal
Suppositories.
Suspensory-bandages.
Suture-instruments.
Syringes, Catarrhal
Syringes, Elastic-bulb
Syringes, Enema
Syringes, Eye
Syringes, Hemorrhoidal
Syringes, Penis
Syringes, Vaginal
Tape-worm, Apparatus for removing
Tenaculums.
Therapeutic applications of heat, cold, air, vacuum.
Tonsils, Instruments for operating on the
Topical applications, Appliances for
Tourniquets.
Trephines.
Trusses.
Uterine-sounds.
Uterine-suppositories.
Vaccinators.
Vacuum-cups.
Vapor-inhalers.
Varicose veins, Bandages for

# CLASS 129.

## TANNING.

Treatment of Hides and Manufacture of Leather.

[*For Enameling and Japanning of Leather, see Class* 91.]

Artificial leather. (*See Class* 91.)
Bark, Extracts of
Bark-leaches.
Bark-mills. (*See Class* 83.)
Bating hides.
Blacking leather, Compositions for
Boarding leather.
Currying-machines.
Currying-tools.
Depilating hides.
Dressing leather.
Finishing leather.
Graining leather.
Hides, Dressing
Leather, Artificial. (*See Class* 91.)
Leather-boarding.
Leather-cloth. (*See Class* 91.)
Leather, Compositions for treating
Leather, Compounds of waste
Leather-dressing
Leather-enameling. (*See Class* 91.)
Leather-finishing.
Leather-graining.
Leather-japanning. (*See Class* 91.)
Leather, Manufacture of
Leather, Water-proofing
Parchment.
Skins, Treatment of
Tanning-machines.
Tanning-materials.
Tanning-processes.
Tawing leather.
Vats, Tanning
Water-proofing leather.
Whitening leather.

# CLASS 130.

## THRASHING.

Machines for Husking, Thrashing, Shelling, Winnowing, and Stacking.

---

Band-cutting machines for thrashers.
Clover-hullers.
Corn-cribs.
Corn-husking gloves.
Corn-husking machines.
Corn-shellers.
Fanning-mills.
Flails.
Fodder stands.
Grain-bins.
Grain-conveyors. *(See Class 57.)*
Grain elevators. *(See Class 57.)*
Grain registers. *(See Class 73.)*
Grain-riddles.
Grain-screens.
Grain-separators.
Grain-thrashers.
Grain-winnowers.
Granaries.
Grain-seed separators.
Hay stackers.
Hullers, Clover
Huskers, Corn
Husking-gloves.
Husking-pins.
Husking stools.
Registers, Grain. *(See Class 73.)*
Stackers.
Straw-carriers.
Straw-stackers.
Thrashing-floors.
Thrashing-machines.
Winnowing-machines.

# CLASS 131.

## TOBACCO.

Processes, Machinery, and Appliances for the Manufacture and Use.

---

Cigar-cases.
Cigarette-fillers.
Cigarettes.
Cigar-holders.
Cigar-making.
Cigar-molds.
Cigar-presses.
Cigars.
Flavoring tobacco.
Meerschaums.
Mouth-pieces for cigars and pipes.
Pipes, Tobacco
Pouches, Tobacco
Smoking-pipes.
Snuff-boxes.
Snuff-sifters.
Tobacco-cartridges.
Tobacco-curing, (*Except Drying, Class* 34.)
Tobacco-cutting.
Tobacco flavoring.
Tobacco-granulating machines.
Tobacco-hooks.
Tobacco packing.
Tobacco-pipes.
Tobacco-pipe stems.
Tobacco-pouches.
Tobacco preparing.
Tobacco-presses. (*See Class* 100.)
Tobacco sheeting.
Tobacco stemming.
Tobacco-stoppers.

# CLASS 132.

## TOILET.

Articles and Appliances for the Toilet and Hair-working.

---

Barbers' appliances.
Bobbins for weaving hair.
Braiding, Hair
Brushes. (*See Class* 15.)
Chignons.
Comb-cleaners.
Combs.
Curling-irons.
Curls.
Dressing-cases.
Frames for weaving hair.
Hair-crimpers.
Hair-curlers.
Hair-cutting machines.
Hair-nets.
Hair-parting instruments.
Hair-pincers.
Hair-pins.
Hair-rolls.
Hair-tongs.
Hair-trimmings.
Hair-tweezers.
Head-nets, Ladies'
Jewel-boxes.
Mirrors. (*See Class* 45.)
Mustache-guard.
Pomatum-cases.
Powder-boxes, Cosmetic
Puffs.
Puffing-irons.
Rats.
Razors. (*See Class* 30.)
Shampooing-appliances.
Shaving-cases.
Shears, Hair-cutting. (*See Class* 30.)
Switches, Hair
Tooth-picks.
Waterfall head-dresses.
Weaving and working human hair.
Wigs.

# CLASS 133.

## TRUNKS.

Traveling and Military Accouterments, *except Mess-kits.*

[*For Manufacture of Metallic Portions of Trunks, see Class* 53, *Section J.*]

Accouterments, Military
Awnings.
Awnings, Detachable
Bags, Carpet
Bags, Mail
Bags, Saddle
Bags, Traveling
Bayonet-frogs.
Bayonet-scabbards.
Boxes, Cartridge
Boxes, Hat
Braces for trunks.
Carpet-bags.
Carpet-bag frames.
Hat-boxes, Leather
Holsters.
Knapsacks.
Knapsack-slings.
Mail-bag fasteners, (*except Locks, Class* 70.)
Mail-bags.
Military accouterments.
Portmanteaus.
Saddle-bags.
Satchel-fastenings.
Satchels.
Shawl-straps.
Slinging-accouterments.
Tents.
Traveling-bags.
Traveling-flasks
Trunk-corners.
Trunk-frames.
Trunk-guards.
Trunk-handles.
Trunk-lid guides.
Trunk-lid supporters.
Trunks.
Trunks and traveling-bags, Special tools for
Trunk-stays.
Valises.

# CLASS 134.

## TUBING AND WIRE.

Manufacture of Wrought-metal and Drawn-metal Tubing and Wire.

---

Barrels, Making gun
Brass tubes, Machines for making
Coiling-pipes.
Coiling-tubes.
Draw-benches.
Drawing-metals.
Draw-plates for tubing.
Draw-rolls for tubing.
Drawing-tubes.
Ferrules.
Expanding-tubes.
Gun-barrels, Making
Iron pipe making, Welded
Lead pipe.
Lead pipe lining.
Lead pipe, Machines for making
Lining cylinders.
Lining gun-barrels.
Lining metallic pipes.
Lining tubes.
Tubes.
Tubes, Bushing
Tubes, Casing
Tubes, Ornamenting
Tubing.
Welded-pipe making.

# CLASS 135.

## UMBRELLAS AND FANS.

Canes, Fans, Parasols, and Umbrellas:

[*For Manufacture of the Metallic Parts of the above, see Class* 53, *Section I.*]

Automatic fans.
Cane and seat combined.
Canes, Ferrules for walking
Canes, Walking
Fan-folding machines.
Fans, Automatic
Fans for chairs.
Fans for sewing-machines.
Handles for canes.
Handles for umbrellas.
Parasols.
Sun-shades.
Umbrella and cane combined.
Umbrella-cases.
Umbrella-covers.
Umbrella-fasteners.
Umbrella-ferrules.
Umbrella-frames.
Umbrella-joints.
Umbrella ribs and stretchers.
Umbrella-runners.
Umbrella-supporters.
Umbrella-tips.
Umbrellas.

# CLASS 136

## VALVES.

Cocks, Faucets, Taps, Valves for all Fluids.

[*For the Manufacture of Water and Gas Fittings, see Class* 53, *Section K.*]

---

Balanced valves.
Ball-cocks.
Beer-taps.
Bottle-faucets.
Check-valves.
Cocks.
Cocks, Blow-off
Cocks, Feed
Cocks, Gage
Cocks, Oil
Cocks, Self-closing
Cocks, Steam
Cocks, Stop
Cocks, Try
Cocks, Water
Cut-off valves.
Escape-valves.
Faucet-filters.
Faucets.
Faucets, Basin
Faucets, Boring
Faucets, Hot-water
Faucets, Molasses
Filter-faucets.
Gates, Molasses
Globe-valves.
Governor-valves.
Hydrants, Cocks for
Hydrant-valves.
Lock-up safety-valves.
Molasses-gates.
Oscillating-valves.
Piston-valves.
Pump-valves.
Puppet-valves.
Rotary-valves.
Safety-valves.
Slide-valves.
Steam-valves.
Stop-cocks.
Stop-valves.
Taps for liquors.
Throttle-valves.
Valve-gear, Steam-engine. (*See Class* 121.)
Valve-regulators.
Valves, Arrangement of
Valves, Automatic
Valves, Back-pressure
Valves, Balance
Valves, Butterfly
Valves, Check
Valves, Clack
Valves, Core
Valves, Cut-off
Valves, Disk
Valve-seats.
Valves, Escape
Valves for gas-engines.
Valves for syringes.
Valves, Globe
Valves, Governor
Valves, Oscillating
Valves, Piston
Valves, Pump
Valves, Puppet
Valves, Rotary
Valves, Safety
Valves, Slide
Valves, Spring-balance
Valves, Steam
Valves, Stop
Valves, Throttle
Valves, Trimmer
Valves, Water-closet
Valve-tubes.
Water-closet valves.
Water-valves.

# CLASS 137.

## WATER-DISTRIBUTION.

Including Well-tubing, Filters, Pipes, and Coupling.

[*For the Manufacture of Water and Gas Fittings, see Class* 53, *Section K.*]

---

Anti-concussion devices for water-pipes.
Artesian-well tubing.
Ash leaches.
Bark leaches.
Barrel washers.
Couplings, Pipe
Driven-wells. (*See Class* 37.)
Filters.
Filters, Globe
Filters, Tube
Filters, Well-tube
Fire-plugs.
Fish traps for water-pipes.
Fountains.
Funnels.
Hose-clamps.
Hose-couplings.
Hose-nozzles.
Hose-pipes.
Hydrant-pipe stops.
Hydrants.
Hydrants, Filtering
Hydrants, Waste for
Irrigating-apparatus.
Irrigating-carts.
Leaches.
Mains, Gas
Mains, Water
Nozzles for fire-engines.
Nozzles for hose.
Nozzles for oil-cups.
Packing for well-tubes.
Pipe-couplings.
Pipe-protectors.
Pipes, Water and gas
Rain-water cut-offs.
Rose-heads.
Sampling tubes.
Street-sprinkling carts.
Street-washers.
Siphons.
Spouting.
Tube-couplings.
Tube-filters.
Tube-wells. (*See Class* 37.)
Tubing, Well. (*See Class* 37.)
Velinches.
Water-tanks for railways, etc.
Wells, Artesian. (*See Class* 37.)
Well-tubes.
Well-tubes, Filters for
Well-tubes, Packing for
Well-tubes, Strainers for

# CLASS 138.

## WATER-WHEELS.

Chutes for water.
Current-wheels.
Forebays.
Gates for water-wheels.
Penstocks.
Tide-wheels.
Turbines.
Water-wheels.

# CLASS 139.

## WEAVING.

Looms, their Mechanisms, Parts, and Products.

[*For Manufacture of Weaving Hardware, see Class 53, Section H.*]

Bags, Woven
Basket-covering for bottles.
Basket-making.
Basket-making machines.
Baskets, Weaving
Belting, Woven
Binding, Woven
Blinds, Woven
Bobbins for shuttles.
Bullion, Machines for making
Carpet fabrics.
Carpeting.
Carpet-lining machines.
Carpet-rag looper.
Chenille-machines.
Circular looms.
Coach-lace.
Corded fabrics.
Crinoline, Weaving
Cut-piled fabrics.
Double-piled fabrics.
Druggets.
Fabrics, Carpet
Fabrics, Corded
Fabrics, Cut-piled
Fabrics, Double-piled
Fabrics, Elastic
Fabrics, Figured
Fabrics, Horse-hair
Fabrics, Piled
Fabrics, Shaped
Fabrics, Straw
Fabrics, Terry
Fabrics, Tubular
Fabrics, Woven
Figured fabrics.
Flannels.
Fringes.
Gauze.
Hair cloth, Machines for making
Harness for looms.
Harness, Machines for making loom
Hats, Weaving
Heddles.
Horse-hair fabrics.
Hose, Weaving
Jacquard apparatus.
Let-off motions for looms.
Looms.
Looms for cross-weaving.
Looms for curvilinear weaving.
Looms for weaving bags.
Looms for weaving baskets.
Looms for weaving belting.
Looms for weaving blinds.
Looms for weaving cane.
Looms for weaving caps.
Looms for weaving carpets.
Looms for weaving cases for bottles.
Looms for weaving corded fabrics.
Looms for weaving counterpanes.
Looms for weaving crinoline.
Looms for weaving cut-piled fabrics.
Looms for weaving double-piled fabrics.
Looms for weaving elastic fabrics.
Looms for weaving figured fabrics.
Looms for weaving fringes.
Looms for weaving hats.
Looms for weaving horse-hair.
Looms for weaving hose.
Looms for weaving ingrain carpets.
Looms for weaving matting.
Looms for weaving metallic tissues.
Looms for weaving piled fabrics.
Looms for weaving ribbons.
Looms for weaving satin.
Looms for weaving scarfs.
Looms for weaving shaped fabrics.
Looms for weaving shawls.
Looms for weaving sieve-cloth.
Looms for weaving silk.
Looms for weaving slats.
Looms for weaving straw fabrics.
Looms for weaving terry fabrics.

Looms for weaving tubular fabrics.
Looms for weaving velvets.
Looms for weaving wire.
Looms for weaving woolen cloth.
Loopers for carpet-rags.
Machines for making welt-bindings.
Matting.
Pattern-cards.
Pattern-chains.
Pattern-cylinders.
Picker-checks.
Picker-motions.
Picker-staves.
Pickers.
Pile-wire actuating-mechanisms.
Pile-wires.
Power-looms.
Reeds for looms.
Reed setter.
Reeds for looms, Machines for making
Shedding-apparatus.
Shuttle-boxes.
Shuttle-boxes, Actuating
Shuttle-checks.
Shuttles.
Shuttle-spindles.
Sieve-cloth, Weaving
Stop-mechanisms for looms.
Take-up motions for looms.
Tape.
Temples, Loom
Warp-beams.
Weavers' harness, Machines for making
Weaving baskets.
Weft-forks.
Wire, Looms for weaving

# CLASS 140.

## WIRE-WORKING.

Cables, Fences, Screens, and Sieves of Wire.

Annealing wire.
Ash-sifters.
Bending wire.
Bird-cages.
Blinds, Woven-wire
Brushes, Wire
Cables for sub-marine telegraphs.
Cables for suspension-bridges.
Cables, Wire
Coal-screens.
Coal-sifters.
Coiling wire.
Corrugating wire.
Covering of wire, wire ropes, and cables.
Draw-plates for wire.
Flour-sieves.
Harness for looms, Making wire
Panels of wire.
Pliers, Wire-cutting
Pointing wire.
Rolling wire.
Rope, Wire
Safes, Wire-work for kitchen
Sand-screens.
Screens of wire.
Sieves of wire.
Sifters.
Sifting-machines.
Skirt-hoops, Covering wire with wire for
Suspension-bridge cables.
Tools for working wire.
Twisting wire.
Wire-annealing.
Wire-bending.
Wire-brushes.
Wire-cables.
Wire-cloth crimping, Machines for
Wire-coiling.
Wire-covering.
Wire dish-covers, Making
Wire dish-stands, Making
Wire-drawing.
Wire fabrics.
Wire-fences, Panels for
Wire-gauze.
Wire heddles for weavers' harness, Making
Wire-making.
Wire-pointing.
Wire-rolling.
Wire ropes.
Wire screens.
Wire shoe-pegs.
Wire sifters.
Wire springs, Making
Wire-straightening.
Wire strainers.
Wire-twisting.
Wire with wire for skirt-hoops, &c., Covering
Wire-working for ornamental fences.

# CLASS 141.

## WOOD SCREWS.

The Article and its Manufacture.

Capping screws.
Coffin-screws, Making
Cutting wood screws.
Screw-blanks, Cutting
Screw-blanks, Feeding
Screw-blanks, Making
Screw-blanks, Threading
Screws, Swaging
Wood screws.
Wood screws, Machines for making

# CLASS 142.

## WOOD-WORKING, 1.

Lathes and Wood-turning.

Axe-handles, Machines for turning
Axles, Machines for turning
Bowls, Machines for turning
Boxes, Machines for turning wooden
Bungs, Machines for turning
Buttons, Machines for turning
Centers for wood-lathes.
Chucks for wood-lathes.
Clothes-pins, Machines for turning
Corks, Machines for polishing
Corks, Machines for turning
Dogs for wood-lathes.
Dowels, Machines for turning
Fluting-lathes, Wood
Frames, Machines for turning round and oval
Gun-stocks, Machines for turning
Handles, Machines for turning
Hat-blocks, Machines for turning
Hubs, Machines for turning
Lasts, Machines for turning
Lathe-dogs for wood-lathes.
Lathes for turning irregular forms.
Lathes, Mandrels for wood
Lathes, Wood
Oars, Machines for turning
Oval frames, Machines for turning
Pegs, Machines for turning
Pins, Machines for turning bedstead and clothes
Plow-handles, Machines for turning
Plugs, Machines for turning
Rests for wood-lathes.
Snaths, Machines for turning
Spools, Machines for boring
Spools, Machines for turning
Spokes, Machines for turning
Tool-rests for wood-lathes.
Tools for wood-lathes.
Tree-nails, Machines for turning
Turning irregular forms.
Wooden boxes, Machines for turning
Wooden pins, Machines for turning
Wooden screws, Machines for turning
Wood-lathes.

# CLASS 143.

## WOOD-WORKING, 2.

Machines for General Work.

[*Except Sawing-machines, Class* 110.]

---

Boring-machines, (wood.)
Carving-machines.
Cutter-heads of planing-machines.
Cutters of molding and planing machines.
Dovetailing machines.
Dressing-machines, (wood.)
Hewing-machines.
Molding-machines, (wood.)
Mortising-machines.
Planing-machines, (wood.)
Polishing-machines, (wood.)
Rabbeting-machines.
Sand-papering machines.
Sawing-machines. (*See Class* 110.)
Saw-mills. (*See Class* 110.)
Shaving-horses.
Slitting timber, Machines for
Slivering-machines.
Tenoning-machines.

# CLASS 144.

## WOOD-WORKING, 3.

Machines for Special Work.

Axle-setters.
Axles, Machines for making wooden
Bark-cutting machines.
Bark, Machines for planing
Bark, Machines for pressing
Barrel-heads.
Barrel-heads, Machines for making
Barrels, Machines for making
Baskets, Machines for making
Basket-splints, Machines for making
Bending wood, Machines for
Blind-slats, Machines for planing
Blind-slats, Machines for tenoning
Blind-slats, Machines for wiring
Blinds, Machines for making
Blind-splints, Machines for making
Blind-stiles, Machines for making
Blind-stiles, Machines for mortising
Blind-stiles, Machines for piercing
Blind-stiles, Machines for spacing
Blocks, Machines for making nautical
Blocks, Machines for making wooden paving
Boxes, Machines for making wooden
Broom-handles, Machines for making
Broom-splints, Machines for making
Brush-backs, Machines for making
Brush-handles, Machines for making
Buckets, Machines for making
Bundling firewood, Machines for
Cane, Machines for polishing
Cane, Machines for splitting
Cane, Machines for stripping
Cane-strippers.
Caning chairs, Instruments for
Carpenters' machines.
Casks, Machines for making
Chair-backs, Machines for dressing
Chair-backs, Machines for rounding
Chair-seats, Machines for boring
Chair-seats, Machines for planing
Chairs, Instruments for caning
Chair-stuff, Machines for scraping
Clap-boards, Machines for gaging
Clap-boards, Machines for planing
Clothes-pins, Machines for making
Clothes-pins, Machines for slotting
Cogs on wooden wheels, Machines for cutting
Cogs, Machines for making
Comb-making machines.
Coopers' machines.
Cutting dyewood.
Dipping-frames for matches.
Doors, Machines for making
Drums, (keg,) Machines for making
Dyewood, Machines for cutting
Eave-troughs, Machines for making wooden
"Excelsior," Machines for making
Fellies, Machines for bending
Fellies, Machines for boring
Fellies, Machines for planing
Fire-wood, Machines for bundling
Fire-wood, Machines for splitting
Frames, Machines for making saw
Frames, Machines for making slate
Gear, Machines for cutting wooden
Girdling trees, Machines for
Grooving lumber, Machines for

Gutters, Machines for making wooden
Handles, Machines for making wooden
Hoops, Machines for bending
Hoops, Machines for cutting
Hoops, Machines for dressing
Hoops, Machines for splitting
Horn, Machines for working in
Horn-molding.
Hubs, Machines for boring
Hubs, Machines for centering
Hubs, Machines for mortising
Hubs, Machines for setting boxes in
Ivory, Machines for working in
Joists, Machines for dressing
Kegs, Machines for making
Lasts, Machines for cutting
Lasts, Machines for making
Lasts, Machines for polishing
Lath, Machines for cutting
Lath, Machines for splitting
Match-blocks, Machines for making
Matches, Dipping frames for
Matches, Machines for pointing
Matches, Machines for making
Matches in frames, Machines for putting
Matches, Machines for rounding
Matches, Machines for splitting
Match-splints, Framing
Match-splints, Machines for making
Nautical-blocks, Machines for making
Oars, Machines for dressing
Oars, Machines for planing
Oars, Machines for riving
Osier-peelers.
Palm-leaves, Machines for splitting
Panels, Rabbeting door
Paving-blocks, Machines for making
Pencils, Machines for making
Pegs, Machines for making
Pegs, Machines for making shoe
Pegs, Machines for pointing
Pegs, Machines for splitting
Picket-pointers.
Pickets, Machines for making
Pins, Machines for cutting clothes
Pins, Machines for slotting clothes
Pipes, Wooden
Plane-stocks, Machines for making
Plow-handles, Machines for bending
Plow-handles, Machines for dressing
Plugs, Machines for cutting wooden
Pointing rails, pickets, matches, &c.
Post-borers.
Railway ties, Machines for dressing
Rake-teeth, Machines for making
Rattan, Machines for cutting
Rattan, Machines for dressing
Rattan, Machines for polishing
Rattan, Machines for reducing
Rattan, Machines for slitting
Rattan, Machines for splitting
Rattan, Machines for stripping
Rings, Machines for cutting wooden
Rossing bark.
Rossing dye-stuffs.
Sand-papering machines.
Sash, Machines for making
Sash, Machines for mortising
Sash, Machines for planing
Scale-board, Working
Sharpeners, Picket and pole
Shingle binders.
Shingles, Machines for jointing
Shingles, Machine for planing
Shingles, Machines for riving
Shoe-pegs, Machines for making
Shoe-soles, Machines for making wooden
Shovel-handles, Machines for making
Skewers, Macnines for making
Slate-frames, Machines for making
Snaths, Machines for bending
Snaths, Machines for finishing
Snaths, Machines for rounding
Splints for baskets, blinds, brooms, and matches.
Splitting firewood, Machines for
Spokes, Machines for driving
Spokes, Machines for planing
Spokes, Machines for polishing

Spokes, Machines for tenoning
Spools, Machines for boring
Staves, Machines for bilging
Staves, Machines for crozing
Staves, Machines for dressing
Staves, Machines for howeling
Staves, Machines for jointing
Staves, Machines for riving
Stumps, Machines for cutting
Table-leaves, Jointing
Tire setters.
Tire tighteners.
Tobacco-pipes, Machines for making wooden
Tooth-picks, Machines for making
Tortoise-shell, Machines for working in
Trays, Machines for making wooden
Treenails, Machines for making
Troughs, Machines for making wooden
Veneering.
Veneers, Machines for cutting
Veneers, Machines for planing
Veneers, Machines for polishing
Veneers, Machines for pressing
Veneers, Machines for straightening
Wainscoting, Making
Washboards, Machines for making
Wedges, Machines for making
Whalebone, Machines for dressing
Whalebone, Machines for splitting
Wheelwrights' machines.
Willow-peeling.
Willow-shaving machines.
Withes, Machines for twisting
Wood-bending machines.
Wood-bundling machines.
Wooden bowls, Machines for making
Wooden trays, Machines for making
Wooden-ware, Machines for making
Wood-gear cutting.
Wood-polishing machines.
Wood-splitting machines.

# CLASS 145.

## WOOD-WORKING, 4.

### Tools.

[*For Manufacture of Wood-working Metallic Tools, see Class* 81, *Section B.*]

---

Adzes.
Annular bits.
Augurs. (Wood.)
Awls. (Wood.)
Ax-helves.
Axes.
Bands for handles.
Bench-clamps.
Bench-dogs.
Benches for carpenters, joiners, &c.
Bench-fittings.
Bench-hooks.
Bench-planes.
Bench-strips.
Bench-vises. (Wood-working.)
Bits, Boring. (Wood.)
Bit-stocks.
Bolt-extractors.
Boring-bits.
Box-scrapers.
Bucks, Saw
Bung-cutters.
Cabinet-makers' tools.
Cabinet-makers' tools, (*Manufacture of, See Class* 81.)
Carpenters' benches.
Carpenters' clamps.
Carpenters' gages.
Carpenters' tools.
Carpenters' tools, (*Manufacture of, See Class* 81.)
Chalk-line holders.
Chamfering-tools.
Chisels, Carpenters'
Chisels, Grafting
Chisels, Joiners'
Chisels, Pruning
Clamps, Carpenters'
Clamps, Joiners'
Clamps, Shipwrights'
Claw-hammers.
Cooper's tools.
Coopers' tools, (*Manufacture of, See Class* 81.)
Countersink-bits.
Cutting-gages.
Deck-scrapers.
Dogs, Bench
Drawing-knives.
Edge-planes. (Wood.)
Expanding-bits.
Extractors, Bolt, nail, spike, tack
Frows.
Gages, Cutting
Gages, Scribing
Gimlets.
Glue hoppers.
Glueing tables.
Gouges.
Hammers, Claw
Handles to tools, Attaching
Handles, Tool, (*except turning.*)
Hatchets.
Hooks, Bench
Horses, Saw
Howeling-tools
Joiners' benches.
Joiners' clamps.
Joiners' tools.
Joiners' tools, (*Manufacture of, See Class* 81.)
Knives, Drawing
Mallets.
Mast-scrapers.
Match-planes.
Miter-boxes.
Molding-planes.
Nail-extractors.
Plane-bits.
Plane-making.
Planes, Carpenters'
Planes, Coopers'
Planes, Joiners'
Planes, Splint
Plane-stocks.
Planking-clamps.
Rasps, Wood

Reaming-bits. (Wood.)
Riving splints, Tools for
Sand-paper holders.
Saws. (*See Class* 110.)
Scrapers, Box
Scrapers, Deck
Scrapers, Mast
Screw-drivers.
Scribing-gages.
Shipwrights' clamps.
Shipwrights' tools.
Shipwrights' tools, (*Manufacture of, See Class* 81.)
Spike-extractors.
Splint-planes.
Spoke-shaves.
Stocks, Bit
Tack-extractors.
Tools for wood-working.
Tool-handles, (*Except turning, Class* 142.)
Tool-holders.
Tool-stocks.
Tools to handles, Attaching
Veneering-tools.
Vises, Carpenters'
Wheelwrights' wood-working tools.

# ALPHABETICAL INDEX OF INVENTIONS.

| SUBJECT. | Name of class—official classification. | No. of class—official classification. | No. of class—subscription classification. |
|---|---|---|---|
| A. | | | |
| Abutments for arches | Bridges | 14 | 21 |
| Accordeons | Music | 84 | 102 |
| Account-books, Binding of | Book-Binding | 11 | 15 |
| Accouterments, Military | Trunks | 133 | 162 |
| Achromatic glasses | Optics | 88 | 157 |
| Acids, Air-tight packages for | Preserving | 99 | 58 |
| Acids, Apparatus for making, (*except Stills, Class* 124) | Chemical Miscellaneous | 23 | 35 |
| Acids, Processes for making | Chemical Miscellaneous | 23 | 35 |
| Acids, Distillation of | Stills | 124 | 47 |
| Acoustic instruments | Surgery | 128 | 156 |
| Acupuncturators | Surgery | 128 | 156 |
| Adding-machines | Measuring-Instruments | 73 | 42 |
| Addressing-machines | Printing | 101 | 117 |
| Address-stamps | Printing | 101 | 146 |
| Adhesion-cars | Railways; 2 | 105 | 121 |
| Adhesive fastenings for paper, &c | Stationery | 120 | 146 |
| Adobes | Clay | 25 | 20 |
| Advertising-cards | Stationery | 120 | 54 |
| Advertising-devices | Stationery | 120 | 54 |
| Advertising-frames | Stationery | 120 | 54 |
| Adzes | Wood-Working; 4 | 145 | 49 |
| Adzes, Manufacture of | Metal-Working; 6 | 81B | 49 |
| Aërated liquids | Aëration | 1 | 1 |
| Aërated liquor apparatus | Aëration | 1 | 1 |
| Aërating dough, Apparatus for | Aëration | 1 | 6 |
| Aërial cars | Pneumatics | 98 | 143 |
| Aërostats | Pneumatics | 98 | 143 |
| Aëro steam-engines | Pneumatics | 98 | 2 |
| Ageing liquors | Beer and Wine | 7 | 19 |
| Agricultural boilers | Stoves | 126 | 59 |
| Agricultural engines | Steam; 3 | 123 | 94 |
| Agricultural furnaces | Stoves | 126 | 59 |
| Agricultural tools of metal, Manufacture of | Metal-Working; 6 | 81B | |
| Air and gas meters | Pneumatics | 98 | 64 |
| Air-blast | Pneumatics | 98 | 12 |
| Air-brakes for car, Steam-actuated | Steam; 1 | 121 | 122 |
| Air-carbureting machines | Gas | 48 | 27 |
| Air-chambers | Pumps | 103 | 118 |
| Air-compressing machines and appliances | Pneumatics | 98 | 12 |
| Air-cooling apparatus | Ice | 62 | 58 |
| Air-cooling processes | Ice | 62 | 58 |
| Air-cushions | Pneumatics | 98 | 62 |
| Air-engines | Pneumatics | 98 | 2 |
| Air-filters | Pneumatics | 98 | 166 |
| Air-guns | Pneumatics | 98 | 12 |
| Air-heating traps for air-engines | Pneumatics | 98 | 2 |
| Air-meters | Pneumatics | 98 | 64 |
| Air-proofing processes | Paint | 91 | 35 |
| Air-pumps | Pneumatics | 98 | 12 |

| SUBJECT. | Name of class—official classification. | No. of class—official classification. | No. of class—subscription classification. |
|---|---|---|---|
| Air-tanks | Pneumatics | 98 | 2 |
| Air-tight cans, jars, and packages | Preserving | 99 | 58 |
| Air-traps for sinks and sewers | Baths and Closets | 4 | 8 |
| Alarm-bells, Manufacture of | Hardware Manufactures | 53B | 10 |
| Alarm-locks | Locks | 70 | 93 |
| Alarms, Bilgewater | Signals | 116 | 137 |
| Alarms, Burglar | Signals | 116 | 24 |
| Alarms, Electro-magnetic | Electricity | 36 | 50 |
| Alarms, Fire | Signals | 116 | 24 |
| Alarms, Fog | Signals | 116 | 138 |
| Alarms for stop-motions of carding-machines | Carding | 19 | 28 |
| Alarms, Gas | Lamps | 67 | 64 |
| Alarms, Grist | Mills | 83 | 57 |
| Alarms, Household fire | Signals | 116 | 24 |
| Alarms, Leak | Signals | 116 | 137 |
| Alarms, Nautical | Signals | 116 | 138 |
| Alarms, Shoal | Signals | 116 | 138 |
| Alarms, Telegraphic | Electricity | 36 | 50 |
| Alarms, Temperature | Signals | 116 | 24 |
| Alarms, Till | Signals | 116 | 24 |
| Alarms, Window | Signals | 116 | 24 |
| Alcohol, Apparatus for making | Stills | 124 | 47 |
| Alcohol, Materials for making | Stills | 124 | 47 |
| Alcoholmeters | Measuring-Instruments | 73 | 47 |
| Albumenizing paper | Photography | 95 | 111 |
| Albums, Binding of | Book-Binding | 11 | 15 |
| Alkalies, Air-tight packages for | Preserving | 99 | 58 |
| Alkalimeters | Measuring-Instruments | 73 | 35 |
| Alloys | Metallurgy | 75 | 99 |
| Almond-peelers | Kitchen Utensils | 65 | 88 |
| Alphabet-blocks | Educational | 35 | 54 |
| Altimeters | Optics | 88 | 157 |
| Altitude-instruments | Optics | 88 | 157 |
| Amalgamators | Ore | 90 | 99 |
| Ambulances | Carriages | 21 | 31 |
| Ammonia-engines | Pneumatics | 98 | 2 |
| Ammunition-cases | Projectiles | 102 | 162 |
| Anæsthesia, Instruments for producing, (*except Refrigerative, Class* 62) | Surgery | 128 | 98 |
| Anæsthetic refrigerators | Ice | 62 | 98 |
| Anchor-balls | Ships; 1 | 114 | 138 |
| Anchors | Ships; 1 | 114 | 138 |
| Anchors, Drift | Ships; 1 | 114 | 138 |
| Anchors for suspension-cables | Bridges | 14 | 21 |
| Anchor-shackles | Ships; 1 | 114 | 138 |
| Anchors, Manufacture of | Metal-Working; 3 | 78 | 60 |
| Anchor-trippers | Ships; 1 | 114 | 138 |
| Andirons | Stoves | 126 | 154 |
| Anemometers | Measuring-Instruments | 73 | 42 |
| Anemoscopes | Measuring-Instruments | 73 | 42 |
| Angle-iron, Bending or straightening | Metal-Working; 1 | 76 | 125 |
| Angle-iron, Rolling | Metal-Working; 5 | 80 | 125 |
| Aniline dyes | Bleaching and Dyeing | 8 | 11 |
| Animals across rivers, Floating | Boats | 9 | 13 |
| Animal from vegetable fiber, Separating | Chemical Miscellaneous | 23 | 35 |
| Animal-black | Sugar | 127 | 155 |
| Animal-catching | Butchering | 17 | 25 |
| Animal-labels | Stationery | 120 | 144 |

| SUBJECT. | Name of class—official classification. | No. of class—official classification. | No. of class—subscription classification. |
|---|---|---|---|
| Animal-powers | Mechanical Powers | 74 | 159 |
| Animals, Preparing food for | Stabling | 119 | 59 |
| Animal-traps | Games and Toys | 46 | 143 |
| Anklets | Boots and Shoes | 12 | 16 |
| Annealing glass | Glass | 4 | 68 |
| Annealing metals | Metallurgy | 75 | 85 |
| Annealing wire | Metallurgy | 75 | 163 |
| Annular bits | Wood-Working; 4 | 145 | 29 |
| Annular boring-bits for metal | Metal-Working; 2 | 77 | 100 |
| Annular saws | Saws | 110 | 130 |
| Annunciators | Signals | 116 | 24 |
| Anti-concussion devices for water-pipes | Water-Distribution | 137 | 167 |
| Anti-fermentation processes and applications | Beer and Wine | 7 | 19 |
| Anti-friction axle-bearings | Carriages | 21 | 31 |
| Anti-friction compounds | Journals and Bearings | 64 | 87 |
| Anti-friction metals | Journals and Bearings | 64 | 87 |
| Anti-friction rollers | Journals and Bearings | 64 | 87 |
| Anti-incrustators, Galvanic | Electricity | 36 | 148 |
| Anti-incrustators, Magnetic | Electricity | 36 | 148 |
| Antiseptics | Chemical Miscellaneous | 23 | 98 |
| Anvils, Manufacture of | Metal-Working; 3 | 78 | 60 |
| Apiaries | Bee-Hives | 6 | 9 |
| Apparel, Paper | Paper Manufactures | 93 | 109 |
| Apple-corers | Kitchen Utensils | 65 | 88 |
| Apple-parers | Kitchen Utensils | 65 | 88 |
| Apple-slicers | Kitchen Utensils | 65 | 88 |
| Aquariums | Fine Arts | 41 | 54 |
| Aqueducts | Hydraulic Engineering | 61 | 82 |
| Archery | Games and Toys | 46 | 143 |
| Arches | Bridges | 14 | 21 |
| Architecture, Naval | Ships; 1 | 114 | 137 |
| Arithmometers | Measuring-Instruments | 73 | 42 |
| Armatures | Electricity | 36 | 50 |
| Armor, Personal | Fire-Arms | 42 | 55 |
| Armor-plates, Bending or straightening | Metal-Working; 1 | 76 | 125 |
| Armor-plates, Rolling | Metal-Working; 5 | 80 | 125 |
| Armor-plating | Ships; 1 | 114 | 137 |
| Armor, Submarine | Hydraulic Engineering | 61 | 82 |
| Arms, Artificial | Artificial Limbs | 3 | 5 |
| Arrastras | Ore | 90 | 99 |
| Arresters | Electricity | 36 | 50 |
| Artesian wells | Stone, Lime, and Cement | 125 | 153 |
| Artesian-well tubing | Water-Distribution | 137 | 51 |
| Articulators | Surgery | 128 | 156 |
| Articulators for fitting dentures | Dental | 32 | 46 |
| Artificial arms | Artificial Limbs | 3 | 5 |
| Artificial ears | Artificial Limbs | 3 | 5 |
| Artificial eyes | Artificial Limbs | 3 | 5 |
| Artificial feet | Artificial Limbs | 3 | 5 |
| Artificial flowers | Fine Arts | 41 | 54 |
| Artificial gems | Glass | 49 | 68 |
| Artificial gums | Dental | 32 | 5 |
| Artificial hands | Artificial Limbs | 3 | 5 |
| Artificial leather | Paint | 91 | 56 |
| Artificial legs | Artificial Limbs | 3 | 5 |
| Artificial marble | Stone, Lime, and Cement | 125 | 153 |
| Artificial palates | Dental | 32 | 46 |
| Artificial slate | Stone, Lime, and Cement | 125 | 153 |
| Artificial stone, (*except Grindstones, Class 51*) | Stone, Lime, and Cement | 125 | 153 |

| SUBJECT. | Name of class—official classification. | No. of class—official classification. | No. of class—sub-scription classification. |
|---|---|---|---|
| Artificial teeth | Dental | 32 | 46 |
| Artillery | Ordnance | 89 | 105 |
| Artillery-carriages | Ordnance | 89 | 105 |
| Artists' appliances, (*except Brushes, Class* 15) | Fine Arts | 41 | 54 |
| Ash-leaches | Water-Distribution | 137 | 167 |
| Ash-pans | Stoves | 126 | 154 |
| Ash-pans of locomotives | Steam; 3 | 123 | 94 |
| Ash-pit covers | Stoves | 126 | 154 |
| Ash-sifters | Wire-Working | 140 | 154 |
| Asphalt and wood pavements combined | Paving | 94 | 110 |
| Asphalt pavements | Paving | 94 | 110 |
| Assaying | Metallurgy | 75 | 99 |
| Assorting yarns and threads | Cordage | 28 | 17 |
| Astatic needles | Electricity | 36 | 50 |
| Astronomical charts | Educational | 35 | 54 |
| Astronomical instruments | Optics | 88 | 157 |
| Astronomical instruments, Electro-magnetic | Electricity | 36 | 50 |
| Astronomical lanterns | Educational | 35 | 54 |
| Atmospheric baths | Surgery | 128 | 156 |
| Atmospheric motors | Pneumatics | 98 | 2 |
| Atomizers | Ice | 62 | 98 |
| Augers for earth-boring | Excavators | 37 | 51 |
| Augers, Manufacture of | Metal-Working; 6 | 81B | 29 |
| Augers. (Wood) | Wood-Working; 4 | 145 | 29 |
| Auricles | Artificial Limbs | 3 | 5 |
| Automata | Games and Toys | 46 | 143 |
| Automatic fans | Umbrellas and Fans | 135 | 165 |
| Automatic presses | Presses | 100 | |
| Automatic valves | Valves | 136 | 152 |
| Awls, Shoemakers' | Boots and Shoes | 12 | 16 |
| Awls. (Wood) | Wood-Working; 4 | 145 | 29 |
| Awnings | Carpentry | 20 | 4 |
| Axe-handles, Machines for turning | Wood-Working; 1 | 142 | 172 |
| Axes | Wood-Working; 4 | 145 | 49 |
| Axes, Manufacture of | Metal-Working; 6 | 81B | 49 |
| Ax-helves | Wood-Working; 4 | 145 | 49 |
| Axle-arms | Carriages | 21 | 31 |
| Axle-boxes, (carriage,) Manufacture of. | Hardware Manufactures | 53C | 31 |
| Axle-boxes, Compositions for | Carriages | 21 | 31 |
| Axle-boxes for carriages | Carriages | 21 | 31 |
| Axle-boxes for railway-cars | Railways; 3 | 106 | 122 |
| Axle-boxes, Making railway-car | Railways; 4 | 107 | 122 |
| Axle, Making locomotive | Railways; 4 | 107 | 122 |
| Axles, Car | Railways; 3 | 106 | 122 |
| Axle-setters | Wood-Working; 3 | 144 | 31 |
| Axles, Lubricators for carriage | Carriages | 21 | 31 |
| Axles, Machines for making wooden | Wood-Working; 3 | 144 | 31 |
| Axles, Machines for turning | Wood-Working; 1 | 142 | 31 |
| Axles, Making railway-car | Railways; 4 | 107 | 122 |
| Axles, Manufacture of carriage | Metal-Working; 3 | 78 | 31 |
| Axle-trees | Carriages | 21 | 31 |
| Azimuth instruments | Optics | 88 | 157 |
| **B.** | | | |
| Baby-jumpers | Furniture | 45 | 62 |
| Baby-walkers | Carriages and Wagons | 21 | 31 |
| Back-bands | Harness | 54 | 72 |

| SUBJECT. | Name of class—official classification. | No. of class—official classification. | No. of class—subscription classification. |
|---|---|---|---|
| Bagasse-dryers | Dryers and Kilns | 34 | 155 |
| Bag-fasteners | Mills | 83 | 57 |
| Baggage-checks | Stationery | 120 | 146 |
| Bag-holders | Mills | 83 | 57 |
| Bag-locks | Locks | 70 | 93 |
| Bag machines, Paper | Paper Manufactures | 93 | 109 |
| Bags, Carpet | Trunks | 133 | 162 |
| Bags, Mail | Trunks | 133 | 162 |
| Bags, Saddle | Trunks | 133 | 162 |
| Bags, Traveling | Trunks | 133 | 162 |
| Bags, Woven | Weaving | 139 | 95 |
| Bait | Fishing | 43 | 143 |
| Bake-pans | Stoves | 126 | 154 |
| Bakers' ovens | Stoves | 126 | 6 |
| Bakers' ovens, Portable | Stoves | 126 | 6 |
| Balanced valves | Valves | 136 | 152 |
| Balance rynds | Mills | 83 | 57 |
| Balances | Measuring-Instruments | 73 | 132 |
| Bale-ties and hoops | Presses | 100 | 7 |
| Baling hay | Presses | 100 | 7 |
| Baling-presses | Presses | 100 | 7 |
| Ballasting | Ships; 1 | 114 | 137 |
| Ball-cocks | Valves | 136 | 167 |
| Balling cord and twine | Cordage | 28 | 17 |
| Balling yarns and threads | Cordage | 28 | 17 |
| Ballistic pendulums | Measuring-Instruments | 73 | 71 |
| Balloons | Pneumatics | 98 | 143 |
| Ballot-boxes | Stationery | 120 | 146 |
| Ball-screws for fire-arms | Fire-Arms | 42 | 55 |
| Balls for cartridges | Projectiles | 102 | 32 |
| Balls for fire-arms | Projectiles | 102 | 32 |
| Balls for ordnance | Projectiles | 102 | 105 |
| Balmorals | Apparel | 2 | 38 |
| Bandages | Surgery | 128 | 156 |
| Bandages, Catamenial | Surgery | 128 | 156 |
| Bandages, Suspensory | Surgery | 128 | 156 |
| Band and cord fasteners | Hose and Belting | 60 | 81 |
| Band-cutting machines for thrashers | Thrashing | 130 | 159 |
| Banding-machines | Cordage | 28 | 17 |
| Banding-spindles | Spinning | 118 | 142 |
| Band-pulleys | Journals and Bearings | 64 | 87 |
| Band-punches | Clasps and Buckles | 24 | 81 |
| Bands | Hose and Belting | 60 | 81 |
| Band-saws | Saws | 110 | 130 |
| Bands for binding grain | Harvesters | 56 | 74 |
| Bands for handles | Wood-Working; 4 | 145 | 29 |
| Bands for pulleys | Hose and Belting | 60 | 81 |
| Band-wheels | Journals and Bearings | 64 | 87 |
| Banjos | Music | 84 | 102 |
| Bank-notes | Fine Arts | 41 | 54 |
| Barbers' appliances | Toilet | 132 | 161 |
| Bark-cutting machines | Wood-Working; 3 | 144 | 92 |
| Bark, Extracts of | Tanning | 129 | 92 |
| Bark-leaches | Water-Distribution | 137 | 167 |
| Bark, Machines for planing | Wood-Working; 3 | 144 | 92 |
| Bark, Machines for pressing | Wood-Working; 3 | 144 | 92 |
| Bark-mills | Mills | 83 | 92 |
| Barley-forks | Harvesters | 56 | 77 |
| Barn-door hangers | Builders' Hardware | 16 | 24 |
| Barn-door rollers | Builders' Hardware | 16 | 24 |

| SUBJECT. | Name of class—official classification. | No. of class—official classification. | No. of class—subscription classification. |
|---|---|---|---|
| Barometers | Measuring-Instruments | 73 | 101 |
| Barrel-cleaners | Aëration | 1 | 19 |
| Barrel-covers | Kitchen Utensils | 65 | 88 |
| Barrel-dryers | Dryers and Kilns | 34 | 19 |
| Barrel-fillers | Aëration | 1 | 167 |
| Barrel-fumigators | Dryers and Kilns | 34 | 19 |
| Barrel-handles | Kitchen Utensils | 65 | 88 |
| Barrel-heads | Wood-Working; 3 | 144 | 40 |
| Barrel-heads, Machines for making | Wood-Working; 3 | 144 | 40 |
| Barrel-heaters | Dryers and Kilns | 34 | 19 |
| Barrel-lining | Paint | 91 | 40 |
| Barrel-pitching | Paint | 91 | 40 |
| Barrel-rollers | Aëration | 1 | 19 |
| Barrels | Wood-Working; 3 | 144 | 40 |
| Barrels impervious to liquids, Rendering | Paint | 91 | 40 |
| Barrels, Machines for making | Wood-Working; 3 | 144 | 40 |
| Barrels, Making gun | Tubing and Wire | 134 | 163 |
| Barrels, Making sheet-metal | Sheet-Metal | 113 | 136 |
| Barrels, Pitching | Paint | 91 | 40 |
| Barrel-stands and skids | Aëration | 1 | 19 |
| Barrels to stocks, Attaching | Fire-Arms | 42 | 55 |
| Barrel-tilters | Aëration | 1 | 19 |
| Barrel-washers | Aëration | 1 | 19 |
| Barrows | Carriages | 21 | 31 |
| Bars and rods, Rolling | Metal-Working; 5 | 80 | 125 |
| Bars, Bending or straightening | Metal-Working; 1 | 76 | 125 |
| Bascule bridges | Bridges | 14 | 21 |
| Base-balls | Games and Toys | 46 | 143 |
| Base-burning fire-place heaters | Stoves | 126 | 154 |
| Base-burning stoves | Stoves | 126 | 154 |
| Basin-clamps | Furniture | 45 | 8 |
| Basins for sewers | Masonry | 72 | 51 |
| Basket-covering for bottles | Weaving | 139 | 169 |
| Basket-making | Weaving | 139 | 169 |
| Basket-making machines | Weaving | 139 | 169 |
| Baskets | Kitchen-Utensils | 65 | 169 |
| Baskets, Car | Railways; 2 | 105 | 121 |
| Baskets for fruit, fish, grain and provisions | Kitchen Utensils | 65 | 169 |
| Baskets, Machines for making | Wood-Working; 3 | 144 | 169 |
| Basket-splints, Machines for making | Wood-Working; 2 | 143 | 174 |
| Baskets, Weaving | Weaving | 139 | 169 |
| Bath-heaters | Stoves | 126 | 8 |
| Bathing-apparatus | Baths and Closets | 4 | 8 |
| Bathing-tubs | Baths and Closets | 4 | 8 |
| Baths | Baths and Closets | 4 | 8 |
| Baths, Douche | Baths and Closets | 4 | 8 |
| Baths, Electro-voltaic | Electricity | 36 | 50 |
| Baths, Medicated | Baths and Closets | 4 | 8 |
| Baths, Shower | Baths and Closets | 4 | 8 |
| Baths, Vapor | Baths and Closets | 4 | 8 |
| Baths, Voltaic | Electricity | 36 | 52 |
| Bating hides | Tanning | 129 | 92 |
| Batons | Games and Toys | 46 | 143 |
| Bats for balls | Games and Toys | 46 | 143 |
| Bats, Forming | Felting and Hats | 38 | 28 |
| Batteries, Floating | Ships; 1 | 114 | 137 |
| Batteries, Submarine | Ships; 1 | 114 | 137 |
| Batteries, Thermo-electric | Electricity | 36 | 50 |
| Batteries, Voltaic | Electricity | 36 | 50 |

| SUBJECT. | Name of class—official classification. | No. of class—official classification. | No. of class—subscription classification. |
|---|---|---|---|
| Batter-machines | Kitchen Utensils | 65 | 88 |
| Batting-machines | Carding | 19 | 28 |
| Bayonet-attachments | Fire-Arms | 42 | 55 |
| Bayonet-frogs | Trunks | 133 | 162 |
| Bayonet-scabbards | Trunks | 133 | 162 |
| Bayonets | Fire-Arms | 42 | 52 |
| Bayonets, Manufacture of | Hardware Manufactures | 53A | 53 |
| Beacons | Signals | 116 | 138 |
| Beacons, Sonorific | Signals | 116 | 138 |
| Beads, Glass | Glass | 49 | 68 |
| Beads, (metallic,) Manufacture of | Hardware Manufactures | 53F | 86 |
| Beaming cloth | Cloth | 26 | 37 |
| Beams of wood, Compound | Carpentry | 20 | 30 |
| Beams, Rolling | Metal-Working; 5 | 80 | 125 |
| Bean-harvesters | Harvesters | 56 | 75 |
| Bean-shellers | Kitchen Utensils | 65 | 88 |
| Bearings for machinery | Journals and Bearings | 64 | 87 |
| Beater-engines | Paper-Making | 92 | 108 |
| Beater-presses | Presses | 100 | 7 |
| Bed-bottoms | Beds | 5 | 62 |
| Bed-bugs, Traps for | Beds | 5 | 62 |
| Bed-clothes clasps | Beds | 5 | 65 |
| Bedding | Beds | 5 | 62 |
| Bed-pans | Surgery | 128 | 156 |
| Bed-plates of rag-engines | Paper-Making | 92 | 108 |
| Bed-quilts | Beds | 5 | 62 |
| Beds | Beds | 5 | 62 |
| Bedstead-canopies | Beds | 5 | 62 |
| Bedstead-cording | Beds | 5 | 62 |
| Bedstead-fastenings | Beds | 5 | 62 |
| Bedstead keys | Beds | 5 | 62 |
| Bedsteads | Beds | 5 | 62 |
| Bedsteads and other furniture combined | Beds | 5 | 62 |
| Bedsteads, Invalid | Beds | 5 | 62 |
| Bedsteads, Sofa | Beds | 5 | 62 |
| Bedsteads, Spring-slats for | Beds | 5 | 62 |
| Bedsteads, Valances for | Beds | 5 | 62 |
| Bedsteads, Ventilating | Beds | 5 | 62 |
| Bedsteads, Wardrobe | Beds | 5 | 62 |
| Bed-wrenches | Beds | 5 | 62 |
| Bee-culture | Bee-Hives | 6 | 9 |
| Bee-feeders | Bee-Hives | 6 | 9 |
| Bee-fumigators | Bee-Hives | 6 | 9 |
| Beer and wine | Beer and Wine | 7 | 19 |
| Beer-coolers | Ice | 62 | 19 |
| Beer-cooling vats | Beer and Wine | 7 | 19 |
| Beer-cups | Kitchen Utensils | 65 | 88 |
| Beer-fountains | Pumps | 103 | 118 |
| Beer-pitchers | Kitchen Utensils | 65 | 88 |
| Beer-preserving vessels | Beer and Wine | 7 | 19 |
| Beer-pumps | Pumps | 103 | 118 |
| Beer-taps | Valves | 136 | 167 |
| Beet-root sugar apparatus | Sugar | 127 | 155 |
| Bell-founding | Casting | 22 | 10 |
| Bell-hanging | Signals | 116 | 24 |
| Bellows-nozzles | Pneumatics | 98 | 12 |
| Bellows | Pneumatics | 98 | 12 |
| Bell-pulls | Builders' Hardware | 16 | 24 |
| Bell-ringing | Signals | 116 | 24 |
| Bells | Signals | 116 | 10 |

| SUBJECT. | Name of class—official classification. | No. of class—official classification. | No. of class—subscription classification. |
|---|---|---|---|
| Bells, Clock | Signals | 116 | 10 |
| Bells, Cow | Harness | 54 | 72 |
| Bells, Dumb | Games and Toys | 46 | 143 |
| Bells, Manufacture of | Hardware Manufactures | 53 B | 10 |
| Bells, Sleigh | Harness | 54 | 72 |
| Bell-straps | Harness | 54 | 72 |
| Belt-clasps | Hose and Belting | 60 | 81 |
| Belt-couplings | Hose and Belting | 60 | 81 |
| Belt-cutting machines | Hose and Belting | 60 | 81 |
| Belt-fasteners | Hose and Belting | 60 | 81 |
| Belting, Compositions for | Hose and Belting | 60 | 81 |
| Belting-Compound | Hose and Belting | 60 | 81 |
| Belting, Modes of making | Hose and Belting | 60 | 81 |
| Belting, Rubber | Hose and Belting | 60 | 81 |
| Belting, Woven | Weaving | 139 | 95 |
| Belt-lacers | Hose and Belting | 60 | 81 |
| Belt-lacing, Machines for cutting | Hose and Belting | 60 | 81 |
| Belt-punches | Clasps and Buckles | 24 | 81 |
| Belts | Hose and Belting | 60 | 81 |
| Belts, Body | Apparel | 2 | 38 |
| Belt-shifters | Journals and Bearings | 64 | 87 |
| Belt-splicing | Hose and Belting | 60 | 81 |
| Belt-tighteners | Journals and Bearings | 64 | 87 |
| Bench-clamps | Wood-Working; 4 | 145 | 29 |
| Bench-dogs | Wood-Working; 4 | 145 | 29 |
| Benches for carpenters, joiners, &c | Wood-Working; 4 | 145 | 29 |
| Bench-fittings | Wood-Working; 4 | 145 | 29 |
| Bench-hooks | Wood-Working; 4 | 145 | 29 |
| Bench-planes | Wood-Working; 4 | 145 | 29 |
| Bench-shears | Cutlery | 30 | 44 |
| Bench-strips | Wood-Working; 4 | 145 | 29 |
| Bench-vises. (Wood-Working) | Wood-Working; 4 | 145 | 29 |
| Bending bars and plates. | Metal-Working; 1 | 76 | 125 |
| Bending rails. | Metal-Working; 1 | 76 | 125 |
| Bending sails | Ships; 1 | 114 | 138 |
| Bending sheet-metal | Sheet-Metal | 113 A | 136 |
| Bending wire | Wire-Working | 140 | 170 |
| Bending wood, Machines for | Wood-Working; 3 | 144 | 176 |
| Berths, Car | Railways; 2 | 105 | 121 |
| Berths, Floating | Ships; 1 | 114 | 137 |
| Berths, Ships' | Ships; 1 | 114 | 137 |
| Béton | Stone, Lime, and Cement | 125 | 153 |
| Béton, Structures of | Masonry | 72 | 4 |
| Beverages, (*except Aërated and Fermented*) | Chemical Miscellaneous | 23 | 98 |
| Beverages, Aërated | Aëration | 1 | 1 |
| Beverages, Fermented | Beer and Wine | 7 | 19 |
| Bidets | Baths and Closets | 4 | 8 |
| Biers | Carriages | 21 | 31 |
| Bilge-boards | Ships; 1 | 114 | 137 |
| Bilgewater-alarms | Signals | 116 | 137 |
| Bilge-water, Discharging | Ships; 1 | 114 | 137 |
| Bilgewater-gages | Measuring-Instruments | 73 | 137 |
| Bill-holders | Stationery | 120 | 146 |
| Bill-hooks | Cutlery | 30 | 44 |
| Billiard-balls | Games and Toys | 46 | 143 |
| Billiard-balls, Compositions for | Chemical Miscellaneous | 23 | 143 |
| Billiard-chalk cups | Games and Toys | 46 | 143 |
| Billiard-cues | Games and Toys | 46 | 143 |
| Billiard-cue tips | Games and Toys | 46 | 143 |
| Billiard-markers | Games and Toys | 46 | 143 |

| SUBJECT. | Name of class—official classification. | No. of class—official classification. | No. of class—subscription classification. |
|---|---|---|---|
| Billiard-table chalk-holders | Games and Toys | 46 | 143 |
| Billiard-table pockets | Games and Toys | 46 | 143 |
| Billiard-tables | Games and Toys | 46 | 143 |
| Binders' boards | Paper Making | 92 | 15 |
| Binders for harvesters | Harvesters | 56 | 74 |
| Binders for sewing-machines | Sewing-Machines | 112 | 134 |
| Binders, Temporary | Stationery | 120 | 15 |
| Bindings for carpet, &c | Cordage | 28 | 17 |
| Binding, Woven | Weaving | 139 | 95 |
| Binnacles | Ships; 1 | 114 | 137 |
| Bins, Meal | Kitchen Utensils | 65 | 88 |
| Bird-cages | Wire-Working | 140 | 170 |
| Bird-houses | Garden and Orchard | 47 | 63 |
| Biscuit-boards | Kitchen Utensils | 65 | 88 |
| Biscuit (clay) ovens | Dryers and Kilns | 34 | 115 |
| Biscuit-cutters | Kitchen Utensils | 65 | 88 |
| Bistouries | Cutlery | 30 | 156 |
| Bits, Boring. (Wood) | Wood-Working; 4 | 145 | 29 |
| Bits, Bridle | Harness | 54 | 72 |
| Bit-stocks | Wood-Working; 4 | 145 | 29 |
| Bitters, Tonic | Chemical Miscellaneous | 23 | 98 |
| Bitting-rigging | Harness | 54 | 144 |
| Bitts, Riding and towing | Ships; 1 | 114 | 137 |
| Black, Animal | Sugar | 127 | 155 |
| Blackboard-rubbers | Educational | 35 | 146 |
| Blackboards | Educational | 35 | 146 |
| Blackboards, Compositions for | Educational | 35 | 146 |
| Black, Bone | Sugar | 127 | 155 |
| Blacking-boxes | Furniture | 45 | 16 |
| Blacking-box holders | Furniture | 45 | 16 |
| Blacking leather, Compositions for | Tanning | 129 | 92 |
| Black, Ivory | Sugar | 127 | 155 |
| Black-leading machines for electrotyping | Printing | 101 | 164 |
| Blacksmiths' forges | Metal-Working; 3 | 78 | 60 |
| Blacksmiths' tools | Metal-Working; 6 | 81A | 60 |
| Blacksmiths' tool-stands | Metal-Working; 3 | 78 | 60 |
| Black-washing molds, Apparatus for | Casting | 22 | 33 |
| Blanching oils, fats, wax, &c | Oil, Fat, and Glue | 87 | 26 |
| Blanks from bars and plates, Cutting | Metal-Working; 4 | 79 | 119 |
| Blast-furnaces | Metallurgy | 75 | 85 |
| Blasting-cartridges | Projectiles | 102 | 124 |
| Blasting-compounds | Gunpowder | 52 | 71 |
| Blasting-fuses | Projectiles | 102 | 124 |
| Blasting-plugs | Projectiles | 102 | 124 |
| Blasting rock | Projectiles | 102 | 124 |
| Blast-nozzles for steam-engines | Steam; 3 | 123 | 94 |
| Blasting | Projectiles | 102 | 124 |
| Blast-machines | Pneumatics | 98 | 12 |
| Blast-regulators | Pneumatics | 98 | 12 |
| Bleaching-materials | Bleaching, Dyeing, &c | 8 | 11 |
| Bleaching-processes | Bleaching, Dyeing, &c | 8 | 11 |
| Bleaching | Bleaching, Dyeing, &c | 8 | 11 |
| Blind-adjusters | Builders' Hardware | 16 | 24 |
| Blind-bridles | Harness | 54 | 72 |
| Blind-fasteners | Builders' Hardware | 16 | 24 |
| Blind-fittings, Manufacture of | Hardware Manufactures | 53B | 24 |
| Blind-folding devices for animals | Harness | 54 | 72 |
| Blind-hinges | Builders' Hardware | 16 | 24 |
| Blind-operators | Builders' Hardware | 16 | 24 |
| Blinds | Carpentry | 20 | 30 |
| Blinds for horses | Harness | 54 | 72 |

| SUBJECT. | Name of class—official classification. | No. of class—official classification. | No. of class—sub-scription classification. |
|---|---|---|---|
| Blind-slats, Machines for planing | Wood-Working; 3 | 144 | 30 |
| Blind-slats, Machines for tenoning | Wood-Working; 3 | 144 | 30 |
| Blind-slats, Machines for wiring | Wood-Working; 3 | 144 | 30 |
| Blinds, Machines for making | Wood-Working; 3 | 144 | 30 |
| Blinds, Metallic | Masonry | 72 | 4 |
| Blind-splints, Machines for making | Wood-Working; 2 | 143 | 174 |
| Blind-staples | Nails | 85 | 30 |
| Blind-stiles, Machines for making | Wood-Working; 3 | 144 | 30 |
| Blind-stiles, Machines for mortising | Wood-Working; 3 | 144 | 30 |
| Blind-stiles, Machines for piercing | Wood-Working; 3 | 144 | 30 |
| Blind-stiles, Machines for spacing | Wood-Working; 3 | 144 | 30 |
| Blinds, Woven | Weaving | 139 | 95 |
| Blinds, Woven-wire | Wire-Working | 140 | 170 |
| Blind, Writing-apparatus for the | Educational | 35 | 146 |
| Blister-spreaders | Surgery | 128 | 156 |
| Blocks, Bushings for | Hoisting | 57 | 79 |
| Blocks for hats and bonnets | Felting and Hats | 38 | 76 |
| Blocks, Machines for making nautical | Wood-Working; 3 | 144 | 174 |
| Blocks, Machines for making wooden paving | Wood-Working; 3 | 144 | 174 |
| Blocks, Pulley | Hoisting | 57 | 79 |
| Blocks, Snatch | Hoisting | 57 | 79 |
| Blocks, Tackle | Hoisting | 57 | 79 |
| Blood, Instruments for infusing | Surgery | 128 | 156 |
| Blood, Instruments for propelling the | Surgery | 128 | 156 |
| Blotters | Stationery | 120 | 146 |
| Blotters, Book | Stationery | 120 | 146 |
| Blotting-pads | Stationery | 120 | 146 |
| Blowers | Pneumatics | 98 | 12 |
| Blowers for furnaces | Pneumatics | 98 | 12 |
| Blowers for organs | Pneumatics | 98 | 12 |
| Blowers, Steam | Pneumatics | 98 | 12 |
| Blowing-apparatus for glass-makers | Pneumatics | 98 | 12 |
| Blowing fur | Felting and Hats | 38 | 76 |
| Blow-pipes | Pneumatics | 98 | 12 |
| Blow-pipes for glass-workers | Pneumatics | 98 | 12 |
| Blubber-presses | Presses | 100 | 26 |
| Boarding leather | Tanning | 129 | 92 |
| Boarding machines, Leather | Tanning | 129 | 92 |
| Boas | Apparel | 2 | 38 |
| Boat-bridges | Boats | 9 | 13 |
| Boat-building | Boats | 9 | 13 |
| Boat-detaching apparatus | Boats | 9 | 13 |
| Boat-hoists | Boats | 9 | 13 |
| Boat-hooks | Boats | 9 | 13 |
| Boats | Boats | 9 | 13 |
| Boats, Canal | Ships; 1 | 114 | 137 |
| Boats, Ferry | Ships; 1 | 114 | 137 |
| Boats, Folding | Boats | 9 | 13 |
| Boats, Ice-breaking | Ships; 1 | 114 | 137 |
| Boats, Life | Boats | 9 | 13 |
| Boats, Metallic | Boats | 9 | 13 |
| Boats, Sectional | Boats | 9 | 13 |
| Boats, Steam | Ships; 1 | 114 | 137 |
| Boats, Surf | Boats | 9 | 13 |
| Boats, Torpedo | Ships; 1 | 114 | 137 |
| Bobbinet-machines | Knitting | 66 | 89 |
| Bobbins | Cordage | 28 | 17 |
| Bobbins | Spinning | 118 | 142 |
| Bobbins for sewing-machines | Sewing-Machines | 112 | 134 |

| SUBJECT. | Name of class—official classification. | No. of class—official classification. | No. of class—subscription classification. |
|---|---|---|---|
| Bobbins for shuttles | Weaving | 138 | 95 |
| Bobbins for braiding hair | Toilet | 132 | 161 |
| Bobbin-winders | Spinning | 118 | 142 |
| Bobbin-winders for sewing-machines | Sewing-Machines | 112 | 134 |
| Bodies of vehicles | Carriages | 21 | 31 |
| Bodkins | Apparel | 2 | 38 |
| Body-conformators | Apparel | 2 | 38 |
| Bog-cutters | Plows | 97 | 114 |
| Boiler-covering | Steam; 2 | 122 | 149 |
| Boiler-feeders | Steam; 2 | 122 | 148 |
| Boiler-flue cleaners | Steam; 2 | 122 | 148 |
| Boiler-flues | Steam; 2 | 122 | 148 |
| Boiler-makers' tools | Metal-Working; 6 | 81A | 148 |
| Boiler-making, Steam | Metal-Working; 3 | 78 | 148 |
| Boilers, Agricultural | Stoves | 126 | 59 |
| Boilers, Domestic | Stoves | 126 | 154 |
| Boiler-setting, Steam | Steam; 2 | 122 | 149 |
| Boilers for heaters | Steam; 2 | 126 | 147 |
| Boilers, Furnace-doors for | Steam; 2 | 122 | 149 |
| Boilers, Furnaces for | Steam; 2 | 122 | 149 |
| Boilers, Locomotive | Steam; 3 | 123 | 94 |
| Boilers, Making sheet-metal domestic | Sheet-Metal | 113B | 136 |
| Boilers of steam-heaters | Steam; 2 | 122 | 147 |
| Boilers, Setting steam | Steam; 2 | 122 | 149 |
| Boilers, Steam | Steam; 2 | 122 | 148 |
| Boiler-tile arrangements | Steam; 2 | 122 | 149 |
| Boiler-tubes, Devices for cleaning | Steam; 2 | 122 | 148 |
| Boiler-tubes, Ferrules for | Steam; 2 | 122 | 148 |
| Boiler-tubes, Tools for scaling | Steam; 2 | 122 | 148 |
| Bolsters for spindles | Spinning | 118 | 142 |
| Bolt-cutters | Metal-Working; 6 | 81A | 100 |
| Bolt-extractors | Wood-Working; 4 | 145 | 29 |
| Bolt-heading machines | Bolts, Nuts, and Rivets | 10 | 14 |
| Bolting meal, &c. | Mills | 83 | 57 |
| Bolt-locks | Builders' Hardware | 16 | 24 |
| Bolt-making machines | Bolts, Nuts, and Rivets | 10 | 14 |
| Bolts, Chain | Builders' Hardware | 16 | 24 |
| Bolts, Flour | Mills | 83 | 57 |
| Bolts for doors, shutters, and blinds | Builders' Hardware | 16 | 24 |
| Bolts, (household,) Manufacture of | Hardware Manufactures | 53B | 24 |
| Bolts, Screw | Bolts, Nuts, and Rivets | 10 | 14 |
| Bolts, Spring | Builders' Hardware | 16 | 24 |
| Bolt-threading machines | Bolts, Nuts, and Rivets | 10 | 14 |
| Bomb-lances | Projectiles | 102 | 143 |
| Bone-black filters | Sugar | 127 | 155 |
| Bone-black furnaces | Sugar | 127 | 155 |
| Bone-black, Processes for reviving | Sugar | 127 | 155 |
| Bone-black, Purifying | Sugar | 127 | 155 |
| Bone-elevators | Surgery | 128 | 156 |
| Bone-mills | Ore | 90 | 3 |
| Bonnet-caps | Felting and Hats | 38 | 38 |
| Bonnet-clasps | Felting and Hats | 38 | 38 |
| Bonnet-frames | Felting and Hats | 38 | 38 |
| Bonnet-fronts | Felting and Hats | 38 | 38 |
| Bonnet-ironing machines | Felting and Hats | 38 | 38 |
| Bonnet-presses | Felting and Hats | 38 | 38 |
| Bonnets | Felting and Hats | 38 | 38 |
| Bonnets for horses | Harness | 54 | 72 |
| Bonnet-shaping machines | Felting and Hats | 38 | 38 |
| Bonnet-stands | Apparel | 2 | 38 |

| SUBJECT. | Name of class—official classification. | No. of class—official classification. | No. of class—sub-scription classification. |
|---|---|---|---|
| Book-binders' presses | Presses | 100 | 15 |
| Book-binders, Temporary | Stationery | 120 | 15 |
| Book-binding materials, processes, styles, tools, and machines | Book-Binding | 11 | 15 |
| Book-cases | Furniture | 45 | 62 |
| Book-clamps | Stationery | 120 | 146 |
| Book-cover fastenings | Book-Binding | 11 | 15 |
| Book-cover protectors | Stationery | 120 | 15 |
| Book-covers | Book-Binding | 11 | 15 |
| Book-covers, Machines for making | Book-Binding | 11 | 15 |
| Book-folding machines | Paper Manufactures | 93 | 117 |
| Book-holders | Stationery | 120 | 146 |
| Book-holders, Portable | Stationery | 120 | 146 |
| Book-markers | Stationery | 120 | 146 |
| Books, Copying | Stationery | 120 | 146 |
| Book-sewing machines | Book-Binding | 11 | 15 |
| Book-shelves | Furniture | 45 | 62 |
| Books in the round, Machines for cutting | Book-Binding | 11 | 15 |
| Books, Manifold-writing | Stationery | 120 | 146 |
| Book-supports | Furniture | 45 | 62 |
| Booms | Ships; 1 | 114 | 138 |
| Booms for logs | Boats | 9 | 13 |
| Boot-boxes of vehicles | Carriages and Wagons | 21 | 31 |
| Boot-channeling machines | Boots and Shoes | 12 | 16 |
| Boot-clamps | Boots and Shoes | 12 | 16 |
| Boot-cleaners | Boots and Shoes | 15 | 16 |
| Boot-crimps | Boots and Shoes | 12 | 16 |
| Boot-grooving machines | Boots and Shoes | 12 | 16 |
| Boot-hooks | Boots and Shoes | 12 | 16 |
| Boot-jacks | Furniture | 45 | 16 |
| Boot-linings | Boots and Shoes | 12 | 16 |
| Boot-patterns | Boots and Shoes | 12 | 16 |
| Boot-racks | Furniture | 45 | 62 |
| Boots and shoes | Boots and Shoes | 12 | 16 |
| Boot-seam rubbers | Boots and Shoes | 12 | 16 |
| Boots for horses | Harness | 54 | 144 |
| Boot-shanks, Manufacture of metallic | Hardware Manufactures | 53G | 16 |
| Boots, Metallic braces for | Boots and Shoes | 12 | 16 |
| Boots, Metallic shanks for | Boots and Shoes | 12 | 16 |
| Boots, Metallic soles for | Boots and Shoes | 12 | 16 |
| Boot-soles | Boots and Shoes | 12 | 16 |
| Boot-stretchers | Boots and Shoes | 12 | 16 |
| Boots, Wooden soles for | Boots and Shoes | 12 | 16 |
| Boot-trees | Boots and Shoes | 12 | 16 |
| Boot-uppers | Boots and Shoes | 12 | 16 |
| Boot-ventilators | Boots and Shoes | 2 | 16 |
| Boring-bits | Wood-Working; 4 | 145 | 29 |
| Boring cannon | Metal-Working; 2 | 77 | 105 |
| Boring cylinders | Metal-Working; 2 | 77 | 100 |
| Boring, Earth | Excavators | 37 | 51 |
| Boring-machines, Metal | Metal-Working; 2 | 77 | 100 |
| Boring-machines, (wood) | Wood-Working; 2 | 143 | ·29 |
| Boring rock | Stone, Lime, and Cement | 125 | 124 |
| Bosom-expanders | Crinoline and Corsets | 29 | 80 |
| Bosom-folders | Apparel | 2 | 38 |
| Bosoms | Apparel | 2 | 38 |
| Bottle-brushes | Brushes and Brooms | 15 | 22 |
| Bottle-cleaners | Aëration and Bottling | 1 | 1 |
| Bottle-envelopes | Aëration and Bottling | 1 | 169 |

| SUBJECT. | Name of class—official classification. | No. of class—official classification. | No. of class—subscription classification. |
|---|---|---|---|
| Bottle-faucets | Valves | 136 | 1 |
| Bottle-fillers | Aëration and Bottling | 1 | 1 |
| Bottle-holders | Kitchen Utensils | 65 | 88 |
| Bottle-holders for glass-making | Glass | 49 | 68 |
| Bottle making, Glass | Glass | 49 | 68 |
| Bottle-racks | Aëration and Bottling | 1 | 1 |
| Bottles, Boxes for transporting | Aëration and Bottling | 1 | 1 |
| Bottles, Dropping | Measuring-Instruments | 73 | 98 |
| Bottles, Graduated medicine | Measuring-Instruments | 73 | 98 |
| Bottles, Mucilage | Stationery | 120 | 146 |
| Bottle-stoppers | Aëration and Bottling | 1 | 1 |
| Bottle-washers | Aëration and Bottling | 1 | 1 |
| Bottling-implements | Aëration and Bottling | 1 | 1 |
| Bottling-machines | Aëration and Bottling | 1 | 1 |
| Bouquet-holders | Garden and Orchard | 47 | 63 |
| Bow-drills | Metal-Working; 2 | 77 | 100 |
| Bow-irons of vehicles | Carriages and Wagons | 21 | 31 |
| Bowls, Machines for turning | Wood-Working; 1 | 142 | 175 |
| Bow-pencils and pens | Drafting | 33 | 97 |
| Bow-string bridges | Bridges | 14 | 21 |
| Bows, Wagon | Carriages and Wagons | 21 | 31 |
| Box-cars | Railways; 2 | 105 | 121 |
| Boxes, Ballot | Stationery | 120 | 42 |
| Boxes, Cartridge | Trunks | 133 | 162 |
| Boxes, Conductors' check | Stationery | 120 | 42 |
| Boxes direct from pulp, Making paper | Paper-Making | 92 | 108 |
| Boxes, Egg-packing | Kitchen Utensils | 65 | 116 |
| Boxes, Fare | Stationery | 120 | 42 |
| Boxes, Fish | Kitchen Utensils | 65 | 63 |
| Boxes, Food | Kitchen Utensils | 65 | 63 |
| Boxes for plants | Garden and Orchard | 47 | 63 |
| Boxes for provisions | Kitchen Utensils | 65 | 63 |
| Boxes for spools and thread | Apparel | 2 | 38 |
| Boxes for transportation, Refrigerating | Ice | 62 | 58 |
| Boxes for vegetables | Kitchen Utensils | 65 | 63 |
| Boxes, Fruit | Kitchen Utensils | 65 | 63 |
| Boxes, Hat | Trunks | 133 | 162 |
| Boxes, Journal | Journals and Bearings | 64 | 87 |
| Boxes, Letter | Stationery | 120 | 146 |
| Boxes, Lunch | Kitchen Utensils | 65 | 63 |
| Boxes, Machines for making wooden | Wood-Working; 3 | 144 | 174 |
| Boxes, Machines for turning wooden | Wood-Working; 1 | 142 | 172 |
| Boxes, Mail-distributing | Stationery | 120 | 146 |
| Boxes, Making sheet-metal | Sheet-Metal | 113 B | 136 |
| Boxes, Match | Lamps and Gas-Fittings | 67 | 90 |
| Boxes of wheel-hubs | Carriages | 21 | 31 |
| Boxes, Paper | Paper Manufactures | 93 | 109 |
| Boxes, Pepper | Kitchen Utensils | 65 | 88 |
| Boxes, Railway-axle | Railways; 3 | 106 | 122 |
| Boxes, Spice | Kitchen Utensils | 65 | 88 |
| Boxes, Ticket | Stationery | 120 | 146 |
| Box machines, Paper | Paper Manufactures | 93 | 109 |
| Box-packing presses | Presses | 100 | 7 |
| Box-plaiting attachments for sewing-machines | Sewing-Machines | 112 | 134 |
| Box-scrapers | Wood-Working; 4 | 145 | 29 |
| Bracelets | Jewelry | 63 | 86 |
| Braces | Apparel | 2 | 38 |
| Braces for carriages, Manufacture of | Hardware Manufactures | 53 C | 31 |
| Braces for trunks | Trunks | 133 | 162 |

| SUBJECT. | Name of class—official classification. | No. of class—official classification. | No. of class—sub-scription classification. |
|---|---|---|---|
| Brackets for eave-troughs | Roofing | 108 | 126 |
| Brackets for roofs | Roofing | 108 | 126 |
| Brackets for staging | Carpentry | 20 | 30 |
| Bracket-shelves | Furniture | 45 | 62 |
| Braid | Cordage | 28 | 17 |
| Braiders for sewing-machines | Sewing-Machines | 112 | 134 |
| Braid-holders | Cordage | 28 | 17 |
| Braiding cords | Cordage | 28 | 17 |
| Braiding, Hair | Toilet | 132 | 161 |
| Braiding-machine carriers | Cordage | 28 | 17 |
| Braiding-machines | Cordage | 28 | 17 |
| Braiding tapes | Cordage | 28 | 17 |
| Braiding whips | Cordage | 28 | 72 |
| Brakes, Car | Railways; 3 | 106 | 122 |
| Brakes, Carriage | Carriages and Wagons | 21 | 31 |
| Brakes for hoisting-machinery | Mechanical Powers | 74 | 79 |
| Brakes for horse-powers | Mechanical Powers | 74 | 159 |
| Brakes for machinery | Mechanical Powers | 74 | 79 |
| Brakes, Wagon | Carriages and Wagons | 21 | 31 |
| Braking-machines | Brakes and Gins | 13 | 78 |
| Bran-dusters | Mills | 83 | 57 |
| Brass-founding | Casting | 22 | 10 |
| Brass tubes, Machines for making | Tubing and Wire | 134 | 163 |
| Brazing | Sheet-Metal | 113 A | 136 |
| Bread and pastry boards | Kitchen Utensils | 65 | 88 |
| Bread-cutters | Kitchen Utensils | 65 | 88 |
| Bread-machines | Kitchen Utensils | 65 | 6 |
| Bread-raising compounds | Chemical Miscellaneous | 23 | 6 |
| Breaking horses, Apparatus for | Harness | 54 | 144 |
| Breakwaters | Hydraulic Engineering | 61 | 82 |
| Breast-bails | Harness | 54 | 72 |
| Breast-cups | Surgery | 128 | 156 |
| Breast-pads | Crinoline and Corsets | 29 | 80 |
| Breast-pins | Jewelry | 63 | 86 |
| Breast-pumps | Surgery | 128 | 156 |
| Breast-slides | Harness | 54 | 72 |
| Breast-straps | Harness | 54 | 72 |
| Breast-strap slides | Harness | 54 | 72 |
| Breech-loading cannon | Ordnance | 89 | 105 |
| Breech-loading small-arms | Fire-Arms | 42 | 18 |
| Breech-straps | Harness | 54 | 72 |
| Brequet-chain hooks | Clasps and Buckles | 24 | 87 |
| Brewery-apparatus | Beer and Wine | 7 | 19 |
| Brewing | Beer and Wine | 7 | 19 |
| Brick-kilns | Dryers and Kilns | 34 | 20 |
| Bricklayers' tools, Manufacture of | Metal-Working; 6 | 31 B | 4 |
| Brick-machines | Clay | 25 | 20 |
| Brick-molds | Clay | 25 | 20 |
| Brick-presses | Clay | 25 | 20 |
| Bricks | Clay | 25 | 20 |
| Bricks, Fire | Clay | 25 | 20 |
| Brick-work of buildings | Masonry | 72 | 4 |
| Bridges | Bridges | 14 | 21 |
| Bridges, Deck | Ships; 1 | 14 | 137 |
| Bridges, Hose | Railway; 1 | 104 | 81 |
| Bridge-signals | Signals | 116 | 21 |
| Bridges, Paddle-box | Ships; 1 | 114 | 137 |
| Bridges, Steam-furnace | Steam; 2 | 122 | 149 |
| Bridles | Harness | 54 | 72 |
| Bridle-winkers | Harness | 54 | 72 |

| SUBJECT. | Name of class—official classification. | No. of class—official classification. | No. of class—sub-scription classification. |
|---|---|---|---|
| Brine-tubs | Butchering | 17 | 25 |
| Bristle-assorting machines | Brushes and Brooms | 15 | 22 |
| Bristle-bunching machines | Brushes and Brooms | 15 | 22 |
| Bristle-cleaning machines | Brushes and Brooms | 15 | 22 |
| Bristle-washing machines | Brushes and Brooms | 15 | 22 |
| Britannia-ware | Sheet-Metal | 113 B | 34 |
| Britannia-ware, Casting-processes for | Casting | 22 | 34 |
| Broadcast sowers | Seeders and Planters | 111 | 133 |
| Broilers | Stoves and Furnaces | 126 | 154 |
| Bronzing-machines | Fine Arts | 41 | 54 |
| Brooches, Manufacture of | Hardware Manufactures | 53 E | 86 |
| Broom-corn presses | Presses | 100 | 22 |
| Broom-handles | Brushes and Brooms | 15 | 22 |
| Broom-handles, Machines for making | Wood-Working; 1 | 142 | 174 |
| Broom-hangers | Furniture | 45 | 62 |
| Broom-heads | Brushes and Brooms | 15 | 22 |
| Brooms | Brushes and Brooms | 15 | 22 |
| Broom-splints, Machines for making | Wood-Working; 2 | 143 | 22 |
| Brush-backs | Brushes and Brooms | 15 | 22 |
| Brush-backs, Machines for making | Wood-Working; 3 | 144 | 22 |
| Brushes, (*all kinds except Wire*) | Brushes and Brooms | 15 | 22 |
| Brushes | Brushes and Brooms | 15 | 22 |
| Brushes, Wire | Wire-Working | 140 | 170 |
| Brush-handles | Brushes and Brooms | 15 | 22 |
| Brush-handles, Machines for making | Wood-Working; 3 | 144 | 22 |
| Brush-heads | Brushes and Brooms | 15 | 22 |
| Brush-making machines | Brushes and Brooms | 15 | 22 |
| Brush-pullers | Hoisting | 57 | 79 |
| Brush-racks | Furniture | 45 | 62 |
| Buckets | Kitchen Utensils | 65 | 88 |
| Buckets, Dinner | Kitchen Utensils | 65 | 63 |
| Buckets, Dumping | Hoisting | 57 | 79 |
| Buckets, Folding | Kitchen Utensils | 65 | 63 |
| Buckets, Hoisting | Hoisting | 57 | 79 |
| Buckets, Machines for making | Wood-Working; 3 | 144 | 40 |
| Buckets, Making sheet-metal | Sheet-Metal | 113 B | 136 |
| Bucking-kiers | Bleaching, Dyeing, &c | 8 | 11 |
| Buckles | Clasps and Buckles | 24 | 23 |
| Buckles, (harness,) Manufacture of | Hardware Manufactures | 53 E | 23 |
| Buckles, Manufacture of | Hardware Manufactures | 53 E | 23 |
| Budding-knives | Metal-Working; 5 | 80 | 63 |
| Buffers | Grinding and Polishing | 51 | 70 |
| Buffers, Car | Railway; 3 | 106 | 122 |
| Buffers for doors | Builders' Hardware | 16 | 24 |
| Buffing-machines | Grinding and Polishing | 51 | 70 |
| Buhl-work | Fine Arts | 41 | 54 |
| Builders' sheet-metal ware | Sheet-Metal | 113 B | 136 |
| Buildings, Moving | Mechanical Powers | 74 | 67 |
| Bulkheads | Ships; 1 | 114 | 137 |
| Bullet-casting | Casting | 22 | 32 |
| Bulletin-boards | Stationery | 120 | 146 |
| Bulletin-boards, Adjustable | Stationery | 120 | 146 |
| Bullet-pressing machines | Hardware Manufactures | 53 A | 32 |
| Bullets | Projectiles | 102 | 32 |
| Bullets, Lubricating | Projectiles | 102 | 32 |
| Bullets, Manufacture of | Hardware Manufactures | 53 A | 32 |
| Bullion, Machines for making | Weaving | 139 | 95 |
| Bull-rings | Harness | 54 | 144 |
| Bumpers, Car | Railways; 3 | 106 | 122 |
| Bumping-posts | Railways; 1 | 104 | 120 |

| SUBJECT. | Name of class—official classification. | No. of class—official classification. | No. of class—subscription classification. |
|---|---|---|---|
| Bundling fire-wood, Machines for | Wood-Working; 3 | 144 | 174 |
| Bundling-presses for yarns and threads | Presses | 100 | 17 |
| Bung-cutters | Wood-Working; 1 | 142 | 175 |
| Bungs | Aëration | 1 | 40 |
| Bungs, Bushes for | Aëration | 1 | 40 |
| Bungs, Extractors for | Aëration | 1 | 40 |
| Bungs, Machines for turning | Wood-Working; 1 | 142 | 172 |
| Buoys | Boats | 9 | 138 |
| Buoy-safes | Boats | 9 | 128 |
| Buoys, Electrical | Electricity | 36 | 138 |
| Burial-cases, Compositions for making | Coffins | 27 | 39 |
| Burial-caskets | Coffins | 27 | 39 |
| Burglar-alarms | Signals | 116 | 24 |
| Bureaus | Furniture | 45 | 62 |
| Burettes | Measuring-Instruments | 73 | 98 |
| Burglar-proof safes | Safes | 109 | 128 |
| Burners | Lamps and Gas-Fittings | 67 | 64 |
| Burning-fluids | Stills | 124 | 47 |
| Burning lime | Stone, Lime, and Cement | 125 | 153 |
| Burnishers | Grinding and Polishing | 51 | 70 |
| Burnishing-machines | Grinding and Polishing | 51 | 70 |
| Burnishing machines, Boot | Boots and Shoes | 12 | 16 |
| Burr-cutters | Metal-Working; 7 | 82 | 100 |
| Burring-machines | Carding | 19 | 28 |
| Burrs, Metallic | Mills | 83 | 57 |
| Bush-extractors | Hoisting | 57 | 79 |
| Bushings for barrels | Aëration | 1 | 40 |
| Bushings for gun-vents | Ordnance | 89 | 105 |
| Busks | Crinoline and Corsets | 29 | 80 |
| Bustles | Crinoline and Corsets | 29 | 80 |
| Butchering appliances | Butchering | 17 | 25 |
| Butter-buckets | Kitchen Utensils | 65 | 45 |
| Butter-dishes | Kitchen Utensils | 65 | 88 |
| Butter-molds | Dairy | 31 | 45 |
| Butter-stampers | Dairy | 31 | 45 |
| Butter-tongs | Kitchen Utensils | 65 | 45 |
| Butter-workers | Dairy | 31 | 45 |
| Button-hole punches | Clasps and Buckles | 24 | 38 |
| Button-hole sewing-machines | Sewing-Machines | 112 | 134 |
| Button-hooks | Clasps and Buckles | 24 | 38 |
| Button-keys or fasteners | Clasps and Buckles | 24 | 23 |
| Button making, Glass | Glass | 49 | 68 |
| Button-rings, Manufacture of | Hardware Manufactures | 53 E | 23 |
| Buttons, (door,) Manufacture of | Hardware Manufactures | 53 B | 24 |
| Buttons, Gilding | Plating | 96 | 113 |
| Buttons, Machines for turning | Wood-Working; 1 | 142 | 172 |
| Buttons, (metallic,) Manufacture of | Hardware Manufactures | 53 E | 23 |
| Buttons of any material, (*except Turning, Class* 142) | Clasps and Buckles | 24 | 23 |
| Buttons to cards, Securing | Claps and Buckles | 24 | 23 |
| Buttons to fabrics, Eyeleting | Clasps and Buckles | 24 | 23 |
| Buttons to fabrics, Riveting | Clasps and Buckles | 24 | 23 |
| **C.** | | | |
| Cabin-door hooks | Builders' Hardware | 16 | 24 |
| Cabinet-hooks, Manufacture of | Hardware Manufactures | 53 D | 24 |
| Cabinet-makers' tools | Wood-Working; 4 | 145 | 49 |
| Cabinet-makers' tools of metal, Manufacture of | Metal-Working; 6 | 81 B | 49 |
| Cabinet-organs | Music | 84 | 106 |

| SUBJECT. | Name of class—official classification. | No. of class—official classification. | No. of class—subscription classification. |
|---|---|---|---|
| Cabinets | Furniture | 16 | 24 |
| Cabins, Floating | Ships; 1 | 114 | 137 |
| Cable-grippers | Ships; 1 | 114 | 138 |
| Cable-hooks | Ships; 1 | 114 | 138 |
| Cable-nippers | Ships; 1 | 114 | 138 |
| Cables for submarine telegraphs | Wire-Working | 140 | 170 |
| Cables for suspension-bridges | Wire-Working | 140 | 170 |
| Cables for vessels | Ships; 1 | 114 | 138 |
| Cable-shackles | Ships; 1 | 114 | 138 |
| Cables, Laying submarine | Hydraulic Engineering | 61 | 82 |
| Cables, Links for | Ships; 1 | 114 | 138 |
| Cables, Securing | Ships; 1 | 114 | 138 |
| Cables, Submarine telegraph | Electricity | 36 | 170 |
| Cables, Suspension-bridge. | Wire-Working | 140 | 170 |
| Cables, Telegraph | Electricity | 36 | 170 |
| Cable-stoppers | Ships; 1 | 114 | 138 |
| Cables, Wire | Wire-Working | 140 | 170 |
| Cabs for railway-cars | Railways; 3 | 106 | 121 |
| Caissons for piers, &c | Hydraulic Engineering | 61 | 82 |
| Cake-cutters | Kitchen Utensils | 65 | 88 |
| Cake-turners | Kitchen Utensils | 65 | 154 |
| Calash-tops for carriages | Carriages | 21 | 31 |
| Calcium lights | Lamps and Gas-Fittings | 67 | 90 |
| Calculating-machines | Measuring-Instruments | 73 | 42 |
| Calculi, Instruments for extracting | Surgery | 128 | 156 |
| Calendars | Stationery | 120 | 146 |
| Calendars, Electrical | Electricity | 36 | 50 |
| Calendering cloth | Cloth | 26 | 37 |
| Calico-printing, Composition for | Bleaching and Dyeing | 8 | 11 |
| Calico-printing machines | Printing | 101 | 11 |
| Calico-printing, Processes for | Bleaching and Dyeing | 8 | 11 |
| Calipers | Draughting | 33 | 97 |
| Calking | Ships; 1 | 114 | 137 |
| Calking-machines | Ships; 1 | 114 | 137 |
| Calks, Boot | Boots and Shoes | 12 | 16 |
| Call-bells | Signals | 116 | 24 |
| Caloric-engines | Pneumatics | 98 | 2 |
| Camels | Ships; 1 | 114 | 82 |
| Cameras | Optics | 88 | 111 |
| Cameras, Photographic | Photography | 95 | 111 |
| Cameras, Solar | Photography | 88 | 111 |
| Camera-stands | Photography | 95 | 111 |
| Camp-beds | Beds | 5 | 62 |
| Camp-cots | Beds | 5 | 62 |
| Camp-kits | Kitchen Utensils | 65 | 88 |
| Camp-stools | Furniture | 45 | 62 |
| Cam-rods | Mechanical Powers | 74 | 67 |
| Cams, Variable | Mechanical Powers | 74 | 67 |
| Canal-boats | Ships; 1 | 114 | 137 |
| Canal-boats, Propulsion of | Ships; 2 | 115 | 139 |
| Canal-hoists | Hydraulic Engineering | 61 | 82 |
| Canal-lining | Hydraulic Engineering | 61 | 82 |
| Canal-lock gates | Hydraulic Engineering | 61 | 82 |
| Canal-locks | Hydraulic Engineering | 61 | 82 |
| Canal-propulsion, apart from the boat. | Hydraulic Engineering | 61 | 82 |
| Canals | Hydraulic Engineering | 61 | 82 |
| Candle-holders | Lamps and Gas-Fitting | 67 | 90 |
| Candle-machinery | Oil, Fat, and Glue | 87 | 26 |
| Candle-making | Oil, Fat, and Glue | 87 | 26 |
| Candle-molds | Oil, Fat, and Glue | 87 | 26 |
| Candle-molds, Compositions for | Oil, Fat, and Glue | 87 | 26 |

| SUBJECT. | Name of class—official classification. | No. of class—official classification. | No. of class—subscription classification. |
|---|---|---|---|
| Candles | Oil, Fat, and Glue | 87 | 26 |
| Candles, Compositions for | Oil, Fat, and Glue | 87 | 26 |
| Candlesticks | Lamps and Gas-Fitting | 67 | 90 |
| Candlesticks, Making sheet-metal | Sheet-Metal | 113 B | 136 |
| Candy-cutters | Sugar | 127 | 155 |
| Candy-making machines | Sugar | 127 | 155 |
| Cane and seat combined | Umbrellas and Fans | 135 | 62 |
| Cane-guns | Fire-Arms | 42 | 55 |
| Cane-harvesters | Harvesters | 56 | 75 |
| Cane-juice, Treatment of | Sugar | 127 | 155 |
| Cane-knives | Cutlery | 30 | 44 |
| Cane, Machines for polishing | Wood-Working; 3 | 144 | 169 |
| Cane, Machines for splitting | Wood-Working; 3 | 144 | 169 |
| Cane, Machines for stripping | Wood-Working; 3 | 144 | 169 |
| Cane-mills | Mills | 83 | 155 |
| Cane-planters | Seeders and Planters | 111 | 133 |
| Canes, Ferrules for walking | Umbrellas and Fans | 135 | 165 |
| Canes, Physicians' | Surgery | 128 | 156 |
| Cane-strippers | Wood-Working; 3 | 144 | 155 |
| Canes, Walking | Umbrellas and Fans | 135 | 165 |
| Caning chairs, Instruments for | Wood-Working; 3 | 144 | 62 |
| Canisters | Kitchen Utensils | 65 | 88 |
| Canister-shot | Projectiles | 102 | 105 |
| Cannons | Ordnance | 89 | 105 |
| Cannon-sights | Ordnance | 89 | 105 |
| Can-openers | Cutlery | 30 | 88 |
| Canopies for bedsteads | Beds | 5 | 62 |
| Cans, Air-tight | Preserving | 99 | 58 |
| Cans for preserving provisions | Preserving | 99 | 28 |
| Cans for slivers | Carding | 19 | 58 |
| Cans, Making sheet-metal | Sheet-Metal | 113 B | 136 |
| Canteens | Kitchen Utensils | 65 | 88 |
| Cant-hooks | Saws | 110 | 130 |
| Can tops, caps, and stoppers | Preserving | 99 | 58 |
| Canvas-stretchers | Fine Arts | 41 | 54 |
| Caoutchouc manufactures | Caoutchouc | 18 | 84 |
| Caoutchouc-printing | Printing | 101 | 117 |
| Cap boxes, Percussion | Projectiles | 102 | 55 |
| Capes | Apparel | 2 | 38 |
| Caponizing devices | Surgery | 128 | 156 |
| Capping nails | Nails | 85 | 103 |
| Capping screws | Nails | 85 | 103 |
| Caps | Felting and Hats | 38 | 38 |
| Cap-spinning machines | Spinning | 118 | 142 |
| Capstans | Hoisting | 57 | 79 |
| Capstans for mole-plows | Hoisting | 57 | 79 |
| Capsules, Bottle | Aëration | 1 | 1 |
| Capsules for cans and bottles, Making | Sheet-Metal | 113B | 136 |
| Capsules, Machines for making medicinal | Chemical Miscellaneous | 23 | 98 |
| Capsules, Medicinal | Chemical Miscellaneous | 23 | 98 |
| Car-axle boxes | Railways; 3 | 106 | 122 |
| Car-axles | Railways; 3 | 106 | 122 |
| Carbonic-acid-gas engines, (motors) | Pneumatics | 98 | 2 |
| Carbonizing-furnaces | Metallurgy | 75 | 99 |
| Carbonizing wood, Furnaces and kilns for | Driers and Kilns | 34 | 61 |
| Carboys | Glass | 49 | 68 |
| Car-brakes | Railways; 3 | 106 | 122 |
| Car-brakes, Automatic | Railways; 3 | 106 | 122 |

| SUBJECT. | Name of class—official classification. | No. of class—official classification. | No. of class—subscription classification. |
|---|---|---|---|
| Car-brakes, Electrical | Electricity | 36 | 122 |
| Car-brake shoes | Railways; 3 | 106 | 122 |
| Car-brakes, Pneumatic | Steam; 1 | 121 | 122 |
| Car-brakes, Steam-actuated | Steam; 1 | 121 | 122 |
| Car-buffers | Railways; 3 | 106 | 122 |
| Car-bumper springs | Railways; 3 | 106 | 122 |
| Carburetors for air or gas | Gas | 48 | 27 |
| Car-cabs | Railways; 3 | 106 | 121 |
| Carcasses | Projectiles | 102 | 105 |
| Car-couches | Railways; 2 | 105 | 121 |
| Car-couplings | Railways; 3 | 106 | 122 |
| Card-board | Paper-Making | 92 | 108 |
| Card-board driers | Driers and Kilns | 34 | 108 |
| Card-calendars | Stationery | 120 | 146 |
| Card-cases | Stationery | 120 | 146 |
| Card-clothing | Carding | 19 | 28 |
| Card-grinding | Grinding and Polishing | 51 | 28 |
| Card-holders | Stationery | 120 | 146 |
| Carding-machines | Carding | 19 | 28 |
| Carding-machines, Alarms for | Carding | 19 | 28 |
| Carding-machines, Cylinders for | Carding | 19 | 28 |
| Carding-machines, Stop-motions for | Carding | 19 | 28 |
| Card-mounts | Fine Arts | 41 | 146 |
| Car-door latches | Locks and Latches | 70 | 121 |
| Car-door locks | Locks and Latches | 70 | 121 |
| Card-racks | Stationery | 120 | 146 |
| Car draw-bars | Railways; 3 | 106 | 122 |
| Car draw-heads | Railways; 3 | 106 | 122 |
| Cards | Stationery | 120 | 146 |
| Cards, Cleaning | Carding | 19 | 28 |
| Cards, Currying | Stabling | 119 | 144 |
| Cards, Playing | Games and Toys | 46 | 143 |
| Cards, Reading | Games and Toys | 46 | 143 |
| Card-strippers | Carding | 19 | 28 |
| Cards, (Wool and Cotton) | Carding | 19 | 28 |
| Cargo-ports | Ships; 1 | 114 | 137 |
| Car-heaters | Stoves and Furnaces | 126 | 154 |
| Car-jacks | Hoisting | 57 | 79 |
| Car-pedestals | Railways; 6 | 106 | 122 |
| Carpenters' benches | Wood-Working; 4 | 145 | 29 |
| Carpenters' clamps | Wood-Working; 4 | 145 | 29 |
| Carpenters' gages | Wood-Working; 4 | 145 | 29 |
| Carpenters' machines | Wood-Working; 3 | 144 | 29 |
| Carpenters' tools | Wood-Working; 4 | 145 | 29 |
| Carpenters' tools of metal, Manufacture of | Metal-Working; 6 | 81B | 49 |
| Carpenters' work of buildings | Carpentry | 20 | 30 |
| Carpet-bag frames | Trunks | 133 | 162 |
| Carpet-bag frames, Manufacture of | Hardware Manufactures | 53J | 162 |
| Carpet-bags | Trunks | 133 | 162 |
| Carpet-cleaners | Furniture | 45 | 62 |
| Carpet fabrics | Weaving | 139 | 56 |
| Carpet-fasteners | Furniture | 45 | 62 |
| Carpet-lining | Weaving | 139 | 95 |
| Carpet-lining machines | Weaving | 139 | 56 |
| Carpet-rag cutter | Cloth | 26 | 95 |
| Carget-rag looper | Weaving | 139 | 95 |
| Carpet-stretchers | Furniture | 45 | 62 |
| Carpeting | Weaving | 139 | 95 |
| Car-platforms | Railways; 3 | 106 | 122 |

| SUBJECT. | Name of class—official classification. | No. of class—official classification. | No. of class—subscription classification. |
|---|---|---|---|
| Carriage-bow joints, Manufacture of | Hardware Manufactures | 53C | 31 |
| Carriage-couplings | Carriages | 21 | 31 |
| Carriage-curtain fastenings | Carriages | 21 | 31 |
| Carriage-door handles, Manufacture of. | Hardware Manufactures | 53C | 31 |
| Carriage-hardware | Carriages | 21 | 31 |
| Carriage-heaters | Stoves and Furnaces | 126 | 154 |
| Carriage-irons | Carriages | 21 | 31 |
| Carriage-jacks | Hoisting | 51 | 79 |
| Carriages | Carriages | 21 | 31 |
| Carriages, Dress-guards for | Carriages | 22 | 31 |
| Carriage-seat fastenings | Carriages | 21 | 31 |
| Carriage-seats | Carriages | 21 | 31 |
| Carriages, Fire-engine | Carriages | 21 | 31 |
| Carriages for artillery | Ordnance | 89 | 105 |
| Carriage-springs | Carriages | 21 | 31 |
| Carriage-springs, Manufacture of | Hardware Manufactures | 53C | 31 |
| Carriage-step protectors | Carriages | 21 | 31 |
| Carriage-steps | Carriages | 21 | 31 |
| Carriage-steps, Manufacture of | Hardware Manufactures | 53C | 31 |
| Carriages, Top-blocks for | Carriages | 21 | 31 |
| Carriage-top braces, Manufacture of | Hardware Manufactures | 53C | 31 |
| Carriage-tops | Carriages | 21 | 31 |
| Carriage-trimmings | Carriages | 21 | 31 |
| Carriage-window frames | Carriages | 21 | 31 |
| Car-replacers | Railways; 1 | 104 | 120 |
| Car-roofs | Railways; 2 | 105 | 121 |
| Cars, Adhesion | Railways; 2 | 105 | 121 |
| Cars, Box | Railways; 2 | 105 | 121 |
| Cars, Cattle | Railways; 2 | 105 | 121 |
| Cars, Coal | Railways; 2 | 105 | 121 |
| Cars, Dummy | Railways; 2 | 105 | 121 |
| Cars, Dumping | Railways; 2 | 105 | 121 |
| Car-seat arm-locks | Locks and Latches | 70 | 121 |
| Car-seats | Railways; 2 | 105 | 121 |
| Cars for transportation, Refrigerating | Ice | 62 | 121 |
| Cars, Freight | Railways; 2 | 105 | 121 |
| Cars, Gravel | Railways; 2 | 105 | 121 |
| Cars, Hand | Railways; 2 | 105 | 121 |
| Cars, Petroleum | Railways; 2 | 105 | 121 |
| Car-springs | Railways; 3 | 106 | 122 |
| Cars, Revolving | Railways; 2 | 105 | 121 |
| Cars, Safety | Railways; 2 | 105 | 121 |
| Cars, Sleeping | Railways; 2 | 105 | 121 |
| Cars, Stock | Railways; 2 | 105 | 121 |
| Cars, Street | Railways; 2 | 105 | 121 |
| Car stake-holders | Railways; 3 | 106 | 122 |
| Car-standards | Railways; 3 | 106 | 122 |
| Cars, Tank | Railways; 2 | 105 | 121 |
| Car-starters | Railways; 3 | 106 | 122 |
| Car-starters, Electrical | Electricity | 36 | 122 |
| Cars, Wrecking | Railways; 2 | 105 | 121 |
| Cartridge-filling machines | Projectiles | 102 | 32 |
| Cartridge-holders | Projectiles | 102 | 32 |
| Cartridge-linings | Projectiles | 102 | 32 |
| Cartridge-loaders | Fire-Arms | 42 | 32 |
| Cartridge-pouches | Projectiles | 102 | 162 |
| Cartridge-priming machines | Projectiles | 102 | 32 |
| Cartridges | Projectiles | 102 | 32 |
| Cartridges, Blasting | Projectiles | 102 | 124 |
| Cartridges, Compositions for | Gunpowder | 52 | 71 |

| SUBJECT. | Name of class—official classification. | No. of class—official classification. | No. of class—subscription classification. |
|---|---|---|---|
| Cartridges for fire-arms | Projectiles | 102 | 32 |
| Cartridges for ordnance | Projectiles | 102 | 105 |
| Cartridge-shell retractors | Fire-Arms | 42 | 18 |
| Cartridge-shells, Manufacture of | Hardware Manufactures | 53A | 32 |
| Cartridges, Metallic | Projectiles | 102 | 32 |
| Cartridges, Priming | Projectiles | 102 | 32 |
| Cartridge-wrappers | Projectiles | 102 | 32 |
| Car-truck frames | Railways; 3 | 106 | 122 |
| Car-trucks | Railways; 3 | 106 | 122 |
| Car-trusses | Railways; 3 | 106 | 122 |
| Carts | Carriages and Wagons | 21 | 31 |
| Cart-saddles | Harness | 54 | 72 |
| Carving-machines | Wood-Working; 2 | 143 | 173 |
| Carving stone | Stone, Lime, and Cement | 125 | 153 |
| Car-wheels | Railways; 3 | 106 | 122 |
| Car-wheels, Making wrought | Railways; 4 | 107 | 122 |
| Car-window fastenings | Builders' Hardware | 16 | 121 |
| Car-windows | Railways; 2 | 105 | 121 |
| Case-hardening compounds | Metallurgy | 75 | 85 |
| Case-hardening processes | Metallurgy | 75 | 85 |
| Caskets, Burial | Coffins | 27 | 39 |
| Casks, Machines for making | Wood-Working; 3 | 144 | 40 |
| Caster-bottles | Kitchen Utensils | 65 | 88 |
| Casters, Furniture | Builders' Hardware | 16 | 24 |
| Casters, Manufacture of | Hardware Manufactures | 53D | 24 |
| Casters, Stove-leg | Stoves and Furnaces | 126 | 154 |
| Caster-stands | Kitchen Utensils | 65 | 88 |
| Casting dental plates | Dental | 32 | 46 |
| Casting glass | Glass | 49 | 68 |
| Casting metal on metal | Casting | 22 | 33 |
| Casting metal rules between rollers | Metal-Working; 5 | 80 | 33 |
| Casting metals, Appliances for | Casting | 22 | 33 |
| Casting metals, Modes of | Casting | 22 | 33 |
| Casting metal under pressure | Casting | 22 | 33 |
| Casting sheet-metal between rollers | Metal-Working; 5 | 80 | 136 |
| Casting soft metals | Casting | 22 | 34 |
| Castings, Tumbling-boxes for | Casting | 22 | 33 |
| Castrating-clamps | Surgery | 128 | 156 |
| Casts, Instruments for obtaining | Fine Arts | 41 | 54 |
| Casts, Processes for obtaining | Fine Arts | 41 | 54 |
| Catamenial sacks | Surgery | 128 | 156 |
| Cataracts for steam-engines | Steam; 1 | 121 | 69 |
| Catarrhal syringes | Surgery | 128 | 156 |
| Catching animals for slaughtering | Butchering | 17 | 25 |
| Catheters | Surgery | 128 | 156 |
| Catoptric lights | Optics | 88 | 90 |
| Cattle-cars | Railways; 2 | 105 | 121 |
| Cattle-feeders | Stabling | 119 | 144 |
| Cattle-gags | Harness | 54 | 144 |
| Cattle-guards | Railways; 1 | 104 | 120 |
| Cattle-leaders | Harness | 54 | 72 |
| Cattle-pens | Stabling | 119 | 144 |
| Cattle-stalls | Stabling | 119 | 144 |
| Cattle-ties | Stabling | 119 | 144 |
| Cauterizing-instruments | Surgery | 128 | 156 |
| Ceilings | Carpentry | 20 | 4 |
| Cellar-floors | Masonry | 72 | 4 |
| Cellars, Draining | Pumps | 103 | 20 |
| Cementation-furnaces | Metallurgy | 75 | 85 |
| Cement, Fire-proof | Stone, Lime, and Cement | 125 | 153 |

| SUBJECT. | Name of class—official classification. | No. of class—official classification. | No. of class—sub-scription classification. |
|---|---|---|---|
| Cement floors | Paving | 94 | 153 |
| Cement for lining pipes | Stone, Lime, and Cement | 125 | 153 |
| Cement, Indurating | Stone, Lime, and Cement | 125 | 153 |
| Cement-kilns | Driers and Kilns | 34 | 48 |
| Cement-mixing machines | Clay | 25 | 153 |
| Cement paving | Paving | 94 | 110 |
| Cement-pipe, Machines for making | Clay | 25 | 20 |
| Cement-pipes | Clay | 25 | 20 |
| Cement-pipes, Coating | Clay | 25 | 20 |
| Cement, Plastic | Stone, Lime, and Cement | 125 | 153 |
| Cements, Earthen | Stone, Lime, and Cement | 125 | 153 |
| Center-boards | Ships; 1 | 114 | 137 |
| Centering for arches | Bridges | 14 | 21 |
| Centering-plates | Metal-Working; 7 | 82 | 100 |
| Centering-tools | Metal-Working; 7 | 82 | 100 |
| Centers for wood-lathes | Wood-Working; 1 | 142 | 172 |
| Centrifugal filters | Sugar | 127 | 155 |
| Centrifugal guns | Ordnance | 89 | 105 |
| Ceramic enamels | Clay | 25 | 115 |
| Ceramic manufactures | Clay | 25 | 115 |
| Chafing-dishes | Stoves and Furnaces | 126 | 154 |
| Chain-making | Metal-Working; 3 | 78 | 60 |
| Chains, (fob and watch,) Manufacture of. | Hardware Manufactures | 53E | 86 |
| Chains, Surveyors' | Optics | 88 | 157 |
| Chain-testers | Presses | 100 | 60 |
| Chair-back sawing-machines | Wood-Working; 3 | 144 | 131 |
| Chair-backs, Machines for dressing | Wood-Working; 3 | 144 | 62 |
| Chair-backs, Machines for rounding | Wood-Working; 3 | 144 | 62 |
| Chair-bottoms | Furniture | 45 | 62 |
| Chairs | Furniture | 45 | 62 |
| Chairs and seats, Head-rests for | Furniture | 45 | 62 |
| Chairs, Barbers' | Furniture | 45 | 62 |
| Chairs, Camp | Furniture | 45 | 62 |
| Chairs, Dentists' | Furniture | 45 | 46 |
| Chair-seats, Machines for boring | Wood-Working; 3 | 144 | 62 |
| Chair-seats, Machines for planing | Wood-Working; 3 | 144 | 62 |
| Chairs, Enema | Furniture | 45 | 156 |
| Chairs, Folding | Furniture | 45 | 62 |
| Chairs for glass-makers | Glass | 49 | 68 |
| Chairs, Instruments for caning | Wood-Working; 3 | 144 | 62 |
| Chairs, Invalid | Furniture | 45 | 62 |
| Chairs, Locomotive | Carriages and Wagons | 21 | 131 |
| Chairs, Making railway | Railways; 4 | 107 | 121 |
| Chairs, Nursery | Furniture | 45 | 62 |
| Chairs, Obstetrical | Furniture | 45 | 156 |
| Chairs, Operating | Furniture | 45 | 156 |
| Chairs, Pew | Furniture | 45 | 62 |
| Chairs, Railway | Railways; 1 | 104 | 121 |
| Chair-stuff, Machines for scraping | Wood-Working; 3 | 144 | 62 |
| Chalk-line holders | Wood-Working; 4 | 145 | 29 |
| Chamber-pots | Baths and Closets | 4 | 8 |
| Chamfering-tools, (Wood) | Wood-Working; 4 | 145 | 173 |
| Chandeliers | Lamps and Gas-Fittings | 67 | 64 |
| Changing speed, Apparatus for | Mechanical Powers | 74 | 67 |
| Channeling-machines | Boots and Shoes | 12 | 16 |
| Channeling stone | Stone, Lime, and Cement | 125 | 153 |
| Charcoal-furnaces for cooking | Stoves and Furnaces | 126 | 154 |
| Charcoal-kilns | Driers and Kilns | 34 | 61 |
| Chart-holders | Educational | 35 | 146 |
| Charts | Educational | 35 | 146 |

| SUBJECT. | Name of class—official classification. | No. of class—official classification. | No. of class—subscription classification. |
|---|---|---|---|
| Chassis for guns | Ordnance | 89 | 105 |
| Checker-boards | Games and Toys | 46 | 143 |
| Check-hooks | Harness | 54 | 72 |
| Check-reins | Harness | 54 | 72 |
| Checks, Baggage | Stationery | 120 | 146 |
| Checks, Conductors' | Stationery | 120 | 146 |
| Check-valves | Valves | 136 | 152 |
| Cheese-covers | Kitchen Utensils | 65 | 88 |
| Cheese-cutters | Kitchen Utensils | 65 | 88 |
| Cheese-grinders | Kitchen Utensils | 65 | 88 |
| Cheese-hoops | Dairy | 31 | 45 |
| Cheese-making | Dairy | 31 | 45 |
| Cheese-presses | Presses | 100 | 45 |
| Cheese-shelves | Dairy | 31 | 45 |
| Cheese-tables | Dairy | 31 | 45 |
| Cheese-turners | Dairy | 31 | 45 |
| Cheese-vats | Dairy | 31 | 45 |
| Chemical apparatus | Chemical Miscellaneous | 23 | 35 |
| Chemical furnaces | Chemical Miscellaneous | 23 | 35 |
| Chemical preparation of food | Preserving | 99 | 58 |
| Chemical processes in sheet-metal making | Metal-Working; 5 | 80 | 35 |
| Chemicals, Air-tight packages for | Preserving | 99 | 35 |
| Chemicals, Preparing | Chemical Miscellaneous | 23 | 35 |
| Chenille-machines | Weaving | 139 | 17 |
| Cherry-stoners | Kitchen Utensils | 65 | 88 |
| Chess | Games and Toys | 46 | 143 |
| Chest-locks | Locks and Latches | 70 | 93 |
| Chest-protectors | Surgery | 128 | 156 |
| Chicken-coops | Stabling | 119 | 116 |
| Chicken-houses | Stabling | 119 | 116 |
| Chicken-nests | Stabling | 119 | 116 |
| Chicken-raising apparatus | Stabling | 119 | 116 |
| Chicken-roosts | Stabling | 119 | 116 |
| Chignons | Toilet | 132 | 161 |
| Children's carriages | Carriages and Wagons | 21 | 31 |
| Children's hand-cars | Carriages and Wagons | 21 | 31 |
| Children's toys | Games and Toys | 46 | 143 |
| Chills | Casting | 22 | 33 |
| Chimney-caps | Stoves and Furnaces | 126 | 154 |
| Chimney-cleaners | Stoves and Furnaces | 126 | 154 |
| Chimney cleaners, Lamp | Kitchen Utensils | 65 | 90 |
| Chimney-cowls | Stoves and Furnaces | 126 | 166 |
| Chimney-pots | Clay | 25 | 115 |
| Chimneys | Stoves and Furnaces | 126 | 4 |
| Chimneys, Lamp | Lamps and Gas-Fittings | 67 | 90 |
| Chimneys, Operating ships' | Ships; 1 | 114 | 138 |
| China-ware | Clay | 25 | 115 |
| Chiroplasts | Music | 84 | 102 |
| Chiropodic instruments | Surgery | 128 | 156 |
| Chisels, Carpenters' | Wood-Working; 4 | 145 | 49 |
| Chisels, Grafting | Wood-Working; 4 | 145 | 63 |
| Chisels, Joiners' | Wood-Working; 4 | 145 | 49 |
| Chisels, Manufacture of | Metal-Working; 6 | 81B | 49 |
| Chisels, Pruning | Wood-Working; 4 | 145 | 63 |
| Chocks, Door | Builders' Hardware | 16 | 24 |
| Chronographs | Horology | 58 | 36 |
| Chronographs, Electro-magnetic | Electricity | 36 | 36 |
| Chronometers | Horology | 58 | 36 |
| Chronoscopes | Horology | 58 | 36 |

| SUBJECT. | Name of class—official classification. | No. of class—official classification. | No. of class—subscription classification. |
|---|---|---|---|
| Chronoscopes, Electro-magnetic | Electricity | 36 | 36 |
| Chucks for metal-lathes | Metal-Working; 7 | 82 | 100 |
| Chucks for wood-lathes | Wood-Working; 1 | 142 | 172 |
| Churn-dashers | Dairy | 31 | 45 |
| Churn-powers | Mechanical Powers | 74 | 45 |
| Churns | Dairy | 31 | 45 |
| Chutes for carrying logs | Hydraulic Engineering | 61 | 82 |
| Chutes for coals | Hydraulic Engineering | 61 | 79 |
| Chutes for water-courses | Hydraulic Engineering | 61 | 82 |
| Cider-making, (*except Presses, Class* 100) | Beer and Wine | 7 | 19 |
| Cider-mills | Presses | 100 | 19 |
| Cider-presses | Presses | 100 | 19 |
| Cigar-cases | Tobacco | 131 | 160 |
| Cigar-dryers | Dryers and Kilns | 34 | 160 |
| Cigarettes | Tobacco | 131 | 160 |
| Cigarette-fillers | Tobacco | 131 | 160 |
| Cigar-holders | Tobacco | 131 | 160 |
| Cigar-lighters | Lamps and Gas-Fittings | 67 | 90 |
| Cigar-making | Tobacco | 131 | 160 |
| Cigar-molds | Tobacco | 131 | 160 |
| Cigar-presses | Tobacco | 131 | 160 |
| Cigar-racks | Furniture | 45 | 160 |
| Cigars | Tobacco | 131 | 160 |
| Cigars, Compositions for lighting | Gunpowder | 52 | 90 |
| Circuit-breakers | Electricity | 36 | 50 |
| Circuit-closers | Electricity | 36 | 50 |
| Circular looms | Weaving | 139 | 95 |
| Circular saws | Saws | 110 | 131 |
| Circumferentors | Optics | 88 | 157 |
| Cistern-covers | Masonry | 72 | 51 |
| Cistern-pumps | Pumps | 103 | 118 |
| Cisterns | Masonry | 72 | 51 |
| Cisterns, Making sheet-metal | Sheet-Metal | 113 B | 136 |
| Cistern-tops | Masonry | 72 | 51 |
| Clamps, Carpenters' | Wood-Working; 4 | 145 | 29 |
| Clamps for molds | Casting | 22 | 33 |
| Clamps for raising buildings | Hoisting | 57 | 79 |
| Clamps for saw-filing, gumming, and sharpening | Grinding and Polishing | 51 | 70 |
| Clamps for sewing harness | Harness | 54 | 72 |
| Clamps for wringers | Laundry | 68 | 91 |
| Clamps, Joiners' | Wood-Working; 3 | 144 | 29 |
| Clamps, Shipwrights' | Wood-Working; 4 | 145 | 137 |
| Clamps, Stitching | Harness | 54 | 72 |
| Clapboard-sawing machines | Wood-Working; 3 | 144 | 131 |
| Clapboards, Machines for gauging | Wood-Working; 3 | 144 | 29 |
| Clapboards, Machines for planing | Wood-Working; 3 | 144 | 173 |
| Clarifying wines, &c | Beer and Wine | 7 | 19 |
| Clarinets | Music | 84 | 102 |
| Clasps | Clasps and Buckles | 24 | 23 |
| Clasps for garments | Clasps and Buckles | 24 | 23 |
| Clasps for skirts | Clasps and Buckles | 24 | 23 |
| Clasps, Manufacture of | Hardware Manufactures | 53 E | 23 |
| Clasps, Rigging | Ships; 1 | 114 | 138 |
| Claw-bars | Mechanical Powers | 74 | 67 |
| Claw-hammers | Wood-Working; 4 | 145 | 60 |
| Clay-mills | Clay | 25 | 20 |
| Clay pipes | Clay | 25 | 20 |
| Clay-pulverizers | Clay | 25 | 20 |
| Clay-screening machines | Clay | 20 | 20 |
| Clay-working tools, Manufacture of | Metal-Working; 6 | 81 B | 20 |

| SUBJECT. | Name of class—official classification. | No. of class—official classification. | No. of class—subscription classification. |
|---|---|---|---|
| Cleading for boilers | Steam; 2 | 122 | |
| Cleading for cylinders | Steam; 2 | 122 | 148 |
| Cleaning carcasses | Butchering | 17 | 25 |
| Cleaning castings | Casting | 22 | 33 |
| Cleaning silk | Silk | 117 | 140 |
| Clearing silk | Silk | 117 | 140 |
| Cleats | Ships; 1 | 114 | 138 |
| Clepsydras | Horology | 58 | 36 |
| Clinchers, Blacksmiths' | Metal-Working; 6 | 81A | 60 |
| Clinch-rings | Clasps and Buckles | 24 | 23 |
| Clipping-tools | Cutlery | 30 | 135 |
| Clipping-tools, Manufacture of | Hardware Manufactures | 53A | 135 |
| Clips, Paper | Stationery | 120 | 146 |
| Cloaks | Apparel | 2 | 38 |
| Cloak-stands | Furniture | 45 | 62 |
| Clock-bells | Signals | 116 | 36 |
| Clock-bells, Manufacture of | Hardware Manufactures | 53B | 36 |
| Clock-cases | Horology | 58 | 36 |
| Clock-keys | Horology | 58 | 36 |
| Clock-pillars, Manufacture of | Hardware Manufactures | 53I | 36 |
| Clocks | Horology | 58 | 36 |
| Clocks, Electrical | Electricity | 36 | 36 |
| Clocks, Illuminated | Horology | 58 | 36 |
| Clod-crushers | Harrows | 55 | 73 |
| Clogs, (metallic,) Manufacture of | Hardware Manufactures | 53 | 16 |
| Closets, Portable | Furniture | 45 | 62 |
| Closets, Revolving | Furniture | 45 | 62 |
| Cloth-crimping machines | Apparel | 2 | 38 |
| Cloth-cutting machines | Cloth | 26 | 38 |
| Cloth-dressing | Cloth | 26 | 37 |
| Cloth-dryers | Cloth | 26 | 37 |
| Clothes-clamps | Beds | 5 | 91 |
| Clothes-dryers | Laundry | 68 | 91 |
| Clothes-frames | Laundry | 68 | 91 |
| Clothes-hooks, Manufacture of | Hardware Manufactures | 53D | 23 |
| Clothes-horses | Laundry | 68 | 91 |
| Clothes-line reels | Laundry | 68 | 91 |
| Clothes-lines | Laundry | 68 | 91 |
| Clothes-pins | Laundry | 68 | 91 |
| Clothes-pins, Machines for making | Wood-Working; 3 | 144 | 175 |
| Clothes-pins, Machines for slotting | Wood-Working; 3 | 144 | 175 |
| Clothes-pins, Machines for turning | Wood-Working; 3 | 144 | 175 |
| Clothes-pins, Manufacture of metallic | Hardware Manufactures | 53E | 24 |
| Clothes-pressing machines | Apparel | 2 | 38 |
| Clothes-sprinklers | Laundry | 68 | 91 |
| Clothes-sticks | Laundry | 68 | 91 |
| Clothes-tongs | Laundry | 68 | 91 |
| Clothes-washers | Laundry | 68 | 91 |
| Clothes wringers | Laundry | 68 | 91 |
| Cloth, Felt | Felting and Hats | 38 | 76 |
| Cloth-finishing | Cloth | 26 | 37 |
| Cloth-folding | Cloth | 26 | 37 |
| Cloth-gassing | Cloth | 26 | 37 |
| Cloth-ironing | Apparel | 2 | 37 |
| Cloth-measuring | Apparel | 2 | 37 |
| Cloth-napping | Apparel | 2 | 37 |
| Cloth-plaiting machines | Apparel | 2 | 37 |
| Cloth-presses | Presses | 100 | 37 |
| Cloth-shearing | Cloth | 26 | 37 |
| Cloth-sponging | Cloth | 26 | 37 |

| SUBJECT. | Name of class—official classification. | No. of class—official classification. | No. of class—subscription classification. |
|---|---|---|---|
| Cloth-steaming | Cloth | 26 | 37 |
| Cloth-stretching | Cloth | 26 | 37 |
| Cloth-teaseling | Cloth | 26 | 37 |
| Cloth-tentering | Cloth | 26 | 37 |
| Clover-headers | Harvesters | 56 | 75 |
| Clover-hullers | Thrashing | 130 | 159 |
| Clover-seed harvesters | Harvesters | 56 | 75 |
| Club feet, Apparatus for | Artificial Limbs | 3 | 156 |
| Clubs, Exercising | Games and Toys | 46 | 143 |
| Clutches, Animal | Butchering | 17 | 25 |
| Clutches for shafting | Journals and Bearings | 64 | 87 |
| Clutches, Friction | Journals and Bearings | 64 | 87 |
| Clutches, Rope | Mechanical Powers | 74 | 87 |
| Clutch-pulleys for sewing-machines | Sewing-Machines | 112 | 134 |
| Coach-lace | Weaving | 139 | 31 |
| Coal-boxes | Stoves and Furnaces | 126 | 154 |
| Coal-breakers | Ore | 90 | 3 |
| Coal-buckets | Stoves and Furnaces | 126 | 154 |
| Coal-cars | Railways; 2 | 105 | 121 |
| Coal-chutes | Hydraulic Engineering | 61 | 79 |
| Coal-driers | Driers and Kilns | 34 | 48 |
| Coal-hods | Stoves and Furnaces | 126 | 154 |
| Coal-hole covers | Paving | 94 | 110 |
| Coal-hole guards | Paving | 94 | 110 |
| Coal-mining | Stone, Lime, and Cement | 125 | 124 |
| Coal-mining buckets | Stone, Lime, and Cement | 125 | 124 |
| Coal, Purification of | Fuel | 44 | 3 |
| Coal-screens | Wire-Working | 140 | 3 |
| Coal-scuttles | Stoves and Furnaces | 126 | 154 |
| Coal-sifters | Wire-Working | 140 | 154 |
| Coal-traps in pavements | Paving | 94 | 110 |
| Coal-washers | Ore | 90 | 3 |
| Coating and lining barrels | Paint | 91 | 107 |
| Coating metallic surfaces | Plating | 96 | 113 |
| Coating or covering yarn | Carding | 19 | 28 |
| Coats | Apparel | 2 | 38 |
| Cock-eyes | Clasps and Buckles | 24 | 23 |
| Cocks | Valves | 136 | 148 |
| Cocks, Blow-off | Valves | 136 | 148 |
| Cocks, Feed | Valves | 136 | 148 |
| Cocks, Gage | Valves | 136 | 148 |
| Cocks, Oil | Valves | 136 | 148 |
| Cocks, Self-closing | Valves | 136 | 148 |
| Cocks, Steam | Valves | 136 | 148 |
| Cocks, Stop | Valves | 136 | 148 |
| Cocks, Try | Valves | 136 | 148 |
| Cocks, Water | Valves | 136 | 148 |
| Cocooneries | Silk | 117 | 140 |
| Cocoon-lodgments | Silk | 117 | 140 |
| Cocoons, Unwinding | Silk | 117 | 140 |
| Coffee-cleaning machines | Mills | 83 | 57 |
| Coffee-cups | Kitchen Utensils | 65 | 88 |
| Coffee-mills | Mills | 83 | 88 |
| Coffee-percolators | Stoves and Furnaces | 126 | 136 |
| Coffee-polishers | Mills | 83 | 57 |
| Coffee-pots | Stoves and Furnaces | 126 | 136 |
| Coffee-roasters | Stoves and Furnaces | 126 | 154 |
| Coffee-urns | Stoves and Furnaces | 126 | 136 |
| Coffer-dams | Hydraulic Engineering | 61 | 82 |
| Coffin-handles | Builders' Hardware | 16 | 39 |

| SUBJECT. | Name of class—official classification. | No. of class—official classification. | No. of class—subscription classification. |
|---|---|---|---|
| Coffin-handles, Manufacture of | Hardware Manufactures | 53D | 39 |
| Coffin-hardware | Builders' Hardware | 16 | 39 |
| Coffin-hardware, Manufacture of | Hardware Manufactures | 53D | 39 |
| Coffins | Coffins | 27 | 39 |
| Coffins, Compositions for covering | Coffins | 27 | 39 |
| Coffins, Compositions for making | Coffins | 27 | 39 |
| Coffin-screws | Coffins | 27 | 39 |
| Coffin-screws, Making | Wood-Screws | 141 | 39 |
| Cog-gearing | Mechanical Powers | 74 | 67 |
| Cogs, Machines for making | Wood-Working; 3 | 144 | 174 |
| Cogs on wooden wheels, Machines for cutting | Wood-Working; 3 | 144 | 174 |
| Coiling or laying slivers | Carding | 19 | 28 |
| Coiling pipes | Tubing and Wire | 134 | 163 |
| Coiling tubes | Tubing and Wire | 134 | 163 |
| Coiling wire | Wire-Working | 140 | 163 |
| Coin-assorters | Measuring-Instruments | 73 | 42 |
| Coin-balances | Measuring-Instruments | 73 | 132 |
| Coin-counting machines | Measuring-Instruments | 73 | 42 |
| Coin detecting, Counterfeit | Measuring-Instruments | 73 | 132 |
| Coin-safes | Safes | 109 | 128 |
| Coir brooms | Brushes and Brooms | 15 | 22 |
| Coke-ovens | Driers and Kilns | 34 | 61 |
| Collapse of boilers, Preventing | Steam; 2 | 122 | 149 |
| Collapsible casks for beer, &c | Aëration and Bottling | 1 | 19 |
| Collar-blocks | Harness | 54 | 72 |
| Collar-fastenings, Manufacture of | Hardware Manufactures | 53E | 72 |
| Collar-fastenings, Shirt | Clasps and Buckles | 24 | 38 |
| Collars, (*except of Paper, Class* 93) | Apparel | 2 | 38 |
| Collar-shells, Manufacture of | Hardware Manufactures | 53F | 72 |
| Collars, Horse | Harness | 54 | 72 |
| Collars, Machines for boxing paper | Paper Manufactures | 93 | 109 |
| Collars, Machines for cutting out paper | Paper Manufactures | 93 | 109 |
| Collars, Machines for embossing paper | Paper Manufactures | 93 | 108 |
| Collars, Machines for folding paper | Paper Manufactures | 93 | 109 |
| Collars, Machines for patching button-holes of paper | Paper Manufactures | 93 | 109 |
| Collars, Machines for planishing paper | Paper Manufactures | 93 | 108 |
| Collars, Machines for punching paper | Paper Manufactures | 93 | 109 |
| Collars, Machines for shaping paper | Paper Manufactures | 93 | 109 |
| Collars, Machines for stuffing | Harness | 54 | 72 |
| Collars, Packing-envelopes for paper | Paper Manufactures | 93 | 109 |
| Collars, Paper | Paper Manufactures | 93 | 109 |
| Collar-straps, Horse | Harness | 54 | 72 |
| Collimators | Optics | 88 | 157 |
| Collodiom | Chemical Miscellaneous | 23 | 35 |
| Coloring metals | Metallurgy | 75 | 99 |
| Color-printing | Printing | 101 | 117 |
| Columns, (*except Iron, Class* 14) | Masonry | 72 | 4 |
| Columns, Iron | Bridges | 14 | 21 |
| Comb-cleaners | Toilet | 132 | 161 |
| Combined reapers and mowers | Harvesters | 56 | 74 |
| Combing-machines | Carding | 19 | 28 |
| Comb-making machines | Wood-Working; 3 | 144 | 161 |
| Combs | Toilet | 132 | 161 |
| Combs, (metallic,) Manufacture of | Hardware Manufactures | 53E | 161 |
| Commodes | Baths and Closets | 4 | 8 |
| Commutators | Electricity | 36 | 50 |
| Compasses, Draughting | Draughting | 33 | 97 |
| Compasses, Mariners' | Electricity | 36 | 50 |

| SUBJECT. | Name of class—official classification. | No. of class—official classification. | No. of class—subscription classification. |
|---|---|---|---|
| Compasses, Surveyors' | Optics | 88 | 50 |
| Composing-machines | Printing | 101 | 117 |
| Composing-sticks | Printing | 101 | 117 |
| Composite-tubing | Clay | 25 | 153 |
| Compositions for belting | Hose and Belting | 60 | 81 |
| Compositions for coating projectiles | Projectiles | 102 | 105 |
| Compositions for elastic tubing | Hose and Belting | 60 | 81 |
| Compositions for hose | Hose and Belting | 60 | 81 |
| Compositions for packing projectiles | Projectiles | 102 | 105 |
| Compositors' appliances | Printing | 101 | 117 |
| Compositors' cases | Printing | 101 | 117 |
| Composts, Fertilizing | Manures | 71 | 96 |
| Composts for fuel | Fuel | 44 | 61 |
| Compressed-air engines | Pneumatics | 98 | 2 |
| Compresses | Surgery | 128 | 156 |
| Compression-casting | Casting | 22 | 33 |
| Compressors for rebaling | Presses | 100 | 7 |
| Computing-tables | Measuring-Instruments | 73 | 42 |
| Concrete pavement | Paving | 94 | 110 |
| Concrete pavement, Devices for laying | Paving | 94 | 110 |
| Concretes | Stone, Lime, and Cement | 125 | 153 |
| Concrete, Structures of | Masonry | 72 | 4 |
| Concussion-fuses | Projectiles | 102 | 105 |
| Condensers | Electricity | 36 | 50 |
| Condensers, Jet | Steam; 2 | 122 | 150 |
| Condensers of steam-engines | Steam; 2 | 122 | 150 |
| Condensers of stills | Stills | 124 | 47 |
| Condensers, Packing-tubes for | Steam; 2 | 122 | 150 |
| Condensers, Surface | Steam; 2 | 122 | 150 |
| Condensers, Tubular | Steam; 2 | 122 | 150 |
| Condensing milk | Preserving Food | 99 | 58 |
| Conductors' check-boxes | Stationery | 120 | 146 |
| Conductors, Electro-telegraphic | Electricity | 36 | 50 |
| Conductors' punches | Clasps and Buckles | 24 | 146 |
| Confectionery | Sugar | 127 | 155 |
| Confectionery-molds | Kitchen Utensils | 65 | 155 |
| Confection-pans | Sugar | 127 | 155 |
| Conformators | Felting and Hats | 38 | 81 |
| Connecting-rods | Mechanical Powers | 74 | 67 |
| Connectors for Voltaic batteries | Electricity | 36 | 50 |
| Construction of vessels | Ships; 1 | 114 | 137 |
| Converting iron into steel | Metallurgy | 75 | 85 |
| Converting motion, Apparatus for | Mechanical Powers | 74 | 67 |
| Cooking by gas | Stoves and Furnaces | 126 | 65 |
| Cooking by lamps | Stoves and Furnaces | 126 | 65 |
| Cooking by lime | Stoves and Furnaces | 126 | 154 |
| Cooking by steam | Stoves and Furnaces | 126 | 154 |
| Cooking-ranges | Stoves and Furnaces | 126 | 154 |
| Cooking-stoves | Stoves and Furnaces | 126 | 154 |
| Cooking-stoves, Base-burning | Stoves and Furnaces | 126 | 154 |
| Cooking-stove shields | Stoves and Furnaces | 126 | 154 |
| Cooling-apparatus | Ice | 62 | 19 |
| Cooling-mixtures | Ice | 62 | 19 |
| Cooling-processes | Ice | 62 | 19 |
| Cooling-wort | Beer and Wine | 7 | 19 |
| Coopers' machines | Wood-Working; 3 | 144 | 40 |
| Coopers' tools | Wood-Working; 4 | 145 | 40 |
| Coopers' tools, Manufacture of | Metal-Working; 6 | 81B | 40 |
| Coping | Masonry | 72 | 4 |
| Copper-furnaces | Metallurgy | 75 | 99 |

| SUBJECT. | Name of class- official classification. | No. of class—official classification. | No. of class—sub-scription classification. |
|---|---|---|---|
| Copper-plate presses | Printing | 101 | 117 |
| Copper-plate printing | Printing | 101 | 117 |
| Cop-tubes | Spinning | 118 | 142 |
| Cop-tubes, Manufacture of | Hardware Manufactures | 53H | 142 |
| Cop-tubes of paper, Machines for making | Spinning | 118 | 142 |
| Copy-books | Educational | 35 | 146 |
| Copy-books | Educational | 35 | 146 |
| Copy-holders | Educational | 120 | 117 |
| Copying-books | Educational | 120 | 146 |
| Copying figures | Fine Arts | 41 | 54 |
| Copying-instruments | Draughting | 33 | 97 |
| Cord and belt fasteners | Hose and Belting | 60 | 81 |
| Cordage | Cordage | 28 | 127 |
| Corded binding, Machine for making | Cordage | 28 | 17 |
| Corded fabrics | Weaving | 139 | 95 |
| Corders for sewing-machines | Sewing-Machines | 112 | 134 |
| Cords | Cordage | 28 | 127 |
| Core-boxes | Casting | 22 | 133 |
| Cores | Casting | 22 | 133 |
| Cores for dikes | Hydraulic Engineering | 61 | 82 |
| Cores, Ventilating | Casting | 22 | 33 |
| Cork hats | Felting and Hats | 38 | 81 |
| Corking-machines | Aëration and Bottling | 1 | 1 |
| Cork-making machines | Wood-Working; 1 | 142 | 176 |
| Cork-presses | Aëration and Bottling | 1 | 1 |
| Cork-pullers | Kitchen Utensils | 65 | 88 |
| Corks, Artificial | Aëration and Bottling | 1 | 1 |
| Corkscrews | Kitchen Utensils | 65 | 88 |
| Corks, Machines for polishing | Wood-Working; 1 | 142 | 176 |
| Corks, Machines for turning | Wood-Working; 1 | 142 | 176 |
| Corn-ball molds | Kitchen Utensils | 65 | 88 |
| Corn brooms | Brushes and Brooms | 15 | 22 |
| Corn-coverers | Plows | 97 | 114 |
| Corn-cribs | Thrashing | 130 | 59 |
| Corn-cutting machines | Harvesters | 56 | 75 |
| Cornets | Music | 84 | 102 |
| Corn-graters | Kitchen Utensils | 65 | 88 |
| Corn-harvesters | Harvesters | 56 | 75 |
| Corn huskers and shellers | Thrashing | 130 | 59 |
| Corn-husking gloves | Thrashing | 130 | 59 |
| Corn-husking machines | Thrashing | 130 | 59 |
| Corn-husk splitters | Brakes and Gins | 13 | 59 |
| Cornice, Metallic | Masonry | 72 | 4 |
| Corning gunpowder | Gunpowder | 52 | 71 |
| Corn-knives | Cutlery | 30 | 44 |
| Corn-mills | Mills | 83 | 58 |
| Corn-poppers | Stoves and Furnaces | 126 | 154 |
| Corn-shellers | Thrashing | 130 | 58 |
| Corn-shocking machines | Harvesters | 56 | 75 |
| Corn-shock tyers | Harvesters | 56 | 75 |
| Corns, Instruments for removing | Surgery | 128 | 156 |
| Corn-stalk choppers | Harrows | 55 | 75 |
| Corpse-coolers | Ice | 62 | 39 |
| Corpse-embalming | Chemical Miscellaneous | 23 | 39 |
| Corrugating machines, Leather | Leather | 69 | 92 |
| Corrugating plates | Metal-Working; 1 | 76 | 136 |
| Corrugating wire | Wire-Working | 140 | 170 |
| Corset-clasps | Crinoline and Corsets | 29 | 38 |
| Corset-fastenings | Crinoline and Corsets | 29 | 38 |

| SUBJECT. | Name of class—official classification. | No. of class—official classification. | No. of class—sub-scription classification. |
|---|---|---|---|
| Corsets | Crinoline and Corsets | 29 | 38 |
| Cosmetics | Chemical Miscellaneous | 23 | 98 |
| Cotton-bale ties | Presses | 100 | 8 |
| Cotton-choppers | Plows | 97 | 43 |
| Cotton-cleaners | Carding | 19 | 28 |
| Cotton-gins | Brakes and Gins | 13 | 41 |
| Cotton-pickers | Harvesters | 56 | 41 |
| Cotton-pickers, (Factory) | Carding | 19 | 28 |
| Cotton-presses | Presses | 100 | 7 |
| Cotton-scrapers | Plows | 97 | 43 |
| Cotton-seed for planting, Preparing | Seeders and Planters | 111 | 133 |
| Cotton-stalk choppers | Harrows | 55 | 43 |
| Cotton-toppers | Harvesters | 56 | 43 |
| Cotton-toppers | Plows | 97 | 43 |
| Couches | Beds | 5 | 62 |
| Couches for railway-cars | Railways; 2 | 105 | 121 |
| Counter-extension apparatus | Artificial Limbs | 3 | 15 |
| Counterpoise gun-carriages | Ordnance | 89 | 105 |
| Counters | Boots and Shoes | 12 | 16 |
| Countersink-bits | Wood-Working; 4 | 145 | 173 |
| Countersink-tools for metal | Metal-Working; 2 | 77 | 100 |
| Counters, Store and shop | Furniture | 45 | 62 |
| Counting-machines | Measuring-Instruments | 73 | 42 |
| Coupling pistols to gun-stocks | Fire-Arms | 42 | 55 |
| Couplings, Band | Hose and Belting | 60 | 87 |
| Couplings, Belt | Hose and Belting | 60 | 81 |
| Couplings, Car | Railways; 3 | 106 | 122 |
| Couplings, Cord | Hose and Belting | 60 | 81 |
| Coupling, Shaft | Journals and Bearings | 64 | 87 |
| Couplings, Pipe | Water Distribution | 137 | 167 |
| Couplings, (pipe,) Manufacture of | Hardware Manufactures | 53K | 167 |
| Couplings, Thill | Carriages and Wagons | 21 | 31 |
| Covering for steam-boilers | Steam; 2 | 122 | 148 |
| Covering of wire, wire ropes, and cables | Wire-Working | 140 | 170 |
| Covering-rollers | Spinning | 118 | 142 |
| Cow-bells, Manufacture of | Hardware Manufactures | 53B | 10 |
| Cow-milkers | Dairy | 31 | 45 |
| Cowls | Stoves and Furnaces | 126 | 166 |
| Cracker-crushers | Kitchen Utensils | 65 | 88 |
| Cracker-machines | Kitchen Utensils | 65 | 88 |
| Cradles | Beds | 5 | 62 |
| Cradles for building vessels | Hydraulic Engineering | 61 | 137 |
| Cradles for launching vessels | Hydraulic Engineering | 61 | 137 |
| Cradles for raising vessels | Hydraulic Engineering | 61 | 137 |
| Cradles, Grain | Harvesters | 56 | 74 |
| Cranberry-gatherers | Garden and Orchard | 47 | 63 |
| Cranes | Hoisting | 57 | 79 |
| Crank-connections | Mechanical Powers | 74 | 67 |
| Crank-motions | Mechanical Powers | 74 | 67 |
| Crank, overcoming the dead-points of | Mechanical Powers | 74 | 67 |
| Crank-pins | Mechanical Powers | 74 | 67 |
| Cranks | Mechanical Powers | 74 | 67 |
| Crank, Substitutes for the | Mechanical Powers | 74 | 67 |
| Crates for fruit, fish, vegetables, &c | Kitchen Utensils | 65 | 63 |
| Cravat-holders | Apparel | 2 | 38 |
| Cravats | Apparel | 2 | 38 |
| Crayons | Fine Arts | 41 | 54 |
| Cream-heaters | Dairy | 31 | 45 |
| Cream-stands | Dairy | 31 | 45 |
| Cream-strainers | Dairy | 31 | 45 |

| SUBJECT. | Name of class—official classification. | No. of class—official classification. | No. of class—sub-scription classification. |
|---|---|---|---|
| Creasers for sewing-machines | Sewing-Machines | 112 | 134 |
| Creasing machines, Leather | Leather | 69 | 92 |
| Creels, Warping | Cordage | 28 | 17 |
| Cressets | Lamps and Gas-Fittings | 67 | 154 |
| Cribbing, Devices to prevent | Harness | 54 | 72 |
| Cribs | Beds | 5 | 62 |
| Crickets for stoves, Heating | Stoves and Furnaces | 126 | 154 |
| Crimping-machines and tools | Boots and Shoes | 12 | 16 |
| Crimping sheet-metal | Sheet-Metal | 113A | 136 |
| Cringles | Ships; 1 | 114 | 138 |
| Crinoline | Crinoline and Corsets | 29 | 38 |
| Crinoline, Weaving | Weaving | 139 | 38 |
| Crochet | Apparel | 2 | 38 |
| Crochet-needles, Making | Needles and Pins | 86 | 104 |
| Croquet | Games and Toys | 46 | 143 |
| Croquet-mallets, &c | Games and Toys | 46 | 143 |
| Crosscut-saws | Saws | 110 | 131 |
| Cross-heads of steam-engines | Steam; 1 | 121 | 150 |
| Crossings for railways, Making | Railways; 4 | 107 | 120 |
| Crossings, Railway | Railways; 1 | 104 | 120 |
| Crossing signals, Railway | Signals | 116 | 120 |
| Crow-bars | Mechanical Powers | 74 | 67 |
| Crown-glass | Glass | 49 | 68 |
| Crucibles, Making | Clay | 25 | 115 |
| Crucibles for metal | Clay | 25 | 99 |
| Crucible-molds | Clay | 25 | 115 |
| Cruet-stands | Kitchen Utensils | 65 | 88 |
| Crumb-removers | Kitchen Utensils | 65 | 88 |
| Cruppers | Harness | 54 | 72 |
| Crupper-loops | Harness | 54 | 72 |
| Crushers, Sugar | Mills | 83 | 155 |
| Crutches | Artificial Limbs | 3 | 156 |
| Cuff-buttons | Clasps and Buckles | 24 | 86 |
| Cuff-buttons, Manufacture of | Hardware Manufactures | 53E | 86 |
| Cuffs | Apparel | 2 | 38 |
| Cuffs, Paper | Paper Manufactures | 93 | 109 |
| Cuff-supporters | Apparel | 2 | 138 |
| Culinary ladles | Kitchen Utensils | 65 | 88 |
| Cultivators | Plows | 97 | 43 |
| Cultivators, Frames for | Plows | 97 | 43 |
| Cultivators, Wheel | Plows | 97 | 43 |
| Cultivator-teeth | Plows | 97 | 43 |
| Culverts | Masonry | 72 | 4 |
| Cupboard-catches | Locks and Latches | 70 | 93 |
| Cupola-furnaces | Metallurgy | 75 | 85 |
| Cupping-instruments | Surgery | 128 | 156 |
| Cups | Kitchen Utensils | 65 | 88 |
| Cups, Medicinal | Surgery | 128 | 156 |
| Curbs for wells | Excavators | 37 | 51 |
| Curbs, Pavement | Paving | 94 | 110 |
| Curculio-traps | Garden and Orchard | 47 | 63 |
| Curd-cutters | Dairy | 31 | 45 |
| Curing meat and fish, (*except Smoking and Drying, Class* 34) | Preserving Food | 99 | 58 |
| Curled hair and its substitutes | Beds | 5 | 62 |
| Curling hat-brims | Felting and Hats | 38 | 76 |
| Curling-irons | Toilet | 132 | 161 |
| Curls | Toilet | 132 | 161 |
| Currency-holders | Stationery | 120 | 146 |
| Current-fenders | Hydraulic Engineering | 61 | 82 |

| SUBJECT. | Name of class—official classification. | No. of class—official classification. | No. of class—subscription classification. |
|---|---|---|---|
| Current-regulators | Electricity | 36 | 50 |
| Current-wheels | Water-Wheels | 138 | 168 |
| Curry-combs | Stabling | 119 | 144 |
| Currying-machines | Tanning | 129 | 92 |
| Currying-tools | Tanning | 129 | 92 |
| Currying-tools | Stabling | 119 | 144 |
| Curtain-fixtures | Furniture | 45 | 62 |
| Curtains | Furniture | 45 | 62 |
| Curved printing-surfaces, Making | Printing | 101 | 117 |
| Cushions for pins, needles, and sewing-work | Apparel | 2 | 38 |
| Cuspadores | Furniture | 45 | 62 |
| Cutlery, Manufacture of | Hardware Manufactures | 53A | 34 |
| Cut nails | Nails | 85 | 104 |
| Cut-off apparatus | Steam; 1 | 121 | 152 |
| Cut-offs, Governor | Governors | 50 | 69 |
| Cut-offs for harvesters | Harvesters | 56 | 74 |
| Cut-off valves | Valves | 136 | 152 |
| Cut-off, Variable | Steam; 1 | 121 | 152 |
| Cut-piled fabrics | Weaving | 139 | 95 |
| Cutter-heads of planing-machines | Wood-Working; 2 | 143 | 173 |
| Cutters, Bolt | Metal-Working; 6 | 81A | 14 |
| Cutters, Feed | Stabling | 119 | 59 |
| Cutters, Harvester | Harvesters | 56 | 74 |
| Cutters of molding and planing machines | Wood-Working; 2 | 143 | 173 |
| Cutters, Paper | Paper Manufactures | 93 | 110 |
| Cutters, Pipe | Metal-Working; 6 | 81A | 100 |
| Cutters, Rod | Metal-Working; 6 | 81A | 100 |
| Cutters, Root | Stabling | 119 | 59 |
| Cutters, Sod | Plows | 97 | 114 |
| Cutters, Straw, hay, &c. | Stabling | 119 | 59 |
| Cutter-stocks for metal-planers | Metal-Working; 7 | 82 | 100 |
| Cutting blanks from bars or plates | Metal-Working; 4 | 79 | 119 |
| Cutting-boards for leather | Boots and Shoes | 12 | 16 |
| Cutting dye-wood | Wood-Working; 3 | 144 | 176 |
| Cutting flock | Felting and Hats | 38 | 76 |
| Cutting fur | Felting and Hats | 38 | 76 |
| Cutting-gages | Wood-Working; 4 | 145 | 173 |
| Cutting glass | Grinding and Polishing | 51 | 68 |
| Cutting jewels | Grinding and Polishing | 51 | 86 |
| Cutting machines, Belt | Hose and Belting | 60 | 81 |
| Cutting-machines, Belt-lacing | Hose and Belting | 60 | 81 |
| Cutting machines, Leather | Leather | 69 | 92 |
| Cutting out garments | Apparel | 2 | 38 |
| Cutting planchets from bars or plates | Metal-Working; 4 | 79 | 119 |
| Cutting slate and stone | Stone, Lime, and Cement | 125 | 153 |
| Cutting-tools for lathes and planers | Metal-Working; 7 | 82 | 100 |
| Cutting wood-screws | Wood-Screws | 141 | 175 |
| Cylinder glass | Glass | 49 | 68 |
| Cylinders, Boring | Metal-Working; 2 | 77 | 100 |
| Cylinders, Steam | Steam; 1 | 121 | 150 |
| Cylindrical saws | Saws | 110 | 130 |
| Cymbals | Music | 84 | 102 |
| **D.** | | | |
| Dairies | Dairy | 31 | 45 |
| Dairy-implements | Dairy | 31 | 45 |
| Dampers for furnaces | Stoves and Furnaces | 126 | 145 |

| SUBJECT. | Name of class—official classification. | No. of class—official classification. | No. of class—subscription classification. |
|---|---|---|---|
| Dampers for musical instruments | Music | 84 | 102 |
| Dampers for stoves | Stoves and Furnaces | 126 | 154 |
| Damping machines, Paper | Printing | 101 | 117 |
| Dams | Hydraulic Engineering | 61 | 82 |
| Dancing figures | Games and Toys | 46 | 143 |
| Darning-lasts | Apparel | 2 | 38 |
| Darning-machines | Apparel | 2 | 89 |
| Darning-machines | Knitting and Netting | 66 | 89 |
| Davits | Boats | 9 | 13 |
| Dead-eyes | Ships; 1 | 114 | 138 |
| Dead-eyes for canal-boats | Ships; 1 | 114 | 138 |
| Dead-lights | Ships; 1 | 114 | 138 |
| Decarbonizing iron | Metallurgy | 75 | 85 |
| Deck-bridges | Ships; 1 | 114 | 137 |
| Deck-covers | Ships; 1 | 114 | 137 |
| Deck-irons | Ships; 1 | 114 | 137 |
| Decks | Ships; 1 | 114 | 137 |
| Deck-scrapers | Wood-Working; 4 | 145 | 137 |
| Decks, Fire-proof | Ships; 1 | 114 | 137 |
| Decks, Water-proof | Ships; 1 | 114 | 137 |
| Decorative gilding | Fine Arts | 41 | 107 |
| Decorative painting | Fine Arts | 41 | 107 |
| Defecators for juices and sirups | Sugar | 127 | 155 |
| Daguerreotype | Photography | 95 | 111 |
| Delineators of grades and routes | Optics | 88 | 157 |
| Dental-compositions | Dental | 32 | 46 |
| Dental-drills | Dental | 32 | 46 |
| Dental-fillings | Dental | 32 | 46 |
| Dental-forceps | Dental | 32 | 46 |
| Dental-instruments | Dental | 32 | 46 |
| Dental-mallets | Dental | 32 | 46 |
| Dental-plates | Dental | 32 | 46 |
| Dental-plates, Casting | Dental | 32 | 46 |
| Dentifrices | Chemical Miscellaneous | 23 | 46 |
| Dentures | Dental | 32 | 46 |
| Deodorizers | Chemical Miscellaneous | 23 | 98 |
| Deodorizing benzole | Stills | 124 | 98 |
| Deodorizing excrements | Manures | 71 | 98 |
| Deodorizing offal | Manures | 71 | 98 |
| Deodorizing vaults | Manures | 71 | 98 |
| Departure and distance instruments | Optics | 88 | 157 |
| Depilating hides | Tanning | 129 | 92 |
| Depilatories | Chemical Miscellaneous | 23 | 92 |
| Depurators | Surgery | 128 | 156 |
| Dermopathic-instruments | Surgery | 128 | 156 |
| Derricks | Hoisting | 57 | 79 |
| Derricks, Floating | Hoisting | 57 | 79 |
| Desk-covers | Furniture | 45 | 62 |
| Desk paper-cutters | Stationery | 120 | 146 |
| Desks | Furniture | 45 | 62 |
| Desiccating grain, food, farina, &c | Dryers and Kilns | 34 | 48 |
| Desulphurizing coal | Metallurgy | 75 | 99 |
| Desulphurizing-furnaces | Metallurgy | 75 | 99 |
| Desulphurizing ores | Metallurgy | 75 | 99 |
| Detaching boats from davits | Boats | 9 | 13 |
| Detaching horses from vehicles | Carriages and Wagons | 21 | 31 |
| Detergents | Chemical Miscellaneous | 23 | 98 |
| Detonating compositions | Gunpowder | 52 | 71 |
| Developing, Photographic | Photography | 95 | 111 |
| Dextrine | Bleaching, Dyeing, &c | 8 | 11 |

| SUBJECT. | Name of class—official classification. | No. of class—official classification. | No. of class—subscription classification. |
|---|---|---|---|
| Dials | Horology | 58 | 36 |
| Dials, Time-piece | Horology | 58 | 36 |
| Diamond-drills | Stone, Lime, and Cement | 125 | 124 |
| Diamond-planers | Stone, Lime, and Cement | 125 | 153 |
| Diamond-saws | Stone, Lime, and Cement | 125 | 153 |
| Diamonds, Glaziers' | Glass | 49 | 68 |
| Diaper-pins | Jewelry | 63 | 86 |
| Diapers | Apparel | 2 | 38 |
| Dibblers, Hand | Seeders and Planters | 111 | 133 |
| Dibbling-machines | Seeders and Planters | 111 | 133 |
| Dice | Games and Toys | 46 | 143 |
| Dice-throwers | Games and Toys | 46 | 143 |
| Die-rolling | Metal-Working; 5 | 80 | 125 |
| Dies for cutting screws | Bolts, Nuts, and Rivets | 10 | 14 |
| Dies for dentists | Dental | 32 | 46 |
| Dies for making articles of cutlery | Hardware Manufacture | 53A | 44 |
| Dies for making spikes and nails | Nails | 85 | 103 |
| Dies for pressing hats | Felting and Hats | 38 | 76 |
| Dies for sheet-metal ware | Sheet-Metal | 113A | 136 |
| Dies for swaging | Metal-Working; 3 | 78 | 131 |
| Die-sinking | Fine Arts | 41 | 143 |
| Die-stocks | Bolts, Nuts, and Rivets | 10 | 14 |
| Differential levers | Mechanical Powers | 74 | 67 |
| Digesters for bones, fats, &c | Stoves and Furnaces | 126 | 26 |
| Digesters for cooking | Stoves and Furnaces | 126 | 154 |
| Diggers, Hand | Garden and Orchard | 47 | 141 |
| Diggers, Peanut | Plows | 97 | 141 |
| Diggers, Potato | Plows | 97 | 114 |
| Diggers, Root | Plows | 97 | 114 |
| Diggers, Thistle | Garden and Orchard | 47 | 141 |
| Digging-forks | Garden and Orchard | 47 | 141 |
| Digging-machines | Plows | 97 | 114 |
| Dikes | Hydraulic Engineering | 61 | 82 |
| Dinner-buckets | Kitchen Utensils | 65 | 136 |
| Dinner-pails | Kitchen Utensils | 65 | 136 |
| Dioptric lights | Optics | 88 | 90 |
| Dippers | Kitchen Utensils | 65 | 136 |
| Dipping-frames for matches | Wood-Working; 3 | 144 | 90 |
| Disabling ordnance, Instruments for | Ordnance | 89 | 105 |
| Dish-covers | Kitchen Utensils | 65 | 88 |
| Dish-cleaners | Kitchen Utensils | 65 | 88 |
| Dish-cloth holders | Kitchen Utensils | 65 | 88 |
| Dish-drainers | Kitchen Utensils | 65 | 88 |
| Dishes | Kitchen Utensils | 65 | 88 |
| Dish-fasteners | Kitchen Utensils | 65 | 88 |
| Dish-heaters | Stoves and Furnaces | 126 | 154 |
| Dish-holders | Kitchen Utensils | 65 | 88 |
| Dish-stands | Kitchen Utensils | 65 | 88 |
| Dish-trays | Kitchen Utensils | 65 | 88 |
| Dish-washers | Kitchen Utensils | 65 | 88 |
| Disinfecting apparatus, (*except Atomizers, Class* 62) | Chemical Miscellaneous | 23 | 35 |
| Disintegrating ores | Ore | 90 | 3 |
| Dislocation apparatus | Artificial Limbs | 3 | 156 |
| Dissected maps | Educational | 35 | 146 |
| Distillation | Stills | 124 | 47 |
| Distributing boxes, Mail | Stationery | 120 | 146 |
| Distributing-machines | Printing | 101 | 117 |
| Ditching-machines | Excavators | 37 | 114 |
| Ditching-plows | Plows | 97 | 114 |

| SUBJECT. | Name of class—official classification. | No. of class—official classification. | No. of class—subscription classification. |
|---|---|---|---|
| Ditching spades and scoops | Garden and Orchard | 47 | 114 |
| Dividers, Measuring | Drafting | 33 | 97 |
| Diving-apparatus | Hydraulic Engineering | 61 | 82 |
| Diving-bells | Hydraulic Engineering | 61 | 82 |
| Dock-holders for horses | Surgery | 128 | 156 |
| Docks, Dry | Hydraulic Engineering | 61 | 82 |
| Docks, Floating | Ships; 1 | 114 | 82 |
| Docks, Graving | Hydraulic Engineering | 61 | 82 |
| Docks, Tidal | Hydraulic Engineering | 61 | 82 |
| Docks, Wet | Hydraulic Engineering | 61 | 82 |
| Doffing apparatus | Carding | 19 | 28 |
| Dog-collars | Harness | 54 | 72 |
| Dog-muzzles | Harness | 54 | 72 |
| Dog-powers | Mechanical Powers | 74 | 159 |
| Dogs, Bench | Wood-Working; 4 | 145 | 174 |
| Dogs for metal-lathes and planers | Metal-Working; 7 | 82 | 100 |
| Dogs for saw-mills | Saws | 110 | 130 |
| Dogs for wood-lathes | Wood-Working; 1 | 142 | 172 |
| Dolls | Games and Toys | 46 | 143 |
| Dolls, Composition for | Chemical Miscellaneous | 23 | 143 |
| Domes | Masonry | 72 | 4 |
| Domestic furnaces | Stoves and Furnaces | 126 | 154 |
| Door-alarms | Signals | 116 | 24 |
| Door-bells | Signals | 116 | 24 |
| Door-bells, Manufacture of | Hardware Manufacture | 53B | 24 |
| Door-bolts | Builders' Hardware | 16 | 24 |
| Door-buttons | Builders' Hardware | 16 | 24 |
| Door-checks | Builders' Hardware | 16 | 24 |
| Door-checks, Manufacture of | Hardware Manufacture | 53B | 24 |
| Door-fasteners | Locks and Latches | 70 | 24 |
| Door-hangers | Carpentry | 20 | 24 |
| Door-hangers | Builders' Hardware | 16 | 24 |
| Door-hinges | Builders' Hardware | 16 | 24 |
| Door-holders | Builders' Hardware | 16 | 24 |
| Door-indicators | Stationery | 120 | 24 |
| Door-knobs | Locks and Latches | 70 | 24 |
| Door-locks | Locks and Latches | 70 | 93 |
| Door-mats | Brushes and Brooms | 15 | 22 |
| Door-operators | Builders' Hardware | 16 | 24 |
| Door-plates | Stationery | 120 | 24 |
| Door-plates, Attaching | Builders' Hardware | 16 | 24 |
| Door-pulls | Builders' Hardware | 16 | 24 |
| Doors | Carpentry | 20 | 30 |
| Door-securers | Builders' Hardware | 16 | 24 |
| Doors for steam-boiler furnaces | Steam; 2 | 122 | 148 |
| Door-sheaves | Builders' Hardware | 16 | 24 |
| Doors, Machines for making | Wood-Working; 2 | 143 | 173 |
| Door-springs | Builders' Hardware | 16 | 24 |
| Door-springs, Manufacture of | Hardware Manufacture | 53 B | 24 |
| Doors, Sliding | Carpentry | 20 | 30 |
| Door-stops | Builders' Hardware | 16 | 24 |
| Door-strips, Manufacture of | Hardware Manufacture | 53 B | 30 |
| Double-piled fabrics | Weaving | 139 | 90 |
| Double-seaming machines for sheet-metal | Sheet-Metal | 113 A | 136 |
| Double shovel-plows | Plows | 97 | 114 |
| Double-trees | Carriages and Wagons | 21 | 31 |
| Doubling-machines | Spinning | 118 | 142 |
| Doubling silk | Silk | 117 | 140 |
| Douche-instruments | Surgery | 128 | 156 |

| SUBJECT. | Name of class—official classification. | No. of class—official classification. | No. of class—sub-scription classification. |
|---|---|---|---|
| Dough-kneaders | Kitchen Utensils | 65 | 88 |
| Dough-machines | Kitchen Utensils | 65 | 88 |
| Dough-raisers | Kitchen Utensils | 65 | 88 |
| Dough-rollers | Kitchen Utensils | 65 | 88 |
| Dovetailing-machines | Wood-Working; 2 | 143 | 30 |
| Dowels, Machines for turning | Wood-Working; 1 | 142 | 175 |
| Drafting-instruments | Drafting | 33 | 97 |
| Drafting-scales | Drafting | 33 | 97 |
| Draftsman's tools and appliances | Drafting | 33 | 97 |
| Draft-regulators for steam-boiler furnaces | Steam; 2 | 122 | 148 |
| Drag-bars for drill-teeth | Seeders and Planters | 111 | 133 |
| Drag-saws | Saws | 110 | 130 |
| Draining cellars | Pumps | 103 | 118 |
| Draining cellars, Automatic devices for | Hydraulic Engineering | 61 | 82 |
| Draining-pipes | Clay | 25 | 20 |
| Draining-plows | Plows | 97 | 114 |
| Draining-tiles | Clay | 25 | 20 |
| Drain-tiles, Machines for, and modes of laying | Hydraulic Engineering | 61 | 20 |
| Drapery-fasteners | Furniture | 45 | 62 |
| Draught of water, Ascertaining | Ships; 1 | 114 | 137 |
| Draught-regulators | Stoves and Furnaces | 126 | 154 |
| Draw-bars for railway-cars | Railways; 3 | 106 | 122 |
| Draw-benches | Tubing and Wire | 134 | 163 |
| Draw-bridges | Bridges | 14 | 21 |
| Draw-bridge signals | Signals | 116 | 158 |
| Draw-plates for tubing | Tubing and Wire | 134 | 163 |
| Draw-plates for wire | Tubing and Wire | 134 | 163 |
| Drawer-locks | Locks and Latches | 70 | 93 |
| Drawer-pulls | Builders' Hardware | 16 | 93 |
| Drawers | Apparel | 2 | 38 |
| Drawers for closets | Furniture | 45 | 62 |
| Drawers for furniture | Furniture | 45 | 62 |
| Draw-gage cutters | Harness | 54 | 72 |
| Drawing and twisting heads and tubes | Spinning | 118 | 142 |
| Drawing-boards | Fine Arts | 41 | 54 |
| Drawing-knives | Wood-Working; 4 | 145 | 49 |
| Drawing liquids | Pumps | 103 | 118 |
| Drawing-machines | Spinning | 118 | 142 |
| Drawing metals | Tubing and Wire | 134 | 163 |
| Drawing-rollers | Spinning | 118 | 142 |
| Drawing-rollers, Cleaners for | Spinning | 118 | 142 |
| Drawing-tables | Furniture | 45 | 62 |
| Drawing-tubes | Tubing and Wire | 134 | 163 |
| Drawing wire | Tubing and Wire | 134 | 163 |
| Draw-rolls for tubing | Tubing and Wire | 134 | 163 |
| Dredges, Oyster | Fishing | 43 | 143 |
| Dredging boxes, Flour | Kitchen Utensils | 65 | 88 |
| Dredging-machines | Excavators | 37 | 82 |
| Dress-cord | Cordage | 28 | 17 |
| Dressing and finishing hosiery | Knitting and Netting | 66 | 89 |
| Dressing-cases | Toilet | 132 | 161 |
| Dressing cloth | Cloth | 26 | 37 |
| Dressing leather | Tanning | 129 | 92 |
| Dressing-machines, (wood) | Wood-Working; 2 | 143 | 173 |
| Dressing slate and stone | Stone, Lime, and Cement | 125 | 153 |
| Dressing yarn, warp, thread, &c. | Cordage | 28 | 17 |
| Dress-pins | Jewelry | 63 | 86 |
| Dress-protectors | Apparel | 2 | 38 |

| SUBJECT. | Name of class—official classification. | No. of class—official classification. | No. of class—sub-scription classification. |
|---|---|---|---|
| Dress-supporters | Apparel | 2 | 80 |
| Drifts | Metal-Working; 2 | 77 | 119 |
| Drill-bits | Metal-Working; 2 | 77 | 100 |
| Drill-gages | Metal-Working; 2 | 77 | 100 |
| Drilling-jigs | Metal-Working; 2 | 77 | 100 |
| Drilling machines, Metal | Metal-Working; 2 | 77 | 100 |
| Drilling metals | Metal-Working; 2 | 77 | 100 |
| Drill-rests, Metal | Metal-Working; 2 | 77 | 100 |
| Drill-rod grabs | Stone, Lime, and Cement | 125 | 124 |
| Drill-rollers | Seeders and Planters | 111 | 133 |
| Drills, Dental | Dental | 32 | 46 |
| Drills, Diamond | Stone, Lime, and Cement | 125 | 124 |
| Drills, Expanding rock | Stone, Lime, and Cement | 125 | 124 |
| Drills for wells | Stone, Lime, and Cement | 125 | 51 |
| Drills, Gages for | Metal-Working; 2 | 77 | 100 |
| Drills, Grain | Seeders and Planters | 111 | 133 |
| Drills, Ratchet | Metal-Working; 2 | 77 | 100 |
| Drills, Rock | Stone, Lime, and Cement | 125 | 124 |
| Drills, Stone | Stone, Lime, and Cement | 125 | 124 |
| Drill-stocks | Metal-Working; 2 | 77 | 100 |
| Drills, Twist | Metal-Working; 2 | 77 | 100 |
| Drills, Watchmakers' | Metal-Working; 2 | 77 | 36 |
| Drill-teeth | Seeders and Planters | 111 | 133 |
| Drinking-cups | Kitchen Utensils | 65 | 88 |
| Drinking-tubes for invalids | Surgery | 128 | 156 |
| Driven-wells | Water-Distribution | 137 | 167 |
| Drop-hammers | Metal-Working; 3 | 78 | 100 |
| Drop-drills | Seeders and Planters | 111 | 133 |
| Dropping-bottles | Measuring-Instruments | 73 | 98 |
| Dropping-platforms for harvesters | Harvesters | 56 | 74 |
| Dropping-tubes | Measuring-Instruments | 73 | 98 |
| Drop-lights | Lamps and Gas-Fittings | 67 | 64 |
| Drop-meters | Measuring-Instruments | 73 | 98 |
| Drop-presses | Metal-Working; 3 | 78 | 100 |
| Drug-crushers | Surgery | 128 | 156 |
| Druggets | Weaving | 139 | 56 |
| Drug-mills | Mills | 83 | 98 |
| Drugs | Chemical Miscellaneous | 23 | 98 |
| Drums | Music | 84 | 102 |
| Drums, Belt | Journals and Bearings | 64 | 87 |
| Drums, (keg,) Machines for making | Wood-Working; 3 | 144 | 40 |
| Dry-docks | Hydraulic Engineering | 61 | 82 |
| Dry-goods packers | Presses | 100 | 7 |
| Drying apparatus, (*except Clothes, Class* 68; *Paper, Class* 92) | Dryers and Kilns | 34 | 48 |
| Drying fruit | Dryers and Kilns | 34 | 48 |
| Drying gunpowder | Gunpowder | 52 | 71 |
| Drying-houses | Dryers and Kilns | 34 | 48 |
| Drying-kilns | Dryers and Kilns | 34 | 48 |
| Drying yarn | Cordage | 28 | 17 |
| Dubbing | Oils, Fats, and Glue | 87 | 26 |
| Dulcimers | Music | 84 | 102 |
| Dumb-bells | Games and Toys | 46 | 143 |
| Dumb-jockeys | Harness | 54 | 72 |
| Dumb-waiters | Hoisting | 57 | 79 |
| Dummies | Steam; 3 | 123 | 94 |
| Dummies for displaying clothing | Apparel | 2 | 38 |
| Dummy-cars | Railways; 2 | 105 | 121 |
| Dumping-buckets | Excavators | 37 | 79 |
| Dumping-cars | Railways; 2 | 105 | 121 |

| SUBJECT. | Name of class—official classification. | No. of class—official classification. | No. of class—subscription classification. |
|---|---|---|---|
| Dumping-carts | Carriages and Wagons | 21 | 31 |
| Dumping-reels for harvesters | Harvesters | 56 | 74 |
| Dumping-tubs and buckets | Hoisting | 57 | 79 |
| Dumping-wagons | Carriages and Wagons | 21 | 31 |
| Dust from cars, Excluding | Pneumatics | 98 | 121 |
| Dust-pans | Furniture | 45 | 62 |
| Dust-receptacles | Furniture | 45 | 62 |
| Dyeing | Bleaching and Dyeing | 8 | 11 |
| Dyeing apparatus | Bleaching and Dyeing | 8 | 11 |
| Dyeing materials | Bleaching and Dyeing | 8 | 11 |
| Dyeing processes | Bleaching and Dyeing | 8 | 11 |
| Dyes | Bleaching and Dyeing | 8 | 11 |
| Dye-vats | Bleaching and Dyeing | 8 | 11 |
| Dye-wood extracts | Bleaching and Dyeing | 8 | 11 |
| Dyewood, Machines for cutting | Wood-Working; 3 | 144 | 176 |
| Dynamite | Gunpowder | 52 | 71 |
| Dynamometers | Measuring-Instruments | 73 | 132 |
| Dyspepsia, Instruments for treating | Surgery | 128 | 156 |
| **E.** | | | |
| Ear, Apparatus for treating the | Surgery | 128 | 156 |
| Ear-drops and rings | Jewelry | 63 | 86 |
| Ear-explorers | Surgery | 128 | 156 |
| Ear, Instruments for treating the | Surgery | 128 | 156 |
| Ears, Artificial | Artificial Limbs | 3 | 5 |
| Earth-boring | Excavators | 37 | 51 |
| Earth-closets | Baths and Closets | 4 | 8 |
| Earthenware | Clay | 25 | 115 |
| Earth-removing machines | Excavators | 37 | 79 |
| Earth-works | Excavators | 37 | 51 |
| Earth-worm traps | Garden and Orchard | 47 | 63 |
| Ear-trumpets | Artificial Limbs | 3 | 156 |
| Ear-tubes | Artificial Limbs | 3 | 156 |
| Easels | Fine Arts | 41 | 54 |
| Eave-troughs | Roofing | 108 | 126 |
| Eave-troughs, Machines for making wooden | Wood-Working; 3 | 144 | 30 |
| Eccentrics | Mechanical Powers | 74 | 87 |
| Edge-planes for soles | Boots and Shoes | 12 | 16 |
| Edge-planes, (wood) | Wood-Working; 4 | 145 | 49 |
| Edge-tools, (not cutlery,) Manufacture of | Metal-Working; 6 | 81B | 49 |
| Edging-irons | Garden and Orchard | 47 | 63 |
| Educational devices | Educational | 35 | 146 |
| Eel-traps | Fishing | 43 | 143 |
| Effervescing preparations for beverages | Aëration and Bottling | 1 | 1 |
| Egg-assorters | Stabling | 119 | 116 |
| Egg-beaters | Kitchen Utensils | 65 | 88 |
| Egg-boilers | Stoves and Furnaces | 126 | 154 |
| Egg-carriers | Stabling | 119 | 116 |
| Egg-cups | Kitchen Utensils | 65 | 88 |
| Egg-hatching apparatus | Stabling | 119 | 116 |
| Egg-packers | Stabling | 119 | 116 |
| Egg-pans | Kitchen Utensils | 65 | 88 |
| Eggs, Preserving | Preserving Food | 99 | 116 |
| Egg-testers | Kitchen Utensils | 65 | 116 |
| Ejectors, Liquid | Pumps | 103 | 118 |
| Elbow-levers for pulls, Manufacture of | Hardware Manufacture | 53B | 24 |
| Electrical apparatus | Electricity | 36 | 50 |

| SUBJECT. | Name of class—official classification. | No. of class—official classification. | No. of class—subscription classification. |
|---|---|---|---|
| Electrical balances | Electricity | 36 | 50 |
| Electrical machines | Electricity | 36 | 50 |
| Electric fuses | Electricity | 36 | 50 |
| Electric lights | Electricity | 36 | 90 |
| Electrizing blood, Instruments for | Electricity | 36 | 156 |
| Electrodes | Electricity | 36 | 50 |
| Electro-magnetic apparatus | Electricity | 36 | 50 |
| Electro-magnetic engines | Electricity | 36 | 50 |
| Electro-magnetic machines | Electricity | 36 | 50 |
| Electro-magnetic meteorological instruments | Electricity | 36 | 50 |
| Electro-magnets | Electricity | 36 | 50 |
| Electrometers | Electricity | 36 | 50 |
| Electronomes | Electricity | 36 | 50 |
| Electrophorus | Electricity | 36 | 50 |
| Electro-plating, Baths and batteries for | Electricity | 36 | 113 |
| Electro-plating | Plating | 96 | 113 |
| Electrotyping | Plating | 96 | 164 |
| Electrotyping, Baths and batteries for | Electricity | 36 | 164 |
| Electrotyping, Producing the matrix for | Printing | 101 | 164 |
| Elevated railways | Railways; 1 | 104 | 120 |
| Elevating scaffolds | Carpentry | 20 | 79 |
| Elevators | Hoisting | 57 | 79 |
| Elevators for bricks | Hoisting | 57 | 79 |
| Elevators for grain | Hoisting | 57 | 79 |
| Elevators for houses | Hoisting | 57 | 79 |
| Elevators for mortar | Hoisting | 57 | 79 |
| Elevators for warehouses | Hoisting | 57 | 79 |
| Elevators, Water | Pumps | 103 | 118 |
| Ellipsographs | Drafting | 33 | 97 |
| Embalming compositions | Chemical Miscellaneous | 23 | 39 |
| Embalming processes | Chemical Miscellaneous | 23 | 39 |
| Embankments, Making and repairing | Hydraulic Engineering | 61 | 82 |
| Embossing glass | Glass | 49 | 68 |
| Embossing hats | Felting and Hats | 38 | 76 |
| Embossing-machines | Fine Arts | 41 | 54 |
| Embossing machines, Leather | Leather | 69 | 92 |
| Embossing-presses | Book-Binding | 11 | 15 |
| Embossing sheet-metal | Sheet-Metal | 113A | 136 |
| Embroidery attachments for sewing-machines | Sewing-Machines | 112 | 134 |
| Emery-paper | Grinding and Polishing | 51 | 70 |
| Emery-wheels | Grinding and Polishing | 51 | 70 |
| Emery-wheels, Compositions for | Grinding and Polishing | 51 | 70 |
| Enameling culinary vessels | Clay | 25 | 68 |
| Enameling-furnaces for metals | Metallurgy | 75 | 100 |
| Enameling hardware | Clay | 25 | 68 |
| Enameling of cloth | Paint | 91 | 107 |
| Enameling of leather | Paint | 91 | 92 |
| Enameling on metal | Fine Arts | 41 | 68 |
| Enameling-ovens | Dryers and Kilns | 34 | 86 |
| Enameling paper | Paper-Making | 92 | 108 |
| Enamels, Vitreous | Glass | 49 | 68 |
| Enchasing | Fine Arts | 41 | 54 |
| Engine-lathes | Metal-Working; 7 | 82 | 100 |
| Engines, Agricultural | Steam; 3 | 123 | 94 |
| Engines, Air | Pneumatics | 98 | 2 |
| Engines, Air-compressing | Pneumatics | 98 | 12 |
| Engines, Ammonia | Pneumatics | 98 | 2 |

| SUBJECT. | Name of class—official classification. | No. of class—official classification. | No. of class—subscription classification. |
|---|---|---|---|
| Engines, Blast | Pneumatics | 98 | 12 |
| Engines, Caloric | Pneumatics | 98 | 2 |
| Engines, Carbonic-acid | Pneumatics | 98 | 2 |
| Engines, Compressed-air | Pneumatics | 98 | 2 |
| Engines, Dummy | Steam; 3 | 123 | 94 |
| Engines for operating rock-drills | Steam; 1 | 121 | 150 |
| Engines for operating steam-hammers | Steam; 1 | 121 | 150 |
| Engines for operating steam-pumps | Steam; 1 | 121 | 150 |
| Engines, Gas | Pneumatics | 98 | 2 |
| Engines, Gunpowder | Pneumatics | 98 | 2 |
| Engines, Locomotive | Steam; 3 | 123 | 94 |
| Engines, Marine steam | Steam; 1 | 121 | 150 |
| Engines, Oscillating | Steam; 1 | 121 | 150 |
| Engines, Portable | Steam; 3 | 123 | 94 |
| Engines, Reciprocating | Steam; 1 | 121 | 150 |
| Engines, Rotary | Steam; 1 | 121 | 150 |
| Engines, Stationary | Steam; 1 | 121 | 150 |
| Engines, Steam | Steam; 1 | 121 | 150 |
| Engines, Steam fire | Steam; 1 | 121 | 118 |
| Engines, Steam plow | Steam; 3 | 123 | 94 |
| Engines, Traction | Steam; 3 | 123 | 94 |
| Engines, Valve-gear for | Steam; 1 | 121 | 152 |
| Engines, Vapor, (except steam) | Pneumatics | 98 | 2 |
| Engravers' clutches | Fine Arts | 41 | 54 |
| Engravers' tools | Fine Arts | 41 | 54 |
| Engravers' vises | Fine Arts | 41 | 54 |
| Engraving, Copper-plate | Fine Arts | 41 | 54 |
| Engraving, Electric apparatus for | Electricity | 36 | 54 |
| Engraving-machines | Fine Arts | 41 | 54 |
| Engraving pantographs | Fine Arts | 41 | 97 |
| Engraving-processes | Fine Arts | 41 | 54 |
| Engraving, Steel-plate | Fine Arts | 41 | 54 |
| Engraving, Wood | Fine Arts | 41 | 54 |
| Envelope-fasteners | Stationery | 120 | 146 |
| Envelope-making machines | Paper Manufacture | 93 | 109 |
| Envelope-openers | Stationery | 120 | 146 |
| Envelopes | Paper Manufacture | 93 | 146 |
| Envelopes and letter-sheets combined | Paper Manufacture | 93 | 146 |
| Eolians | Music | 84 | 102 |
| Epaulets | Apparel | 2 | 38 |
| Epergnes | Kitchen Utensils | 65 | 88 |
| Eprouvettes | Measuring-Instruments | 73 | 71 |
| Equalizing-bars | Carriages and Wagons | 21 | 31 |
| Erasers | Stationery | 120 | 146 |
| Escapements | Horology | 58 | 36 |
| Escapements, Electrical | Electricity | 36 | 50 |
| Escape-pipes for steam-engines | Steam; 3 | 123 | 150 |
| Escape-valves | Valves | 136 | 152 |
| Etching | Fine Arts | 41 | 54 |
| Eudiometers | Measuring-Instruments | 73 | 101 |
| Evaporators for saccharine juices | Sugar | 127 | 155 |
| Evaparators for salines | Sugar | 127 | 129 |
| Evaporators for vegetable extracts | Sugar | 127 | 155 |
| Evening cloth | Cloth | 26 | 37 |
| Excavators | Excavators | 37 | 51 |
| "Excelsior," Machines for making | Wood-Working; 2 | 143 | 174 |
| Exercising devices | Games and Toys | 46 | 143 |
| Exhaust-fans | Pneumatics | 98 | 12 |
| Exhaust for millstones | Mills | 83 | 57 |
| Exhaust-mechanism for steam-engines | Steam; 3 | 123 | 150 |

| SUBJECT. | Name of class—official classification. | No. of class—official classification. | No. of class—subscription classification. |
|---|---|---|---|
| Exhaust-nozzles for steam-engines | Steam; 3 | 123 | 94 |
| Exhaust-pipes for steam-engines | Steam; 3 | 123 | 150 |
| Expanding-bits | Wood-Working; 4 | 145 | 173 |
| Expanding-drills for metal | Metal-Working; 2 | 77 | 100 |
| Expanding-tubes | Metal-Working; 6 | 81 | 163 |
| Expansion-joints for steam-engines | Steam; 1 | 121 | 150 |
| Explosive balls | Projectiles | 102 | 32 |
| Explosive projectiles | Projectiles | 102 | 105 |
| Ex-sections, Apparatus for | Surgery | 128 | 156 |
| Extension-apparatus for fractures | Artificial Limbs | 3 | 156 |
| Extension-ladders | Carpentry | 20 | 30 |
| Extinguishers | Lamps and Gas-Fittings | 67 | 90 |
| Extractors, Bolt, nail, spike, tack | Wood-Working; 4 | 145 | 29 |
| Eye, Apparatus for treating the | Surgery | 128 | 156 |
| Eye-cups | Surgery | 128 | 156 |
| Eye-glasses | Optics | 88 | 157 |
| Eye, Instruments for treating the | Surgery | 128 | 156 |
| Eyeleting-machines | Clasps and Buckles | 24 | 23 |
| Eyeleting-stamps | Clasps and Buckles | 24 | 23 |
| Eyelet-presses | Clasps and Buckles | 24 | 23 |
| Eyelet-punches | Clasps and Buckles | 24 | 23 |
| Eyelets | Clasps and Buckles | 24 | 23 |
| Eyelets, Manufacture of | Hardware Manufacture | 53E | 23 |
| Eye-protectors | Surgery | 128 | 156 |
| Eyes, Artificial | Artificial Limbs | 3 | 5 |
| Eyes for harness, Manufacture of | Hardware Manufacture | 53H | 72 |
| Eye-shades | Surgery | 128 | 156 |
| **F.** | | | |
| Fabrics, Carpet | Weaving | 139 | 56 |
| Fabrics, Corded | Weaving | 139 | 95 |
| Fabrics, Cut-piled | Weaving | 139 | 95 |
| Fabrics, Double-piled | Weaving | 139 | 95 |
| Fabrics, Elastic | Weaving | 139 | 95 |
| Fabrics, Figured | Weaving | 139 | 95 |
| Fabrics for re-use of fiber, Tearing up | Brakes and Gins | 13 | 78 |
| Fabrics, Horse-hair | Weaving | 139 | 95 |
| Fabrics, Piled | Weaving | 139 | 95 |
| Fabrics, Printing | Printing | 101 | 11 |
| Fabrics, Shaped | Weaving | 139 | 95 |
| Fabrics, Straw | Weaving | 139 | 95 |
| Fabrics, Terry | Weaving | 139 | 95 |
| Fabrics, Tubular | Weaving | 139 | 95 |
| Fabrics, Woven | Weaving | 139 | 95 |
| Facing compositions | Casting | 22 | 33 |
| Facing molds, Machines for | Casting | 22 | 33 |
| Facing slate and stone | Stone, Lime, and Cement | 125 | 153 |
| Fagots and piles for rolling | Metal-Working; 5 | 80 | 125 |
| Fan-blowers | Pneumatics | 98 | 12 |
| Fan-folding machines | Umbrellas and Fans | 135 | 165 |
| Fanning-mills | Thrashing | 130 | 159 |
| Fans, Automatic | Umbrellas and Fans | 135 | 165 |
| Fans for chairs | Umbrellas and Fans | 135 | 165 |
| Fans for sewing-machines | Umbrellas and Fans | 135 | 165 |
| Fare-boxes | Stationery | 120 | 42 |
| Farm implements of metal, Manufacture of | Metal-Working; 6 | 81B | 49 |
| Farriers' tools | Metal-Working; 6 | 81A | 60 |
| Fasteners for bands, belts, and cords | Hose and Belting | 60 | 81 |

| SUBJECT. | Name of class—official classification. | No. of class—official classification. | No. of class—subscription classification. |
|---|---|---|---|
| Fasteners, Paper | Clasps and Buckles | 24 | 146 |
| Fastenings, Upholstery | Furniture | 45 | 62 |
| Fats, Treatment of | Oils, Fats, and Glue | 87 | 26 |
| Faucet-filters | Valves | 136 | 167 |
| Faucets | Valves | 136 | 167 |
| Faucets, Basin | Valves | 136 | 167 |
| Faucets, Boring | Valves | 136 | 167 |
| Faucets, Hot-water | Valves | 136 | 167 |
| Faucets, Manufacture of | Hardware Manufacture | 53J | |
| Faucets, Molasses | Valves | 136 | 167 |
| Feather-dressers | Apparel | 2 | 62 |
| Feather-renovators | Dryers and Kilns | 34 | 62 |
| Feed-bags | Harness | 54 | 72 |
| Feed-devices for saw-mills | Saws | 110 | 130 |
| Feed-devices for sewing-machines | Sewing-Machines | 112 | 134 |
| Feeders, Paper | Paper Manufacture | 93 | 117 |
| Feeders, Steam-boiler | Steam; 2 | 122 | 148 |
| Feeding apparatus for carding-machines | Carding | 19 | 28 |
| Feed-mixers | Stabling | 119 | 59 |
| Feed-racks | Stabling | 119 | 144 |
| Feed-regulators for mills | Mills | 83 | 57 |
| Feed-troughs | Stabling | 119 | 144 |
| Feed-water apparatus | Steam; 2 | 122 | 148 |
| Feed-water heaters for steam-boilers | Steam; 2 | 122 | 148 |
| Feed-water regulators | Steam; 2 | 122 | 148 |
| Feet, Artificial | Artificial Limbs | 3 | 5 |
| Fellers for sewing-machines | Sewing-Machines | 112 | 134 |
| Fellies | Carriages and Wagons | 21 | 31 |
| Fellies, Attaching | Carriages and Wagons | 21 | 31 |
| Fellies, Expanding | Carriages and Wagons | 21 | 31 |
| Fellies, Machines for bending | Wood-Working; 3 | 144 | 176 |
| Fellies, Machines for boring | Wood-Working; 3 | 144 | 173 |
| Fellies, Machines for planing | Wood-Working; 3 | 144 | 173 |
| Felling trees, Machines for | Saws | 110 | 131 |
| Felly-clamps | Carriages and Wagons | 21 | 31 |
| Felly-plates, Manufacture of | Hardware Manufacture | 53C | 31 |
| Felt | Felting and Hats | 38 | 76 |
| Felted fabrics | Felting and Hats | 38 | 76 |
| Felt-hats | Felting and Hats | 38 | 76 |
| Felting cloth | Felting and Hats | 38 | 37 |
| Felting hats | Felting and Hats | 38 | 76 |
| Felting, Materials for | Felting and Hats | 38 | 76 |
| Felting yarn | Felting and Hats | 38 | 17 |
| Fence-posts | Fences | 39 | 52 |
| Fences | Fences | 39 | 52 |
| Fences, Flood | Fences | 39 | 52 |
| Fences, Portable | Fences | 39 | 52 |
| Fences, Wire | Fences | 39 | 52 |
| Fenders for roofs | Roofing | 108 | 30 |
| Fenders, Furniture | Furniture | 45 | 62 |
| Fermentation-checks | Beer and Wine | 7 | 19 |
| Fermentation-guards for barrels | Aëration and Bottling | 1 | 19 |
| Fermented liquors, Cooling | Aëration and Bottling | 7 | 19 |
| Fermented liquors, Preserving | Aëration and Bottling | 7 | 19 |
| Ferrules | Tubing and Wire | 134 | 163 |
| Ferrules, Manufacture of | Hardware Manufacture | 53I | 163 |
| Ferry-boat locks | Ships; 1 | 114 | 137 |
| Ferry-boat platforms | Hydraulic Engineering | 61 | 137 |
| Ferry-boats | Ships; 1 | 114 | 137 |

| SUBJECT. | Name of class—official classification. | No. of class—official classification. | No. of class—sub-scription classification. |
|---|---|---|---|
| Fertilizer-dryers | Dryers and Kilns | 34 | 96 |
| Fertilizer-mixers | Manures | 71 | 96 |
| Fertilizers. (*See Class* 71) | | | 96 |
| Fertilizers, Manufacture of | Manures | 71 | 96 |
| Fertilizer-sowers | Seeders and Planters | 111 | 133 |
| Fertilizers, Packaging | Manures | 71 | 96 |
| Fertilizers, Treating | Manures | 71 | 96 |
| Fetters | Harness | 54 | 72 |
| Fetters for animals | Harness | 54 | 72 |
| Fibers of plants, Separating | Brakes and Gins | 13 | 78 |
| Fids | Ships; 1 | 114 | 138 |
| Fifes | Music | 84 | 102 |
| Fifth-wheels | Carriages and Wagons | 21 | 31 |
| Fifth-wheels, Manufacture of | Hardware Manufacture | 53C | 31 |
| Figured fabrics | Weaving | 139 | 95 |
| File-blanks | Files | 40 | 53 |
| File-chisels | Files | 40 | 53 |
| File-cutters | Files | 40 | 53 |
| File-cutting machines | Files | 40 | 53 |
| File-hammers | Files | 40 | 53 |
| Filers, Paper | Stationery | 120 | 146 |
| Files | Files | 40 | 53 |
| Files, Machines for making | Files | 40 | 53 |
| Files, Redressing | Files | 40 | 53 |
| Files, Sharpening | Files | 40 | 53 |
| Files, Tempering | Metallurgy | 75 | 53 |
| Filigree glass | Glass | 49 | 68 |
| Filing-machines | Grinding and Polishing | 51 | 70 |
| Filling for safes | Safes | 109 | 128 |
| Filter and cooler combined | Ice | 62 | 58 |
| Filter-faucets | Valves | 136 | 167 |
| Filtering-presses | Presses | 100 | 19 |
| Filters | Water-Distribution | 137 | 167 |
| Filters for medicinal preparations | Surgery | 128 | 156 |
| Filters, Globe | Water-Distribution | 137 | 167 |
| Filters, Tube | Water-Distribution | 137 | 167 |
| Filters, Well-tube | Water-Distribution | 137 | 167 |
| Finger-guards | Apparel | 2 | 38 |
| Finger-rings | Jewelry | 63 | 86 |
| Finger-rings, Manufacture of | Hardware Manufacture | 53E | 86 |
| Fingers, Harvester | Harvesters | 56 | 74 |
| Finishing cloth | Cloth | 26 | 37 |
| Finishing leather | Tanning | 129 | 92 |
| Finishing stone | Stone, Lime, and Cement | 125 | 153 |
| Fire-alarms, Household | Signals | 116 | 158 |
| Fire-alarm telegraphs | Electricity | 36 | 50 |
| Fire-alarm telegraphs, Signal-boxes for | Electricity | 36 | 50 |
| Fire-annihilators, Gas-generating | Aëration and Bottling | 1 | 35 |
| Fire-arms | Fire-Arms | 42 | 55 |
| Fire-arms, Converting | Fire-Arms | 42 | 55 |
| Fire-arms, Safety-guards for | Fire-Arms | 42 | 55 |
| Fire-back bricks | Clay | 25 | 20 |
| Fire-backs | Stoves and Furnaces | 126 | 154 |
| Fire-bars | Steam; 2 | 122 | 149 |
| Fire-bricks | Clay | 25 | 20 |
| Fire-dogs | Stoves and Furnaces | 126 | 154 |
| Fire-engines, Chemical | Aëration and Bottling | 1 | 35 |
| Fire-engines, Hydraulic | Pumps | 103 | 118 |
| Fire-engines, Steam | Steam; 1 | 121 | 118 |
| Fire-escapes | Carpentry | 20 | 30 |

| SUBJECT. | Name of class—official classification. | No. of class—official classification. | No. of class—subscription classification. |
|---|---|---|---|
| Fire-extinguishers for boilers | Steam; 2 | 122 | 149 |
| Fire-extinguishers on ships | Ships; 1 | 114 | 138 |
| Fire-fenders | Stoves and Furnaces | 126 | 154 |
| Fire-grates | Stoves and Furnaces | 126 | 154 |
| Fire-irons | Stoves and Furnaces | 126 | 154 |
| Fire-kindlers | Stoves and Furnaces | 126 | 154 |
| Fire-kindling compounds | Fuel | 44 | 154 |
| Fire-ladders | Carpentry | 20 | 30 |
| Fire-lighters | Stoves and Furnaces | 126 | 154 |
| Fire-place heaters, Base-burning | Stoves and Furnaces | 126 | 154 |
| Fire-places | Stoves and Furnaces | 126 | 154 |
| Fire-plates for stoves | Stoves and Furnaces | 126 | 154 |
| Fire-plugs | Water-Distribution | 137 | 167 |
| Fire-pokers | Stoves and Furnaces | 126 | 154 |
| Fire-pots for stoves | Stoves and Furnaces | 126 | 154 |
| Fire-proofing compounds | Paint | 91 | 107 |
| Fire-proofing processes | Paint | 91 | 35 |
| Fire-proofing wood, fabrics, &c | Chemical Miscellaneous | 23 | 35 |
| Fire-proof building | Masonry | 72 | 4 |
| Fire-proof floors | Masonry | 72 | 4 |
| Fire-proof safes | Safes | 109 | 128 |
| Fire-proof shutters | Masonry | 72 | 4 |
| Fire-proof walls | Masonry | 72 | 4 |
| Fire-proof walls, Portable | Masonry | 72 | 4 |
| Fire regulators for steam-boilers | Steam; 2 | 122 | 149 |
| Fire-screens | Furniture | 45 | 62 |
| Fire-shovels | Stoves and Furnaces | 126 | 154 |
| Fire-tests for hydrocarbons | Stills | 124 | 47 |
| Fire-tiles | Clay | 25 | 20 |
| Fire-tongs | Stoves and Furnaces | 126 | 154 |
| Fire-wood, Machines for bundling | Wood-Working; 3 | 144 | 176 |
| Fire-wood, Machines for splitting | Wood-Working; 3 | 144 | 174 |
| Fish-bars for railways, Making | Railways; 4 | 107 | 120 |
| Fish-bars, Railway | Railways; 1 | 104 | 120 |
| Fish, Cans for packing | Preserving Food | 99 | 58 |
| Fish-carrying boxes | Preserving Food | 99 | 63 |
| Fish-curing and drying | Dryers and Kilns | 34 | 58 |
| Fish-decoys | Fishing | 43 | 143 |
| Fish-flakes | Dryers and Kilns | 34 | 58 |
| Fish-hooks | Fishing | 43 | 143 |
| Fish-hooks, Manufacture of | Hardware Manufacture | 53I | 104 |
| Fishing | Fishing | 43 | 143 |
| Fishing-floats | Fishing | 43 | 143 |
| Fishing-line reels | Fishing | 43 | 143 |
| Fishing-lines | Cordage | 28 | 143 |
| Fishing-line sinkers | Fishing | 43 | 143 |
| Fishing-line swivels | Fishing | 43 | 143 |
| Fishing-nets | Fishing | 43 | 143 |
| Fishing-nets, Making | Knitting and Netting | 66 | 89 |
| Fishing-rods | Fishing | 43 | 143 |
| Fishing-seines | Fishing | 43 | 143 |
| Fish-spears | Fishing | 43 | 143 |
| Fish-oil, Treatment of | Oils, Fats, and Glue | 87 | 26 |
| Fish-scalers | Kitchen Utensils | 65 | 88 |
| Fish-traps for water-pipes | Water-Distribution | 137 | 167 |
| Fish-traps | Fishing | 43 | 143 |
| Fissure-needles | Surgery | 128 | 156 |
| Fitting garments | Apparel | 2 | 38 |
| Flageolets | Music | 84 | 102 |
| Flags | Signals | 116 | 158 |

| SUBJECT. | Name of class—official classification. | No. of class—official classification. | No. of class—sub-scription classification. |
|---|---|---|---|
| Flails | Thrashing | 130 | 159 |
| Flannels | Weaving | 139 | 95 |
| Flask-clamps | Casting | 22 | 33 |
| Flasks, Molders' | Casting | 22 | 33 |
| Flasks, Powder | Projectiles | 102 | 55 |
| Flat-iron heaters | Stoves and Furnaces | 126 | 91 |
| Flat-strippers | Carding | 19 | 28 |
| Flavoring tobacco | Tobacco | 131 | 160 |
| Flax-brakes | Brakes and Gins | 13 | 78 |
| Flax-pullers | Harvesters | 56 | 75 |
| Flax-rotting | Brakes and Gins | 13 | 78 |
| Flax-scutchers | Brakes and Gins | 13 | 78 |
| Fleece-tables and tyers | Presses | 100 | 135 |
| Fleece-tyers | Stabling | 119 | 135 |
| Flesh-brushes | Brushes and Brooms | 15 | 22 |
| Flexible gas-tubing | Hose and Belting | 60 | 64 |
| Float-valves | Valves | 136 | 152 |
| Floating-batteries | Ships; 1 | 114 | 138 |
| Floating-berths | Ships; 1 | 114 | 137 |
| Floating-bridges | Ships; 1 | 114 | 137 |
| Floating-cabins | Ships; 1 | 114 | 137 |
| Floating-derricks | Hoisting | 57 | 79 |
| Floating-docks | Ships; 1 | 114 | 82 |
| Floating-safes | Boats | 9 | 128 |
| Floating-wharves | Ships; 1 | 114 | 82 |
| Floats, Boot | Boots and Shoes | 12 | 16 |
| Floats for steam-boilers | Steam; 2 | 122 | 148 |
| Floats, Plasterers | Masonry | 72 | 4 |
| Flock-grinding | Felting and Hats | 38 | 78 |
| Flocking fabrics | Felting and Hats | 38 | 37 |
| Flocking hats | Felting and Hats | 38 | 76 |
| Flocking paper | Paper-Making | 92 | 108 |
| Flood-fences | Fences | 39 | 52 |
| Flood-gates | Hydraulic Engineering | 61 | 82 |
| Floor-cloth | Paint | 91 | 56 |
| Floor cloth, Paper | Paper Manufacture | 93 | 108 |
| Floor-cloth printing | Printing | 101 | 56 |
| Floor-covering, Slatted | Brushes and Brooms | 15 | 56 |
| Floors | Carpentry | 20 | 30 |
| Floors, Cement | Paving | 94 | 110 |
| Floors, Fire-proof | Masonry | 72 | 4 |
| Floors of stone or concrete | Masonry | 72 | 110 |
| Floors, Wooden | Carpentry | 20 | 30 |
| Flour-bolts | Mills | 83 | 57 |
| Flour-chests | Kitchen Utensils | 65 | 88 |
| Flour-coolers | Mills | 83 | 57 |
| Flour-dredgers | Kitchen Utensils | 65 | 88 |
| Flour-dryers | Dryers and Kilns | 34 | 48 |
| Flour-mills | Mills | 83 | 57 |
| Flour-packers | Presses | 100 | 57 |
| Flour-scoops | Kitchen Utensils | 65 | 88 |
| Flour-sieves | Wire-Working | 140 | 88 |
| Flower-boxes | Garden and Orchard | 47 | 63 |
| Flower-pots | Garden and Orchard | 47 | 63 |
| Flower-pots, (clay) | Clay | 25 | 115 |
| Flowers, Artificial | Fine Arts | 41 | 54 |
| Flowers, Preserving | Fine Arts | 41 | 54 |
| Flower-stands | Garden and Orchard | 47 | 63 |
| Flue-boilers | Steam; 2 | 122 | 148 |
| Flue-cleaners | Steam; 2 | 122 | 148 |

| SUBJECT. | Name of class—official classification. | No. of class—official classification. | No. of class—subscription classification. |
|---|---|---|---|
| Flues, Boiler | Steam; 2 | 122 | 148 |
| Flue-scrapers | Steam; 2 | 122 | 148 |
| Fluids, Writing | Chemical Miscellaneous | 23 | 146 |
| Flumes for logs | Hydraulic Engineering | 61 | 82 |
| Flushing-apparatus | Hydraulic Engineering | 61 | 82 |
| Flutes | Music | 84 | 102 |
| Fluting-irons | Laundry | 68 | 91 |
| Fluting-lathes, Wood | Wood-Working; 1 | 142 | 172 |
| Fluting-machines | Apparel | 2 | 38 |
| Fluxing-sand for glass | Glass | 49 | 68 |
| Flyer-guides, Manufacture of | Hardware Manufacture | 53H | 142 |
| Flyers | Spinning | 118 | 142 |
| Flyers for printing-presses | Printing | 101 | 117 |
| Fly-frames | Spinning | 118 | 142 |
| Flying-bridges | Bridges | 14 | 21 |
| Flying-machines | Pneumatics | 98 | 143 |
| Fly-nets | Harness | 54 | 72 |
| Fly-nets, Machines for making leather | Harness | 54 | 72 |
| Fly-traps | Furniture | 45 | 143 |
| Fly-wheels | Mechanical Powers | 74 | 67 |
| Foam-collectors for steam-boilers | Steam; 2 | 122 | 148 |
| Fob-chains | Jewelry | 63 | 86 |
| Fodder-stands | Thrashing | 130 | 144 |
| Fog-alarms | Signals | 116 | 138 |
| Fog-bells | Signals | 116 | 138 |
| Folding boats | Boats | 9 | 13 |
| Folding or tying yarn | Cordage | 28 | 17 |
| Folding sheet-metal | Sheet-Metal | 113A | 136 |
| Fomentations, Modes of applying | Surgery | 128 | 156 |
| Food, Cans, jars, and air-tight packages for | Preserving Food | 99 | 58 |
| Food, Chemical preparation of | Preserving Food | 99 | 58 |
| Food for animals, Preparing | Stabling | 119 | 59 |
| Food, Hermetically sealed packages for | Preserving Food | 99 | 58 |
| Food, Preservation of | Preserving Food | 99 | 58 |
| Foot-bellows | Pneumatics | 98 | 12 |
| Foot-lifters for horses | Stabling | 119 | 144 |
| Foot-pads for horses | Harness | 54 | 144 |
| Foot-rests for horses | Metal-Working; 6 | 81A | 60 |
| Foot-rot curatives | Chemical Miscellaneous | 23 | 98 |
| Footsteps for spindles | Spinning | 118 | 142 |
| Foot-stools | Furniture | 45 | 62 |
| Foot-stoppers for horses | Surgery | 128 | 144 |
| Foot-stoves | Stoves and Furnaces | 126 | 154 |
| Foot-warmers | Stoves and Furnaces | 126 | 154 |
| Forage-rations | Stabling | 119 | 59 |
| Forceps, Artery | Surgery | 128 | 156 |
| Forceps, Calculi | Surgery | 128 | 156 |
| Forceps, Dental | Dental | 32 | 56 |
| Forceps, Obstetrical | Surgery | 128 | 156 |
| Forceps, Polypus | Surgery | 128 | 156 |
| Forcing-pits | Garden and Orchard | 47 | 63 |
| Forebays | Water-Wheels | 138 | 168 |
| Forge-bellows | Pneumatics | 98 | 12 |
| Forges, Portable | Metal-Working; 3 | 78 | 60 |
| Forging | Metal-Working; 3 | 78 | 60 |
| Forks, Culinary | Cutlery | 30 | 44 |
| Forks, Digging | Garden and Orchard | 47 | 141 |
| Forks, Flesh | Cutlery | 30 | 44 |
| Forks, Grain | Harvesters | 56 | 77 |

| SUBJECT. | Name of class—official classification. | No. of class—official classification. | No. of class—subscription classification. |
|---|---|---|---|
| Forks, Hand hay | Garden and Orchard | 47 | 77 |
| Forks, Horse hay | Harvesters | 56 | 77 |
| Forks, Manufacture of | Hardware Manufactures | 53 A | 44 |
| Forks, Manure | Garden and Orchard | 47 | 77 |
| Forks, Table | Cutlery | 30 | 44 |
| Forming bats | Felting and Hats | 38 | 76 |
| Fortifications | Masonry | 72 | 51 |
| Foundations of stone or concrete | Masonry | 72 | 4 |
| Foundry, Appliances of the | Casting | 22 | 33 |
| Fountain-levers | Pumps | 103 | 118 |
| Fountains | Water-Distribution | 137 | 167 |
| Fountains, Beer | Pumps | 103 | 118 |
| Fowls, Coops, nests, and roosts for | Stabling | 119 | 116 |
| Fractures, Apparatus for reducing | Artificial Limbs | 3 | 156 |
| Frames, Advertising | Stationery | 120 | 54 |
| Frames for measuring and fitting garments | Apparel | 2 | 38 |
| Frames for plants | Garden and Orchard | 47 | 63 |
| Frames for tambour-work and quilting | Apparel | 2 | 38 |
| Frames for weaving hair | Toilet | 132 | 161 |
| Frames, Hanging picture | Furniture | 45 | 62 |
| Frames, Machines for making saw | Wood-Working; 3 | 144 | 131 |
| Frames, Machines for making slate | Wood-Working; 3 | 144 | 174 |
| Frames, Machines for turning round and oval | Wood-Working; 1 | 142 | 172 |
| Frames, Numeral | Educational | 35 | 42 |
| Frames of wood | Carpentry | 20 | 30 |
| Frames, Picture | Fine Arts | 41 | 54 |
| Freezing-liquids | Ice | 62 | 83 |
| Freezing-mixtures | Ice | 62 | 83 |
| Freight-cars | Railways; 2 | 105 | 121 |
| Friction-clutches | Journals and Bearings | 64 | 67 |
| Friction-couplings | Journals and Bearings | 64 | 67 |
| Friction-primers | Projectiles | 102 | 105 |
| Frilling | Apparel | 2 | 38 |
| Fringe | Cordage | 28 | 17 |
| Fringe-making machines | Cordage | 28 | 17 |
| Fringe-twisters | Cordage | 28 | 17 |
| Fringing attachments for sewing-machines | Sewing-Machines | 112 | 134 |
| Frit | Glass | 49 | 68 |
| Frit-furnaces | Glass | 49 | 68 |
| Frogs for railways, Making | Railways; 4 | 107 | 120 |
| Frogs, Railway | Railways; 1 | 104 | 120 |
| Frosting glass | Grinding and Polishing | 51 | 68 |
| Frows | Wood-Working; 4 | 145 | 40 |
| Fruit-assorters | Garden and Orchard | 47 | 63 |
| Fruit-baskets | Kitchen Utensils | 65 | 63 |
| Fruit-boxes | Kitchen Utensils | 65 | 63 |
| Fruit-cans | Preserving Food | 99 | 58 |
| Fruit-cars | Preserving Food | 99 | 58 |
| Fruit cellars and houses | Preserving Food | 99 | 58 |
| Fruit-chambers | Preserving Food | 99 | 58 |
| Fruit-crates | Kitchen Utensils | 65 | 63 |
| Fruit-dryers | Dryers and Kilns | 34 | 48 |
| Fruit-frames | Garden and Orchard | 47 | 63 |
| Fruit-gatherers | Garden and Orchard | 47 | 63 |
| Fruit-jars | Preserving Food | 99 | 58 |
| Fruit-knives | Cutlery | 30 | 44 |
| Fruit-mashers | Kitchen Utensils | 65 | 88 |

| SUBJECT. | Name of class—official classification. | No. of class—official classification. | No. of class—sub-scription classification. |
|---|---|---|---|
| Fruit-packers | Presses | 100 | 63 |
| Fruit-panniers | Kitchen Utensils | 65 | 63 |
| Fruit-peelers | Kitchen Utensils | 65 | 88 |
| Fruit-presses | Presses | 100 | 19 |
| Fruit-ripeners | Garden and Orchard | 47 | 63 |
| Fruits, Chemical preparation of | Preserving Food | 99 | 58 |
| Fruit-separators | Garden and Orchard | 47 | 63 |
| Fruit-strainers | Kitchen Utensils | 65 | 88 |
| Frying-pans | Stoves and Furnaces | 126 | 154 |
| Fuel, Apparatus for burning wet | Steam; 2 | 122 | 149 |
| Fuel, Artificial | Fuel | 44 | 61 |
| Fuel | Fuel | 44 | 61 |
| Fuel-dryers | Dryers and Kilns | 34 | 48 |
| Fuel-feeders for steam-boiler furnaces | Steam; 2 | 122 | 149 |
| Fulling | Felting and Hats | 38 | 37 |
| Fulling-mills | Felting and Hats | 38 | 37 |
| Fulminates | Gunpowder | 52 | 71 |
| Fumigators for insects | Garden and Orchard | 47 | 63 |
| Funnels | Water-Distribution | 137 | 167 |
| Fur, Articles of | Apparel | 2 | 38 |
| Fur-blowing | Felting and Hats | 38 | 76 |
| Fur-cutting | Felting and Hats | 38 | 76 |
| Fur from pelts, Cutting | Felting and Hats | 38 | 76 |
| Fur hats | Felting and Hats | 38 | 76 |
| Furling sails | Ships; 1 | 114 | 138 |
| Furnace-doors for steam-boiler furnaces | Steam; 2 | 122 | 149 |
| Furnace-grates for steam-boiler furnaces | Steam; 2 | 122 | 149 |
| Furnace-lining | Metallurgy | 75 | 99 |
| Furnaces, Bath | Stoves and Furnaces | 126 | 154 |
| Furnaces, Blast | Metallurgy | 75 | 85 |
| Furnaces, Boiling | Metallurgy | 75 | 85 |
| Furnaces, Bone-black | Sugar | 127 | 155 |
| Furnaces, Catalan | Metallurgy | 75 | 85 |
| Furnaces, Charcoal | Stoves and Furnaces | 126 | 154 |
| Furnaces, Chemical | Chemical Miscellaneous | 23 | 35 |
| Furnaces, Copper | Metallurgy | 75 | 99 |
| Furnaces, Cupola | Metallurgy | 75 | 85 |
| Furnaces, Desulphurizing | Metallurgy | 75 | 95 |
| Furnaces, Domestic | Stoves and Furnaces | 126 | 154 |
| Furnaces for blanks | Metal-Working; 3 | 78 | 60 |
| Furnaces for reducing metals | Metallurgy | 75 | 99 |
| Furnaces for roasting ores | Metallurgy | 75 | 99 |
| Furnaces for skelps | Metal-Working; 3 | 78 | 60 |
| Furnaces for tires | Metal-Working; 3 | 78 | 60 |
| Furnaces for treatment of iron | Metallurgy | 75 | 85 |
| Furnaces, Glass | Glass | 49 | 68 |
| Furnaces, Heating | Stoves and Furnaces | 126 | 154 |
| Furnaces, Hot-air | Stoves and Furnaces | 126 | 154 |
| Furnaces, Hot-blast | Metallurgy | 75 | 85 |
| Furnaces, Iron | Metallurgy | 75 | 85 |
| Furnaces, Lead | Metallurgy | 75 | 99 |
| Furnaces, Locomotive | Steam; 3 | 123 | 94 |
| Furnaces, Portable | Stoves and Furnaces | 126 | 154 |
| Furnaces, Puddling | Metallurgy | 75 | 85 |
| Furnaces, Salt | Sugar | 127 | 129 |
| Furnaces, Smelting | Metallurgy | 75 | 85 |
| Furnaces, Steam-boiler | Steam; 2 | 122 | 149 |
| Furnaces, Sugar | Sugar | 127 | 155 |
| Furnaces, Zinc | Metallurgy | 75 | 99 |

| SUBJECT. | Name of class—official classification. | No. of class—official classification. | No. of class—subscription classification. |
|---|---|---|---|
| Furniture-casters | Builders' Hardware | 16 | 24 |
| Furniture, Feet for legs of | Furniture | 45 | 62 |
| Furniture-hooks | Builders' Hardware | 16 | 62 |
| Furniture-knobs | Builders' Hardware | 16 | 62 |
| Furniture-pads | Furniture | 45 | 62 |
| Furniture, Printers' | Printing | 101 | 117 |
| Furniture-springs | Furniture | 45 | 62 |
| Fur-picking | Felting and Hats | 38 | 76 |
| Furrowing-machines | Plows | 97 | 114 |
| Furrowing-stone | Stone, Lime, and Cement | 125 | 153 |
| Fur, Treating | Felting and Hats | 38 | 76 |
| Fuse-making machines | Cordage | 28 | 17 |
| Fuses, Blasting | Projectiles | 102 | 105 |
| Fuses, Concussion | Projectiles | 102 | 105 |
| Fuses, Firing | Projectiles | 102 | 105 |
| Fuses, Safety | Projectiles | 102 | 105 |
| Fuses, Tape | Projectiles | 102 | 105 |
| Fuses, Time | Projectiles | 102 | 105 |
| Fusible plugs for steam-boilers | Steam; 2 | 122 | 148 |
| G. | | | |
| Gaff-jaws | Ships; 1 | 114 | 138 |
| Gaffs | Ships; 1 | 114 | 138 |
| Gage-cocks | Valves | 136 | 148 |
| Gage-knives | Harness | 54 | 72 |
| Gage-pins for tympan-sheets | Printing | 101 | 117 |
| Gages, Barrel-filling | Measuring-Instruments | 73 | 167 |
| Gages, Bilgewater | Measuring-Instruments | 73 | 137 |
| Gages, Cutting | Wood-Working; 4 | 145 | 29 |
| Gages for cutting garments | Apparel | 2 | 38 |
| Gages for sawing-machines | Saws | 110 | 130 |
| Gages for sewing-machines | Sewing-Machines | 112 | 134 |
| Gages for steam-boilers | Measuring-Instruments | 73 | 148 |
| Gages, Measuring | Measuring-Instruments | 73 | 132 |
| Gages, Pressure | Measuring-Instruments | 73 | 148 |
| Gages, Rain | Measuring-Instruments | 73 | 101 |
| Gages, Scribing | Wood-Working; 4 | 145 | 29 |
| Gages, Sliding | Measuring-Instruments | 73 | 132 |
| Gages, Steam | Measuring-Instruments | 73 | 148 |
| Gages, Vacuum | Measuring-Instruments | 73 | 148 |
| Gaging-instruments | Measuring-Instruments | 73 | 132 |
| Gaging silk | Silk | 117 | 140 |
| Gag-runners | Harness | 54 | 72 |
| Gags for animals | Harness | 54 | 144 |
| Gaiterettes, (of cloth) | Apparel | 2 | 38 |
| Gaiters, Leather | Boots and Shoes | 12 | 16 |
| Galley-rests | Printing | 101 | 117 |
| Galleys | Printing | 101 | 117 |
| Galleys, Ships' | Ships; 1 | 114 | 137 |
| Gallows for suspending carcasses | Butchering | 17 | 25 |
| Galoshes | Boots and Shoes | 12 | 16 |
| Galvanic appliances | Electricity | 36 | 50 |
| Galvanic attachments | Electricity | 36 | 50 |
| Galvanizing metals | Plating | 96 | 113 |
| Galvanometers | Electricity | 36 | 50 |
| Galvanoscopes | Electricity | 36 | 50 |
| Gambrels | Butchering | 17 | 25 |
| Game-bags | Games and Toys | 46 | 143 |
| Game-boards | Games and Toys | 46 | 143 |

| SUBJECT. | Name of class—official classification. | No. of class—official classification. | No. of class—sub-scription classification. |
|---|---|---|---|
| Game-counters | Measuring-Instruments | 73 | 143 |
| Games | Games and Toys | 46 | 143 |
| Gang-plows | Plows | 97 | 114 |
| Garbage-boxes in pavement | Paving | 94 | 110 |
| Garbage-sinks | Masonry | 72 | 110 |
| Gardeners' basket-supports | Garden and Orchard | 47 | 63 |
| Gardeners' implements of metal, Manufacture of | Metal-Working; 6 | 81B | 141 |
| Garden-forks | Garden and Orchard | 47 | 141 |
| Garden-hoes | Garden and Orchard | 47 | 141 |
| Garden-implements, (*except Shears, Class* 30.) | Garden and Orchard | 47 | 63 |
| Garden-pots | Clay | 25 | 63 |
| Garden-tiles | Clay | 25 | 63 |
| Garden-walks, Machines for scraping | Plows | 97 | 114 |
| Garment-fasteners | Clasps and Buckles | 24 | 23 |
| Garments, (*except of Paper, Class* 93) | Apparel | 2 | 38 |
| Garments, Gages for cutting | Apparel | 2 | 38 |
| Garments, Machines for cutting out | Apparel | 2 | 38 |
| Garments, Systems of measuring and cutting | Apparel | 2 | 38 |
| Garments, Patterns for | Apparel | 2 | 38 |
| Garters | Apparel | 2 | 38 |
| Gasaliers | Lamps and Gas-Fittings | 67 | 64 |
| Gas-brackets | Lamps and Gas-Fittings | 67 | 64 |
| Gas-burners | Lamps and Gas-Fittings | 67 | 64 |
| Gas-burners, Manufacture of | Hardware Manufactures | 53J | 64 |
| Gas-carburetors | Gas | 48 | 27 |
| Gas-checks for breech-loading ordnance | Ordnance | 89 | 105 |
| Gas-compressing machines | Pneumatics | 98 | 12 |
| Gas-engines | Pneumatics | 98 | 2 |
| Gas-engines, Electrical apparatus for | Electricity | 36 | 2 |
| Gases, Methods and processes for obtaining | Gas | 48 | 66 |
| Gas-fitters' tools | Metal-Working; 6 | 81A | 64 |
| Gas-fittings | Lamps and Gas-Fittings | 67 | 64 |
| Gas-heaters | Stoves and Furnaces | 126 | 65 |
| Gaskets for steam-joints | Steam; 1 | 121 | 151 |
| Gas-lighters | Lamps and Gas-Fittings | 67 | 64 |
| Gas-lighting by electricity | Electricity | 36 | 64 |
| Gas-making apparatus | Gas | 48 | 66 |
| Gas manufacture | Gas | 48 | 66 |
| Gas-meters | Pneumatics | 98 | 64 |
| Gas-meters, Registering devices of | Measuring-Instruments | 73 | 64 |
| Gas-pumps | Pneumatics | 98 | 12 |
| Gas-purifiers | Gas | 48 | 66 |
| Gas-regulators | Lamps and Gas-Fittings | 67 | 64 |
| Gas-retorts | Gas | 48 | 66 |
| Gassing yarns, threads, &c | Cordage | 28 | 17 |
| Gas-stoves | Stoves and Furnaces | 126 | 65 |
| Gas-tubing | Paint | 91 | 64 |
| Gate-hinges | Builders' Hardware | 16 | 24 |
| Gate-latches | Locks and Latches | 70 | 93 |
| Gate-posts | Fences | 39 | 52 |
| Gates | Fences | 39 | 52 |
| Gate-saws | Saws | 110 | 130 |
| Gates, Canal-lock | Hydraulic Engineering | 61 | 82 |
| Gates, Flood | Hydraulic Engineering | 61 | 82 |
| Gates for water-wheels | Water-Wheels | 138 | 168 |

| SUBJECT. | Name of class—official classification. | No. of class—official classification. | No. of class—subscription classification. |
|---|---|---|---|
| Gates, Molasses | Valves | 136 | 167 |
| Gates, Railway | Railways; 1 | 104 | 120 |
| Gates, Sluice | Hydraulic Engineering | 61 | 82 |
| Gatherers for sewing-machines | Sewing-Machines | 112 | 134 |
| Gaveling-attachments for harvesters | Harvesters | 56 | 74 |
| Gaveling-forks | Harvesters | 56 | 74 |
| Gauze | Weaving | 139 | 95 |
| Gazogenes | Aëration and Bottling | 1 | 1 |
| Gear-cutting machines | Metal-Working; 7 | 82 | 100 |
| Gearing, Arrangements of | Mechanical Powers | 74 | 67 |
| Gearing, Harvester | Harvesters | 56 | 74 |
| Gear, Machines for cutting wooden | Wood-Working; 3 | 144 | 174 |
| Gelatine | Oil, Fat, and Glue | 87 | 26 |
| Gems, Artificial | Glass | 49 | 68 |
| Gems, Setting | Jewelry | 63 | 86 |
| Generators, Steam | Steam; 2 | 122 | 148 |
| Geometrical constructors | Drafting | 33 | 97 |
| Gibs for pitmen, &c | Mechanical Powers | 74 | 67 |
| Gig-mills | Cloth | 26 | 37 |
| Gig-trees | Harness | 54 | 72 |
| Gilding | Plating | 96 | 113 |
| Gilding, Leaf | Fine Arts | 41 | 54 |
| Gimlets | Wood-Working; 4 | 145 | 29 |
| Gimp | Knitting and Netting | 66 | 17 |
| Gimp-machines | Cordage | 28 | 17 |
| Gins, Cotton | Brakes and Gins | 13 | 41 |
| Girder-bar, Rolling | Metal-Working; 5 | 80 | 125 |
| Girders for bridges, floors, roofs | Bridges | 14 | 21 |
| Girders of iron or wood | Bridges | 14 | 21 |
| Girdling trees, Machines for | Wood-Working; 3 | 144 | 29 |
| Glass-blowing | Glass | 49 | 68 |
| Glass-cementing | Glass | 49 | 68 |
| Glass-cleaners | Kitchen Utensils | 65 | 88 |
| Glass, Compositions for | Glass | 49 | 68 |
| Glass-cutting machine | Grinding and Polishing | 51 | 68 |
| Glass-cutting processes | Grinding and Polishing | 51 | 68 |
| Glass furnaces | Glass | 49 | 68 |
| Glass-gilding | Plating | 96 | 54 |
| Glass-grinding | Grinding and Polishing | 51 | 68 |
| Glass-house pots | Glass | 49 | 68 |
| Glass-houses | Glass | 49 | 68 |
| Glass-molding | Glass | 49 | 68 |
| Glass-platinizing | Plating | 96 | 68 |
| Glass-polishing | Grinding and Polishing | 51 | 68 |
| Glass pots | Glass | 49 | 68 |
| Glass-pressing | Glass | 49 | 68 |
| Glass, Printing | Printing | 101 | 54 |
| Glass-roughening | Grinding and Polishing | 51 | 68 |
| Glass-silvering | Plating | 96 | 68 |
| Glass-working tools, Manufacture of | Metal-Working; 6 | 81B | 68 |
| Glazed fabrics | Bleaching and Dyeing | 8 | 11 |
| Glazes for pottery, bricks, &c | Clay | 25 | 115 |
| Glaziers' diamonds | Glass | 49 | 68 |
| Glaziers' diamonds, Setting | Glass | 49 | 68 |
| Glaziers' diamonds, Substitutes for | Glass | 49 | 68 |
| Glaziers' points, Making | Sheet-Metal | 113B | 68 |
| Glazing gunpowder | Gunpowder | 52 | 71 |
| Glazing paper | Paper-Making | 92 | 108 |
| Globes, Astronomical | Educational | 35 | 146 |
| Globes, Geographical | Educational | 35 | 146 |

| SUBJECT. | Name of class—official classification. | No. of class—official classification. | No. of class—sub-scription classification. |
|---|---|---|---|
| Globe-valves | Valves | 136 | 152 |
| Globule-machines | Surgery | 128 | 98 |
| Glossing silk | Silk | 117 | 140 |
| Glove-buttoners | Clasps and Buckles | 24 | 38 |
| Glove-buttons | Clasps and Buckles | 24 | 38 |
| Glove-buttons, Manufacture of | Hardware Manufactures | 53E | 23 |
| Glove-fasteners | Clasps and Buckles | 24 | 38 |
| Glove-fastenings | Clasps and Buckles | 24 | 38 |
| Glove-making | Apparel | 2 | 38 |
| Gloves | Apparel | 2 | 38 |
| Glove-stretching machines | Apparel | 2 | 38 |
| Glove-turning machines | Apparel | 2 | 38 |
| Glucose, Manufacture of | Sugar | 127 | 155 |
| Glue-cutting machines | Oil, Fat, and Glue | 87 | 26 |
| Glue-dryers | Dryers and Kilns | 34 | 26 |
| Glue-hoppers | Wood-Working; 4 | 145 | 29 |
| Glueing machines, Leather | Leather | 69 | 92 |
| Glueing tables | Wood-Working; 4 | 145 | 29 |
| Glue-making | Oil, Fat, and Glue | 87 | 26 |
| Glue-pots | Stoves and Furnaces | 126 | 29 |
| Glue | Oil, Fat, and Glue | 87 | 26 |
| Glycerine | Oil, Fat, and Glue | 87 | 26 |
| Gold-beating | Metal-Working; 3 | 78 | 136 |
| Gold-leaf condensers | Dental | 32 | 46 |
| Gold pens, Manufacture of | Hardware Manufactures | 53 I | 136 |
| Gold-washers | Ore | 90 | 3 |
| Gongs | Signals | 116 | 24 |
| Goniometers | Measuring-Instruments | 73 | 132 |
| Gouges | Wood-Working; 4 | 145 | 49 |
| Governor cut-off motions | Governors | 50 | 69 |
| Governors | Governors | 50 | 69 |
| Governor-valves | Valves | 136 | 152 |
| Grabs, Drill-rod | Stone, Lime, and Cement | 125 | 124 |
| Grading | Excavators | 37 | 51 |
| Grading-scrapers | Excavators | 37 | 51 |
| Graduated glasses | Measuring-Instruments | 73 | 132 |
| Graduated measures | Measuring-Instruments | 73 | 132 |
| Graduated rules | Drafting | 33 | 97 |
| Grafting-knives | Cutlery | 30 | 63 |
| Grafting-machines | Garden and Orchard | 47 | 63 |
| Grain-bagging | Mills | 83 | 57 |
| Grain-binders | Harvesters | 56 | 74 |
| Grain-bins | Thrashing | 130 | 59 |
| Grain-bruisers | Mills | 83 | 57 |
| Grain-cars | Railways; 2 | 105 | 121 |
| Grain-cleaners for mills | Mills | 83 | 57 |
| Grain-conveyors | Hoisting | 57 | 79 |
| Grain-cradles | Harvesters | 56 | 49 |
| Grain-dampeners | Mills | 83 | 57 |
| Grain-distributors for mills | Mills | 83 | 57 |
| Grain-drills | Seeders and Planters | 111 | 133 |
| Grain-dryers | Dryers and Kilns | 34 | 48 |
| Grain-elevators | Hoisting | 57 | 79 |
| Grain-feeders | Mills | 83 | 57 |
| Grain for distillation, Preparing | Stills | 124 | 47 |
| Grain for grinding, Preparing | Mills | 83 | 57 |
| Grain-forks | Harvesters | 56 | 77 |
| Grain-bags | Mills | 83 | 57 |
| Grain-harvesters | Harvesters | 56 | 74 |
| Graining gunpowder | Gunpowder | 52 | 71 |

| SUBJECT. | Name of class—official classification. | No. of class—official classification. | No. of class—subscription classification. |
|---|---|---|---|
| Graining leather | Tanning | 129 | 92 |
| Graining-tools and appliances | Fine Arts | 41 | 107 |
| Grain-levelers in vessels | Ships; 1 | 114 | 138 |
| Grain, Manufacture of sugar from | Sugar | 127 | 155 |
| Grain-mills | Mills | 83 | 57 |
| Grain-rakes | Harvesters | 56 | 77 |
| Grain-registers | Measuring-Instruments | 73 | 159 |
| Grain-riddles | Thrashing | 130 | 159 |
| Grain-screens | Thrashing | 130 | 159 |
| Grain-seed separators | Thrashing | 130 | 159 |
| Grain-spouts for elevators | Hoisting | 57 | 79 |
| Grain-separators | Thrashing | 130 | 159 |
| Grain-spouts for mills | Mills | 83 | 57 |
| Grain-tallies | Measuring-Instruments | 73 | 159 |
| Grain-thrashers | Thrashing | 130 | 159 |
| Grain-tollers | Mills | 83 | 57 |
| Grain-weighers | Measuring-Instruments | 73 | 132 |
| Grain-winnowers | Thrashing | 130 | 159 |
| Granaries | Thrashing | 130 | 59 |
| Grape-presses | Presses | 100 | 19 |
| Grape-shot | Projectiles | 102 | 105 |
| Grape-sugar, Manufacture of | Sugar | 127 | 155 |
| Grape-trellises | Garden and Orchard | 47 | 63 |
| Grape-vine protectors | Garden and Orchard | 47 | 63 |
| Grapnels | Ships; 1 | 114 | 138 |
| Grappling-irons | Ships; 1 | 114 | 138 |
| Grass-burners | Garden and Orchard | 47 | 63 |
| Grass-harvesters | Harvesters | 56 | 75 |
| Grate-bars for steam-boilers | Steam; 2 | 122 | 149 |
| Graters, Corn | Kitchen Utensils | 65 | 88 |
| Graters, Spice | Kitchen Utensils | 65 | 88 |
| Grates for stoves | Stoves and Furnaces | 126 | 154 |
| Grate-tippers | Stoves and Furnaces | 126 | 154 |
| Gratings for windows | Carpentry | 20 | 4 |
| Gravel-cars | Railways; 2 | 105 | 121 |
| Gravel-heaters | Stoves and Furnaces | 126 | 126 |
| Gravers | Fine Arts | 41 | 54 |
| Graving-docks | Hydraulic Engineering | 61 | 82 |
| Grease-cups | Journals and Bearings | 64 | 87 |
| Grease from slush, Separating | Oil, Fat, and Glue | 87 | 26 |
| Green-houses | Garden and Orchard | 47 | 63 |
| Grenades | Projectiles | 102 | 105 |
| Griddles | Stoves and Furnaces | 126 | 154 |
| Gridirons | Stoves and Furnaces | 126 | 154 |
| Grinding flock | Felting and Hats | 38 | 76 |
| Grinding glass | Grinding and Polishing | 51 | 68 |
| Grinding-machines | Grinding and Polishing | 51 | 70 |
| Grinding-materials | Grinding and Polishing | 51 | 70 |
| Grinding-plates for mills | Mills | 83 | 57 |
| Grinding stone | Grinding and Polishing | 51 | 153 |
| Grindstones | Grinding and Polishing | 51 | 70 |
| Grindstones, Artificial | Grinding and Polishing | 51 | 70 |
| Grindstones, Hanging | Grinding and Polishing | 51 | 70 |
| Grist alarms | Mills | 83 | 57 |
| Grommets | Ships; 1 | 114 | 138 |
| Grooming apparatus | Stabling | 119 | 144 |
| Groove-cutting machines | Metal-Working; 7 | 82 | 100 |
| Grooving lumber, Machines for | Wood-Working; 2 | 143 | 173 |
| Grooving sheet-metal | Sheet-Metal | 113 A | 136 |
| Grubbing-harrows | Harrows | 55 | 73 |

| SUBJECT. | Name of class—official classification. | No. of class—official classification. | No. of class—subscription classification. |
|---|---|---|---|
| Grubbing-hoes | Garden and Orchard | 47 | 141 |
| Grubbing-machines | Harrows | 55 | 73 |
| Guano, Treatment of | Manures | 71 | 96 |
| Guard-fingers for harvesters | Harvesters | 56 | 74 |
| Guards for locks of fire-arms | Fire-Arms | 42 | 55 |
| Guards for triggers of fire-arms | Fire-Arms | 42 | 55 |
| Guards for windows | Carpentry | 20 | 30 |
| Guards, Railway | Railways; 1 | 104 | 120 |
| Guides for sewing-machines | Sewing-Machines | 112 | 134 |
| Guitars | Music | 84 | 102 |
| Gumming-saws | Metal-Working; 4 | 79 | 131 |
| Gum resins, Treatment of | Paint | 91 | 107 |
| Gums, Artificial | Dental | 32 | 46 |
| Gun-barrels, Making | Tubing and Wire | 134 | 163 |
| Gun-carriages | Ordnance | 89 | 105 |
| Gun-carriages, Counterpoise | Ordnance | 89 | 105 |
| Gun-carriages, Making of iron | Metal-Working; 3 | 78 | 105 |
| Gun-chassis | Ordnance | 89 | 105 |
| Gun-cotton, Manufacture of | Gunpowder | 52 | 71 |
| Gun-covers | Fire-Arms | 42 | 55 |
| Gun-hammers | Fire-Arms | 42 | 55 |
| Gun-locks | Fire-Arms | 42 | 55 |
| Gun-locks, Manufacture of | Hardware Manufactures | 53A | 55 |
| Gun-platforms | Ordnance | 89 | 105 |
| Gunpowder, Compositions for | Gunpowder | 52 | 71 |
| Gunpowder, Corning | Gunpowder | 52 | 71 |
| Gunpowder, Drying | Gunpowder | 52 | 71 |
| Gunpowder-engines | Pneumatics | 98 | 2 |
| Gunpowder, Glazing | Gunpowder | 52 | 71 |
| Gunpowder, Graining | Gunpowder | 52 | 71 |
| Gunpowder-mills | Mills | 83 | 71 |
| Gunpowder-polishing | Gunpowder | 52 | 71 |
| Gunpowder-presses | Presses | 100 | 71 |
| Gunpowder | Gunpowder | 52 | 71 |
| Gun-scrapers | Fire-Arms | 42 | 55 |
| Gun-stocks | Fire-Arms | 42 | 55 |
| Gun-stocks, Machines for turning | Wood-Working; 1 | 142 | 172 |
| Gun-wads | Projectiles | 102 | 105 |
| Gun-wipers | Fire-Arms | 42 | 55 |
| Gun-worms | Fire-Arms | 42 | 55 |
| Gutta-percha manufactures | Caoutchouc | 18 | 84 |
| Gutter-clearers | Paving | 94 | 110 |
| Gutters, Pavement | Paving | 94 | 110 |
| Gutters for roofs | Roofing | 108 | 126 |
| Gutters, Machines for making wooden | Wood-Working; 3 | 144 | 173 |
| Gutters, Making sheet-metal | Sheet-Metal | 113B | 136 |
| Gutter-tubes, Making sheet-metal | Sheet-Metal | 113B | 136 |
| Gymnastic apparatus | Games and Toys | 46 | 143 |
| **H.** | | | |
| Hackles | Brakes and Gins | 13 | 78 |
| Hackling corn-husks | Brakes and Gins | 13 | 78 |
| Hackling-machines | Brakes and Gins | 13 | 78 |
| Hair-brooms | Brushes and Brooms | 15 | 22 |
| Hair-brushes | Brushes and Brooms | 15 | 161 |
| Hair-cleaning | Butchering | 17 | 22 |
| Hair cloth, Machines for making | Weaving | 139 | 95 |
| Hair cloth, Machines for trimming | Cloth | 26 | 37 |
| Hair-crimpers | Toilet | 132 | 161 |

| SUBJECT. | Name of class—official classification. | No. of class—official classification. | No. of class—sub-scription classification. |
|---|---|---|---|
| Hair-curlers | Toilet | 132 | 161 |
| Hair-cutting machines | Toilet | 132 | 161 |
| Hair-dyes | Chemical Miscellaneous | 23 | 161 |
| Hair from fur, Separating | Felting and Hats | 38 | 76 |
| Hair from skins, Pulling | Felting and Hats | 38 | 76 |
| Hair from skins, Separating | Felting and Hats | 38 | 76 |
| Hair-nets | Toilet | 132 | 161 |
| Hair-parting instruments | Toilet | 132 | 161 |
| Hair-pickers | Brakes and Gins | 13 | 78 |
| Hair-pincers | Toilet | 132 | 161 |
| Hair-pins | Toilet | 132 | 104 |
| Hair-pins, Making | Needles and Pins | 86 | 104 |
| Hair-restoratives | Chemical Miscellaneous | 23 | 161 |
| Hair-rolls | Toilet | 132 | 161 |
| Hair-tongs | Toilet | 132 | 161 |
| Hair-trimmings | Toilet | 132 | 161 |
| Hair-tweezers | Toilet | 132 | 161 |
| Halters | Harness | 54 | 72 |
| Hame-clips | Harness | 54 | 72 |
| Hame-fasteners | Harness | 54 | 72 |
| Hames | Harness | 54 | 72 |
| Hames, Manufacture of | Hardware Manufactures | 53F | 72 |
| Hammering machines, Leather | Leather | 69 | 92 |
| Hammers, Claw | Wood-Working; 4 | 145 | 29 |
| Hammers, Drop | Metal-Working; 3 | 78 | 60 |
| Hammers, Gun | Fire-Arms | 42 | 55 |
| Hammers, Hydraulic | Metal-Working; 3 | 78 | 60 |
| Hammers, Millstone | Stone, Lime, and Cement | 125 | 57 |
| Hammers, Pneumatic | Metal-Working; 3 | 78 | 60 |
| Hammers, Power | Metal-Working; 3 | 78 | 60 |
| Hammers, Safety-guards for gun | Fire-Arms | 42 | 55 |
| Hammers, Steam | Metal-Working; 3 | 78 | 60 |
| Hammers, Steam-engines for | Steam; 1 | 121 | 150 |
| Hammers, Water | Metal-Working; 3 | 78 | 60 |
| Hammocks | Beds | 5 | 62 |
| Hampers | Kitchen Utensils | 65 | 169 |
| Hand-barrows | Carriages and Wagons | 21 | 31 |
| Hand-bearers | Carriages and Wagons | 21 | 31 |
| Hand-cars | Railways; 2 | 105 | 121 |
| Hand-cuffs, Locks for | Locks and Latches | 70 | 93 |
| Hand-drills for grain | Seeders and Planters | 111 | 133 |
| Hand-grenades | Projectiles | 102 | 105 |
| Hand-hammers | Metal-Working; 6 | 81A | 60 |
| Hand-holes for boilers and tanks | Steam; 2 | 122 | 148 |
| Handles for canes | Umbrellas and Fans | 135 | 165 |
| Handles for tea and coffee pots | Stoves and Furnaces | 126 | 136 |
| Handles for umbrellas | Umbrellas and Fans | 135 | 165 |
| Handles, Machines for making wooden | Wood-Working; 3 | 144 | 174 |
| Handles, Machines for turning | Wood-Working; 1 | 142 | 175 |
| Handles, Tool, (*except Turning*) | Wood-Working; 4 | 145 | 29 |
| Handles to tools, Attaching | Wood-Working; 4 | 145 | 29 |
| Hand-mirrors | Furniture | 45 | 62 |
| Hand-planters | Seeders and Planters | 111 | 133 |
| Hand-plows | Plows | 97 | 114 |
| Hands, Artificial | Artificial Limbs | 3 | 5 |
| Hand-saws | Saws | 110 | 131 |
| Hands for clocks and watches | Horology | 58 | 36 |
| Hand-stamps | Printing | 101 | 146 |
| Hand-trucks | Carriages and Wagons | 21 | 31 |
| Hand-vises | Metal-Working; 6 | 81A | 60 |

| SUBJECT. | Name of class—official classification. | No. of class—official classification. | No. of class—subscription classification. |
|---|---|---|---|
| Hangers for clothing | Apparel | 2 | 38 |
| Hangers for lamps | Lamps and Gas-Fittings | 67 | 90 |
| Hangers for shafting | Journals and Bearings | 64 | 67 |
| Hanging-baskets | Garden and Orchard | 47 | 63 |
| Hanging carriage-bodies | Carriages and Wagons | 21 | 31 |
| Hanging grindstones | Grinding and Polishing | 51 | 70 |
| Hanging looking-glasses | Furniture | 45 | 62 |
| Hanging picture-frames | Furniture | 45 | 62 |
| Hangings for doors, gates, and sash | Builders' Hardware | 16 | 24 |
| Hanging-shelves | Furniture | 45 | 62 |
| Hanks for sails | Ships; 1 | 114 | 138 |
| Harbor defenses | Hydraulic Engineering | 61 | 82 |
| Harbor obstructions, Removing | Hydraulic Engineering | 61 | 82 |
| Harbors | Hydraulic Engineering | 61 | 82 |
| Harbors, Obstructing | Hydraulic Engineering | 61 | 82 |
| Hardening-machines for hats | Felting and Hats | 38 | 76 |
| Hard rubber | Caoutchouc | 18 | 84 |
| Harmoniums | Music | 84 | 106 |
| Harness | Harness | 54 | 72 |
| Harness-chains | Harness | 54 | 72 |
| Harness, Detaching | Harness | 54 | 72 |
| Harness for looms | Weaving | 139 | 95 |
| Harness for looms, Making wire | Wire-Working | 140 | 95 |
| Harness-hooks | Harness | 54 | 72 |
| Harness, Machines for making loom | Weaving | 139 | 95 |
| Harness-makers' tools | Harness | 54 | 72 |
| Harness-mountings | Harness | 54 | 72 |
| Harness-oiling machines | Harness | 54 | 72 |
| Harness-pads | Harness | 54 | 72 |
| Harness-reins, Machines for rounding | Harness | 54 | 72 |
| Harness-swivels | Harness | 54 | 72 |
| Harness-wardrobes | Harness | 54 | 72 |
| Harpoons | Fishing | 43 | 143 |
| Harpoons, Gun | Projectiles | 102 | 105 |
| Harps | Music | 84 | 102 |
| Harrows | Harrows | 55 | 73 |
| Harrows, Rotary | Harrows | 55 | 73 |
| Harrow-teeth | Harrows | 55 | 73 |
| Harvester-gearing | Harvesters | 56 | 74 |
| Harvesters, Bean | Harvesters | 56 | 75 |
| Harvesters, Binders for | Harvesters | 56 | 74 |
| Harvesters, Cane | Harvesters | 56 | 75 |
| Harvesters, Clover-seed | Harvesters | 56 | 75 |
| Harvesters, Corn | Harvesters | 56 | 75 |
| Harvesters, Cutters for | Harvesters | 56 | 74 |
| Harvesters, Dropping-platforms for | Harvesters | 56 | 74 |
| Harvesters, Dumping-reels for | Harvesters | 56 | 74 |
| Harvesters, Fingers for | Harvesters | 56 | 74 |
| Harvesters, Gaveling-attachments for | Harvesters | 56 | 74 |
| Harvesters, Gearing for | Harvesters | 56 | 74 |
| Harvesters, Grain | Harvesters | 56 | 74 |
| Harvesters, Grass | Harvesters | 56 | 75 |
| Harvesters, Grinding-apparatus for | Grinding and Polishing | 51 | 74 |
| Harvesters, Guard-fingers for | Harvesters | 56 | 74 |
| Harvesters, Hemp | Harvesters | 56 | 75 |
| Harvesters, Pitmen for | Mechanical Powers | 74 | 74 |
| Harvesters, Rakes for | Harvesters | 56 | 74 |
| Harvesters, Reels for | Harvesters | 56 | 74 |
| Harvesters, Track-clearers for | Harvesters | 56 | 74 |
| Harvesting-bags | Harvesters | 56 | 74 |

| SUBJECT. | Name of class—official classification. | No. of class—official classification. | No. of class—sub-scription classification. |
|---|---|---|---|
| Hasp-locks | Locks and Latches | 70 | 93 |
| Hat and coat hooks | Builders' Hardware | 16 | 24 |
| Hat and coat racks | Furniture | 45 | 62 |
| Hat-blocks, Machines for turning | Wood-Working; 1 | 142 | 175 |
| Hat-bodies | Felting and Hats | 38 | 76 |
| Hat-bodies, Forming | Felting and Hats | 38 | 76 |
| Hat-bodies, Hardening | Felting and Hats | 38 | 76 |
| Hat-bodies, Napping | Felting and Hats | 38 | 76 |
| Hat-blocking machines | Felting and Hats | 38 | 76 |
| Hat-blocks | Felting and Hats | 38 | 76 |
| Hat-boxes, Leather | Trunks | 133 | 162 |
| Hat-brushing machines | Felting and Hats | 38 | 76 |
| Hatchets | Wood-Working; 4 | 145 | 49 |
| Hatchways, Devices for operating | Carpentry | 20 | 30 |
| Hatchways, Safety-guards for | Carpentry | 20 | 30 |
| Hat-conformators | Felting and Hats | 38 | 76 |
| Hat-finishing | Felting and Hats | 38 | 76 |
| Hat-frames | Felting and Hats | 38 | 76 |
| Hat-hooks, Manufacture of | Hardware Manufactures | 53D | 24 |
| Hat-ironing machines | Felting and Hats | 38 | 76 |
| Hat-lining | Felting and Hats | 38 | 76 |
| Hat-measures | Felting and Hats | 38 | 76 |
| Hat-napping | Felting and Hats | 38 | 76 |
| Hat-planking | Felting and Hats | 38 | 76 |
| Hat-pressing | Felting and Hats | 38 | 76 |
| Hat-printing machines | Printing | 101 | 76 |
| Hat-protectors | Felting and Hats | 38 | 76 |
| Hats | Felting and Hats | 38 | 76 |
| Hat-shaping machines | Felting and Hats | 38 | 76 |
| Hat-shrinking machines | Felting and Hats | 38 | 76 |
| Hat-sizing machines | Felting and Hats | 38 | 76 |
| Hat-stiffening compositions | Felting and Hats | 38 | 76 |
| Hat-stiffening machines | Felting and Hats | 38 | 76 |
| Hat-stretching machines | Felting and Hats | 38 | 76 |
| Hat supporters | Felting and Hats | 38 | 76 |
| Hats, weaving | Weaving | 139 | 95 |
| Hatters' kettles | Felting and Hats | 38 | 76 |
| Hat-tips | Felting and Hats | 38 | 76 |
| Hat-ventilators | Felting and Hats | 38 | 76 |
| Haversacks | Kitchen Utensils | 65 | 162 |
| Hay-caps | Harvesters | 56 | 77 |
| Hay-elevators | Hoisting | 57 | 79 |
| Hay-elevator hook | Hoisting | 57 | 79 |
| Hay-forks | Garden and Orchard | 47 | 77 |
| Hay-forks, Horse | Harvesters | 56 | 77 |
| Hay-knives | Cutlery | 30 | 49 |
| Hay-loaders for wagons | Carriages and Wagons | 21 | 79 |
| Hay-loading derricks | Hoisting | 57 | 79 |
| Hay-presses | Presses | 100 | 7 |
| Hay-racks for feeding | Stabling | 119 | 144 |
| Hay-racks for wagons | Carriages and Wagons | 21 | 31 |
| Hay rakers and cockers | Harvesters | 56 | 77 |
| Hay rakers and loaders | Harvesters | 56 | 77 |
| Hay-rakes, Hand | Garden and Orchard | 47 | 77 |
| Hay-rakes, Horse | Harvesters | 56 | 77 |
| Hay-spreaders | Harvesters | 56 | 77 |
| Hay-stackers | Thrashing | 130 | 77 |
| Hay-tedders | Harvesters | 56 | 77 |
| Hay-unloaders | Hoisting | 57 | 79 |
| Hawks | Masonry | 72 | 4 |

| SUBJECT. | Name of class—offici classification. | No. of class—official classification. | No. of class—subscription classification. |
|---|---|---|---|
| Hawse-pipes | Ships; 1 | 114 | 137 |
| Hawser-clamps | Ships; 1 | 114 | 138 |
| Hawsers | Cordage | 28 | 17 |
| Head-blocks for saw-mills | Saws | 110 | 130 |
| Head-coverings | Felting and Hats | 38 | 76 |
| Head-dresses | Felting and Hats | 38 | 76 |
| Heading-machines, Grain | Harvesters | 56 | 74 |
| Head-lights, Locomotive | Lamps and Gas-Fittings | 67 | 90 |
| Head-nets, Ladies' | Toilet | 132 | 161 |
| Head-protectors for horses | Harness | 54 | 72 |
| Head-rests | Furniture | 45 | 62 |
| Head-rests, Photographic | Photography | 95 | 111 |
| Heads for nails and screws | Builders' Hardware | 16 | 24 |
| Hearses | Carriages and Wagons | 21 | 39 |
| Heaters | Stoves and Furnaces | 126 | 154 |
| Heaters, Boilers of steam | Steam; 2 | 122 | 148 |
| Heaters, Electrical | Electricity | 36 | 50 |
| Heaters for stoves and ranges, Making sheet-metal | Sheet-Metal | 113 B | 136 |
| Heating-furnaces | Stoves and Furnaces | 126 | 154 |
| Heaters, Hot-water | Stoves and Furnaces | 126 | 147 |
| Heaters, Railway-car | Stoves and Furnaces | 126 | 154 |
| Heaters, Steam | Stoves and Furnaces | 126 | 147 |
| Heaters, Street-car | Stoves and Furnaces | 126 | 154 |
| Heating-irons | Stoves and Furnaces | 126 | 154 |
| Heating skates, boots, &c | Stoves and Furnaces | 126 | 154 |
| Heat-regulators | Stoves and Furnaces | 126 | 154 |
| Heddles | Weaving | 139 | 95 |
| Hedge-benders | Garden and Orchard | 47 | 52 |
| Hedge-fasteners | Garden and Orchard | 47 | 52 |
| Hedge-planters | Garden and Orchard | 47 | 52 |
| Hedge-setters | Garden and Orchard | 47 | 52 |
| Hedge-trimming machines | Harvesters | 56 | 52 |
| Heel and toe caps, Materials for | Boots and Shoes | 12 | 16 |
| Heel-caps | Boots and Shoes | 12 | 16 |
| Heel-fastenings | Boots and Shoes | 12 | 16 |
| Heel-irons | Boots and Shoes | 12 | 16 |
| Heel-plates | Boots and Shoes | 12 | 16 |
| Heel-polishing machines | Boots and Shoes | 12 | 16 |
| Heel-protectors | Boots and Shoes | 12 | 16 |
| Heels | Boots and Shoes | 12 | 16 |
| Heel-shaves | Boots and Shoes | 12 | 16 |
| Heels, Machines for cutting | Boots and Shoes | 12 | 16 |
| Heels, Machines for dressing | Boots and Shoes | 12 | 16 |
| Heel-stiffeners | Boots and Shoes | 12 | 16 |
| Heel-trimming machines | Boots and Shoes | 12 | 16 |
| Helices, Electrical | Electricity | 36 | 50 |
| Heliometers | Optics | 88 | 157 |
| Heliostats | Optics | 88 | 157 |
| Heliotropes | Optics | 88 | 157 |
| Heliotypes | Photography | 95 | 111 |
| Hemmers for sewing-machines | Sewing-Machines | 112 | 134 |
| Hemmers, Hand | Apparel | 2 | 38 |
| Hemorrhage, Instruments for arresting | Surgery | 128 | 156 |
| Hemp-brakes | Brakes and Gins | 13 | 78 |
| Hemp-hackling | Brakes and Gins | 13 | 78 |
| Hemp-harvesters | Harvesters | 56 | 75 |
| Hemp-pullers | Harvesters | 56 | 75 |
| Hemp-rotting | Brakes and Gins | 13 | 78 |
| Hen's-nests | Stabling | 119 | 116 |

| SUBJECT. | Name of class—official classification. | No. of class—official classification. | No. of class—subscription classification. |
|---|---|---|---|
| Herbariums | Fine Arts | 41 | 54 |
| Hermetically sealed packages for provisions | Preserving Food | 99 | 58 |
| Hernia, Instruments for treating | Surgery | 128 | 156 |
| Hewing-machines | Wood-Working; 3 | 144 | 173 |
| Hewing stone | Stone, Lime, and Cement | 125 | 153 |
| Hickory brooms | Brushes and Brooms | 15 | 22 |
| Hides, Dressing | Tanning | 129 | 92 |
| High-pressure alarms for steam-boilers | Steam; 2 | 122 | 148 |
| Hinges | Builders' Hardware | 16 | 24 |
| Hinges, Blind | Builders' Hardware | 16 | 24 |
| Hinges, Door | Builders' Hardware | 16 | 24 |
| Hinges, Gate | Builders' Hardware | 16 | 24 |
| Hinges, Loose-joint | Builders' Hardware | 16 | 24 |
| Hinges, Manufacture of | Hardware Manufactures | 53 B | 24 |
| Hinges, Self-shutting | Builders' Hardware | 16 | 24 |
| Hinges, Shutter | Builders' Hardware | 16 | 24 |
| Hinges, Spring | Builders' Hardware | 16 | 24 |
| Hitching-devices | Stabling | 119 | 144 |
| Hitching-posts | Stabling | 119 | 144 |
| Hobby-horses | Games and Toys | 46 | 143 |
| Hods | Masonry | 72 | 4 |
| Hod-unloaders | Hoisting | 57 | 79 |
| Hoeing-machines | Plows | 97 | 114 |
| Hoes | Garden and Orchard | 47 | 141 |
| Hoes, Horse | Plows | 97 | 114 |
| Hog-elevators | Butchering | 17 | 25 |
| Hogs' nose-trimmers | Cutlery | 30 | 44 |
| Hog-pens | Stabling | 119 | 144 |
| Hog-rings | Harness | 54 | 144 |
| Hog-slaughtering apparatus | Butchering | 17 | 25 |
| Hog-troughs | Stabling | 119 | 144 |
| Hoisting-apparatus | Hoisting | 57 | 79 |
| Hoisting-buckets | Hoisting | 57 | 79 |
| Hoisting-machines | Hoisting | 57 | 79 |
| Hold-backs for vehicles | Carriages | 21 | 31 |
| Holsters | Trunks | 133 | 162 |
| Hominy-machines | Mills | 83 | 57 |
| Hones | Grinding and Polishing | 51 | 70 |
| Hones, Magnetic | Electricity | 36 | 70 |
| Honey-boxes for bee-hives | Bee-Hives | 6 | 9 |
| Hoods, Ladies' | Felting and Hats | 38 | 38 |
| Hoof-spreaders | Metal-Working; 6 | 81A | 60 |
| Hook-buttons | Clasps and Buckles | 24 | 23 |
| Hook-buttons, Implements for setting | Clasps and Buckles | 24 | 38 |
| Hooks | Clasps and Buckles | 24 | 23 |
| Hooks and eyes | Clasps and Buckles | 24 | 104 |
| Hooks and eyes, Manufacture of | Hardware Manufactures | 53E | 104 |
| Hooks, Bench | Wood-Working; 4 | 145 | 29 |
| Hooks, Coat and hat | Builders' Hardware | 16 | 24 |
| Hooks, Door | Builders' Hardware | 16 | 24 |
| Hooks for cam-rods | Mechanical Powers | 74 | 67 |
| Hooks for watch-chains | Clasps and Buckles | 24 | 23 |
| Hooks, Furniture | Builders' Hardware | 16 | 24 |
| Hooks, Mousing | Clasps and Buckles | 24 | 23 |
| Hooks, Shutter | Builders' Hardware | 16 | 24 |
| Hooks, Snap | Clasps and Buckles | 24 | 23 |
| Hooks, Swivel | Clasps and Buckles | 24 | 23 |
| Hoop-iron, Rolling | Metal-Working; 5 | 80 | 125 |
| Hoop-locks | Presses | 100 | 7 |

| SUBJECT. | Name of class—official classification. | No. of class—official classification. | No. of class—subscription classification. |
|---|---|---|---|
| Hoops for skirts | Crinoline and Corsets | 29 | 80 |
| Hoops for skirts, Coating or covering | Crinoline and Corsets | 29 | 80 |
| Hoop-skirt formers | Crinoline and Corsets | 29 | 80 |
| Hoop-skirts | Crinoline and Corsets | 29 | 80 |
| Hoop-skirts, Machines for making | Crinoline and Corsets | 29 | 80 |
| Hoops | Wood-Working; 3 | 144 | 40 |
| Hoops, Machines for bending | Wood-Working; 3 | 144 | 40 |
| Hoops, Machines for cutting | Wood-Working; 3 | 144 | 40 |
| Hoops, Machines for dressing | Wood-Working; 3 | 144 | 40 |
| Hoops, Machines for sawing | Wood-Working; 3 | 144 | 40 |
| Hoops, Machines for splitting | Wood-Working; 3 | 144 | 40 |
| Hoop-strainers | Presses | 100 | 7 |
| Hop-backs | Beer and Wine | 7 | 19 |
| Hop-extracts | Beer and Wine | 7 | 19 |
| Hop-frames | Garden and Orchard | 47 | 63 |
| Hop-kilns | Dryers and Kilns | 34 | 48 |
| Hopper-boys | Mills | 83 | 57 |
| Hopper-shoes for mills | Mills | 83 | 57 |
| Hop-pole extractors | Fences | 39 | 63 |
| Hop-presses | Presses | 100 | 7 |
| Hops, Treatment of | Beer and Wine | 7 | 19 |
| Hop-strippers | Garden and Orchard | 47 | 63 |
| Horizons, Artificial | Optics | 88 | 157 |
| Horn, Artificial | Chemical Miscellaneous | 23 | 35 |
| Horn-knobs, Cattle | Harness | 54 | 144 |
| Horn, Machines for working in | Wood-Working; 3 | 144 | 176 |
| Horn-molding | Wood-Working; 3 | 144 | 176 |
| Horological instruments | Horology | 58 | 36 |
| Horse-blankets | Harness | 54 | 144 |
| Horse-brushes | Brushes and Brooms | 15 | 22 |
| Horse-hair fabrics | Weaving | 139 | 95 |
| Horse-hair hats | Felting and Hats | 38 | 76 |
| Horse-hoes | Plows | 97 | 114 |
| Horse-powers | Mechanical Powers | 74 | 159 |
| Horse-powers, Endless chains for | Mechanical Powers | 74 | 159 |
| Horse-powers, Links for | Mechanical Powers | 74 | 159 |
| Horse-powers, Treads for | Mechanical Powers | 74 | 159 |
| Horse-protectors | Harness | 54 | 144 |
| Horseshoe-bar, Rolling | Metal-Working; 5 | 80 | 125 |
| Horseshoe-bars | Horseshoes | 59 | 60 |
| Horseshoe bars, Machines for bending | Horseshoes | 59 | 60 |
| Horseshoe bars, Machines for shaping | Horseshoes | 59 | 60 |
| Horseshoe blanks, Making | Horseshoes | 59 | 60 |
| Horseshoe-machines | Horseshoes | 59 | 60 |
| Horseshoe-nails | Nails | 85 | 104 |
| Horseshoe-nails, Machines for making | Nails | 85 | 104 |
| Horseshoes | Horseshoes | 59 | 60 |
| Horseshoes, Calks for | Horseshoes | 59 | 60 |
| Horseshoes, Pads for | Horseshoes | 59 | 60 |
| Horses, Saw | Wood-Working; 4 | 145 | 131 |
| Horse-taming | Harness | 54 | 144 |
| Horse-training apparatus | Harness | 54 | 144 |
| Horse-washers | Stabling | 119 | 144 |
| Horse-yokes | Harness | 54 | 144 |
| Hose | Hose and Belting | 60 | 81 |
| Hose-carriages | Carriages and Wagons | 21 | 31 |
| Hose-bridges | Railways; 1 | 121 | 81 |
| Hose-clamps | Water-Distribution | 137 | 81 |
| Hose, Compositions for | Hose and Belting | 60 | 81 |
| Hose-couplings | Water-Distribution | 137 | 81 |

| SUBJECT. | Name of class—official classification. | No. of class—official classification. | No. of class—subscription classification. |
|---|---|---|---|
| Hose, Leather | Hose and Belting | 60 | 81 |
| Hose, Modes of making | Hose and Belting | 60 | 81 |
| Hose-nozzles | Water Distribution | 137 | 81 |
| Hose-pipes | Water Distribution | 137 | 81 |
| Hose-protectors | Hose and Belting | 60 | 81 |
| Hose, Rubber | Hose and Belting | 60 | 81 |
| Hose, Weaving | Weaving | 139 | 95 |
| Hosiery | Knitting and Netting | 66 | 89 |
| Hosiery-ironing machines | Apparel | 2 | 89 |
| Hot-air engines | Pneumatics | 98 | 2 |
| Hot-blast engines | Pneumatics | 98 | 12 |
| Hot-air furnaces | Stoves and Furnaces | 126 | 154 |
| Hot-air furnace-doors | Stoves and Furnaces | 126 | 154 |
| Hot-bed frames | Garden and Orchard | 47 | 63 |
| Hot-closets for stoves | Stoves and Furnaces | 126 | 154 |
| Hotel-annunciators | Signals | 116 | 24 |
| Hot-houses | Garden and Orchard | 47 | 63 |
| Hot-presses for fabrics and paper | Presses | 100 | 37 |
| Hot-water heaters | Stoves and Furnaces | 126 | 147 |
| Hour-glasses | Horology | 58 | 36 |
| House-fronts, Iron | Masonry | 72 | 4 |
| House-fronts, Stone-facing | Masonry | 72 | 4 |
| Houses | Carpentry | 20 | 30 |
| Houses, Construction of | Carpentry | 20 | 30 |
| Houses, Iron, (*except Trusses*) | Masonry | 72 | 4 |
| Houses, Portable | Carpentry | 20 | 30 |
| Housewives | Apparel | 2 | 38 |
| Housings for harness | Harness | 54 | 72 |
| Howeling-tools | Wood-Working; 3 | 144 | 40 |
| Hub-bands, Manufacture of | Hardware Manufactures | 53 C | 31 |
| Hubs, Carriage | Carriages and Wagons | 21 | 31 |
| Hubs, Machines for boring | Wood-Working; 3 | 144 | 31 |
| Hubs, Machines for centering | Wood-Working; 3 | 144 | 31 |
| Hubs, Machines for mortising | Wood-Working; 3 | 144 | 31 |
| Hubs, Machines for setting boxes in | Wood-Working; 3 | 144 | 31 |
| Hubs, Machines for turning | Wood-Working; 1 | 142 | 172 |
| Hullers, Clover | Thrashing | 130 | 159 |
| Hulling-machines | Mills | 83 | 57 |
| Hurdles for silk-winders | Silk | 117 | 140 |
| Husk-brooms | Brushes and Brooms | 15 | 22 |
| Huskers, Corn | Thrashing | 130 | 59 |
| Husking-gloves | Thrashing | 130 | 59 |
| Husking-pins | Thrashing | 130 | 59 |
| Husking-stools | Thrashing | 130 | 59 |
| Husk-splitting | Brakes and Gins | 13 | 78 |
| Hydrant-pipe stops | Water-Distribution | 137 | 167 |
| Hydrants | Water-Distribution | 137 | 167 |
| Hydrants, Cocks for | Valves | 136 | 167 |
| Hydrants, Filtering | Water-Distribution | 137 | 167 |
| Hydrants, Waste for | Water-Distribution | 137 | 167 |
| Hydrant-valves | Valves | 136 | 167 |
| Hydraulic buffers | Pumps | 103 | 118 |
| Hydraulic cement | Stone, Lime, and Cement | 125 | 153 |
| Hydraulic elevators | Pumps | 103 | 79 |
| Hydraulic engines | Pumps | 103 | 118 |
| Hydraulic jacks | Pumps | 103 | 79 |
| Hydraulic mains | Gas | 48 | 66 |
| Hydraulic meters | Pumps | 103 | 118 |
| Hydraulic mining | Hydraulic Engineering | 61 | 82 |
| Hydraulic motors | Pumps | 103 | 118 |

| SUBJECT. | Name of class—official classification. | No. of class—official classification. | No. of class—subscription classification. |
|---|---|---|---|
| Hydraulic presses | Pumps | 103 | 118 |
| Hydraulic propellers | Ships; 2 | 115 | 139 |
| Hydraulic rams | Pumps | 103 | 118 |
| Hydro-carbon furnaces for steam-boilers | Steam; 2 | 122 | 149 |
| Hydro-carbons, Manufacture of | Stills | 124 | 47 |
| Hydro-electric machines | Electricity | 36 | 50 |
| Hydrometers | Measuring-Instruments | 73 | 132 |
| Hydrostatic balances | Measuring-Instruments | 73 | 132 |
| Hygrometers | Measuring-Instruments | 73 | 101 |
| Hypodermic syringes | Surgery | 128 | 156 |
| Hypsometers | Measuring-Instruments | 73 | 101 |
| **I.** | | | |
| Ice-boats | Ships; 1 | 114 | 137 |
| Ice-breakers | Ships; 1 | 114 | 137 |
| Ice-cars | Carriages and Wagons | 21 | 121 |
| Ice-chairs and carriages | Carriages and Wagons | 21 | 31 |
| Ice-cream freezers | Ice | 62 | 58 |
| Ice-creepers | Boots and Shoes | 12 | 16 |
| Ice-crushers | Ice | 62 | 58 |
| Ice-cutters | Ice | 62 | 58 |
| Ice-elevators | Hoisting | 57 | 79 |
| Ice-houses | Ice | 62 | 58 |
| Ice-levelers | Ice | 62 | 83 |
| Ice-making machines | Ice | 62 | 83 |
| Ice manufacture | Ice | 62 | 83 |
| Ice-picks | Ice | 62 | 58 |
| Ice-plows | Ice | 62 | 58 |
| Ice-runs | Hoisting | 57 | 79 |
| Ice-safes | Ice | 62 | 58 |
| Ice-sandals | Boots and Shoes | 12 | 16 |
| Ice-saws | Saws | 110 | 58 |
| Ice-shavers | Ice | 62 | 58 |
| Ice-smoothers | Ice | 62 | 83 |
| Illuminated clocks | Horology | 58 | 36 |
| Illuminating-reflectors | Lamps and Gas-Fittings | 67 | 90 |
| Illuminated signs | Stationery | 120 | 54 |
| Illuminators, Stove | Stoves and Furnaces | 126 | 154 |
| Imitation leather | Paint | 91 | 107 |
| Inclined planes for canals | Hydraulic Engineering | 61 | 82 |
| Inclinometers | Measuring-Instruments | 73 | 157 |
| Incrustation in steam-boilers, Preventing | Steam; 2 | 122 | 148 |
| Incrustation in steam-boilers, Removing | Steam; 2 | 122 | 148 |
| Incubators | Stabling | 119 | 116 |
| Indelible pencils | Stationery | 120 | 146 |
| Indexing | Book-Binding | 11 | 15 |
| India rubber | Caoutchouc | 18 | 84 |
| Indicators, Automatic station | Measuring-Instruments | 73 | 120 |
| Indicators, Automatic time | Horology | 58 | 36 |
| Indicators, Door | Stationery | 120 | 146 |
| Indicators in cars, Automatic street | Measuring-Instruments | 73 | 42 |
| Indicators, Leak | Measuring-Instruments | 73 | 137 |
| Indicators, Lee-way | Measuring-Instruments | 73 | 137 |
| Indicators, Registering devices of steam-pressure | Measuring-Instruments | 73 | 148 |

| SUBJECT. | Name of class—official classification. | No. of class—official classification. | No. of class—subscription classification. |
|---|---|---|---|
| Indicators, Station | Stationery | 120 | 146 |
| Indicators, Street | Stationery | 120 | 146 |
| Indicators, Time, (*non-automatic*) | Stationery | 120 | 36 |
| Indicators, Weather | Measuring-Instruments | 73 | 101 |
| Induction-apparatus | Electricity | 36 | 50 |
| Incendiary compounds | Gunpowder | 52 | 71 |
| Incendiary shells | Projectiles | 102 | 105 |
| Inclined planes | Excavators | 37 | 51 |
| Incrustation of glass | Glass | 49 | 68 |
| Inhalers | Surgery | 128 | 156 |
| Injectors for steam-boilers | Steam; 2 | 122 | 148 |
| Injectors, Medical | Surgery | 128 | 156 |
| Inking-apparatus | Printing | 101 | 117 |
| Inking-rollers, Compositions for | Printing | 101 | 117 |
| Ink, Printers' | Paint | 91 | 117 |
| Inkstands | Stationery | 120 | 146 |
| Ink, Writing | Chemical Miscellaneous | 23 | 146 |
| Inlaying | Fine Arts | 41 | 54 |
| Inlets to sewers | Paving | 94 | 110 |
| Insect-destroyers | Garden and Orchard | 47 | 63 |
| Insect-destroying compounds | Chemical Miscellaneous | 23 | 63 |
| Insect-fumigators | Garden and Orchard | 47 | 63 |
| Insect-lamps | Garden and Orchard | 47 | 63 |
| Insect-traps | Garden and Orchard | 47 | 63 |
| Insect-traps, Household | Furniture | 45 | 63 |
| Insertable saw-teeth | Saws | 110 | 131 |
| Insoles | Boots and Shoes | 12 | 16 |
| Insulated wire | Electricity | 36 | 50 |
| Insulators | Electricity | 36 | 50 |
| Interfering, Devices to prevent | Harness | 54 | 144 |
| Intermittent gearing | Mechanical Powers | 74 | 67 |
| Iron and wood pavements combined | Paving | 94 | 110 |
| Iron blocks, Nautical | Ships; 1 | 114 | 137 |
| Iron bridges | Bridges | 14 | 21 |
| Iron-clad vessels | Ships; 1 | 114 | 137 |
| Iron columns | Bridges | 14 | 21 |
| Iron facing for houses | Masonry | 72 | 4 |
| Iron filings, Separators for | Electricity | 36 | 50 |
| Iron frames, Ships' | Ships; 1 | 114 | 137 |
| Iron fronts for houses | Masonry | 72 | 4 |
| Iron girders | Bridges | 14 | 21 |
| Iron houses | Masonry | 72 | 4 |
| Iron houses, Portable | Masonry | 72 | 4 |
| Ironing-apparatus | Hydraulic Engineering | 61 | 91 |
| Ironing cloth | Apparel | 2 | 37 |
| Ironing-machines, Tailors' and hosiery | Apparel | 2 | 38 |
| Ironing-tables | Laundry | 68 | 91 |
| Iron, Manufacture of | Metallurgy | 75 | 85 |
| Iron pavements | Paving | 94 | 110 |
| Iron piers | Bridges | 14 | 21 |
| Iron pipe making, Welded | Tubing and Wire | 134 | 163 |
| Iron posts | Bridges | 14 | 21 |
| Iron ribs, Ships' | Ships; 1 | 114 | 137 |
| Irons, Carriage | Carriages | 21 | 31 |
| Irons, Carriage-top | Carriages | 21 | 31 |
| Irons, Setting carriage | Carriages | 21 | 31 |
| Iron trusses | Bridges | 14 | 21 |
| Iron tunnels | Hydraulic Engineering | 61 | 82 |
| Iron vessels | Ships; 1 | 114 | 137 |
| Irrigating-apparatus | Water-Distribution | 137 | 82 |

| SUBJECT. | Name of class—official classification. | No. of class—official classification. | No. of class—subscription classification. |
|---|---|---|---|
| Irrigating-apparatus | Hydraulic Engineering | 61 | 82 |
| Irrigating-carts | Water-Distribution | 137 | 31 |
| Irrigators for nose, vagina, &c | Surgery | 128 | 156 |
| Isinglass | Oil, Fat, and Glue | 87 | 26 |
| Italian irons | Laundry | 68 | 91 |
| Ivory, Artificial | Chemical Miscellaneous | 23 | 35 |
| Ivory, Machines for working in | Wood-Working; 3 | 144 | 173 |
| J. | | | |
| Jacks, Carriage | Hoisting | 57 | 79 |
| Jacks for hats, Rounding | Felting and Hats | 38 | 76 |
| Jacks, Hydraulic | Pumps | 103 | 118 |
| Jacks, Lifting | Hoisting | 57 | 79 |
| Jacks, Pegging | Boots and Shoes | 12 | 16 |
| Jack-spinning | Spinning | 118 | 142 |
| Jacks, Railway-car | Hoisting | 57 | 79 |
| Jacks, Screw | Hoisting | 57 | 79 |
| Jacks, Wagon | Hoisting | 57 | 79 |
| Jacquard apparatus | Weaving | 139 | 95 |
| Jails | Masonry | 72 | 4 |
| Japanning cloth | Paint | 91 | 107 |
| Japanning leather | Paint | 91 | 107 |
| Japanning wood | Paint | 91 | 107 |
| Japanning | Paint | 91 | 107 |
| Jars, Earthen | Clay | 25 | 115 |
| Jars for fruit, meat, &c | Preserving Food | 99 | 58 |
| Jars, Self-sealing | Preserving Food | 99 | 58 |
| Jaws for gaffs | Ships; 1 | 114 | 138 |
| Jewel-boxes | Toilet | 132 | 86 |
| Jewelers' tools | Metal-Working; 6 | 81A | 86 |
| Jeweling watches, Tools for | Metal-Working; 6 | 81A | 36 |
| Jewelry, Manufacture of | Hardware Manufactures | 53E | 86 |
| Jewels, Cutting | Grinding and Polishing | 51 | 86 |
| Jiggers | Ore | 90 | 3 |
| Joiners' benches | Wood-Working; 4 | 145 | 29 |
| Joiners' clamps | Wood-Working; 4 | 145 | 29 |
| Joiners' tools | Wood-Working; 4 | 145 | 29 |
| Joiners' tools, Manufacture of | Metal-Working; 6 | 81B | 29 |
| Joiners' work of buildings | Carpentry | 20 | 30 |
| Joints for timbers | Carpentry | 20 | 30 |
| Joints, Railway | Railways; 1 | 104 | 120 |
| Joists | Carpentry | 20 | 30 |
| Joists, Machines for dressing | Wood-Working; 2 | 143 | 173 |
| Journal-bearings, Compositions for | Journals and Bearings | 64 | 87 |
| Journal-bearings | Journals and Bearings | 64 | 87 |
| Journal-boxes | Journals and Bearings | 64 | 87 |
| Journal-box lining | Journals and Bearings | 64 | 87 |
| Journals | Journals and Bearings | 64 | 87 |
| K. | | | |
| Kaolin, Purifying | Clay | 25 | 115 |
| Keels | Ships; 1 | 114 | 137 |
| Keelsons | Ships; 1 | 114 | 137 |
| Keepers for door-locks | Locks and Latches | 70 | 93 |
| Kegs, Machines for making | Wood-Working; 3 | 144 | 40 |
| Key-boards for musical instruments | Music | 84 | 112 |
| Key-boards, Telegraphic | Electricity | 36 | 50 |
| Key-fasteners | Locks and Latches | 70 | 93 |

| SUBJECT. | Name of class—official classification. | No. of class—official classification. | No. of class—subscription classification. |
|---|---|---|---|
| Key-hole guards | Locks and Latches | 70 | 93 |
| Key-hole guides | Locks and Latches | 70 | 93 |
| Key-rings | Clasps and Buckles | 24 | 93 |
| Key-rings, Manufacture of | Hardware Manufactures | 53E | 93 |
| Keys | Locks and Latches | 70 | 93 |
| Keys, Clock | Horology | 58 | 36 |
| Keys, Dental | Dental | 32 | 46 |
| Keys, (door,) Manufacture of | Hardware Manufactures | 53B | 93 |
| Key-seat cutting-machines | Metal-Working; 7 | 82 | 100 |
| Keys for musical instruments | Music | 84 | 112 |
| Keys, Telegraphic | Electricity | 36 | 50 |
| Keys, Watch | Horology | 58 | 36 |
| Kilns, Brick | Dryers and Kilns | 34 | 20 |
| Kilns, Charcoal | Dryers and Kilns | 34 | 48 |
| Kilns for drying grain, peat, &c | Dryers and Kilns | 34 | 48 |
| Kilns for glass | Glass | 49 | 68 |
| Kilns, Hop | Dryers and Kilns | 34 | 19 |
| Kilns, Lime | Dryers and Kilns | 34 | 153 |
| Kilns, Malt | Dryers and Kilns | 34 | 19 |
| Kilns, Pottery | Dryers and Kilns | 34 | 115 |
| Kindling arrangements for stoves | Stoves and Furnaces | 126 | 154 |
| Kindling-baskets | Stoves and Furnaces | 126 | 154 |
| Kindling-materials | Fuel | 44 | 154 |
| Kindling-packages | Fuel | 44 | 154 |
| King-bolts, Manufacture of | Hardware Manufactures | 53C | 31 |
| Kitchen furniture, Making sheet-metal | Sheet-Metal | 113B | 136 |
| Kitchen-safes | Kitchen Utensils | 65 | 88 |
| Kitchen-sinks | Baths and Closets | 4 | 8 |
| Knapsacks | Trunks | 133 | 162 |
| Knapsack-slings | Trunks | 133 | 162 |
| Kneading-machines | Kitchen Utensils | 65 | 88 |
| Knee-caps | Harness | 54 | 72 |
| Knee-formers | Ships; 1 | 114 | 137 |
| Knee-guards for horses | Harness | 54 | 72 |
| Knees | Ships; 1 | 114 | 137 |
| Knee-swells for organs | Music | 84 | 106 |
| Knife-cleaners | Kitchen Utensils | 65 | 88 |
| Knife-racks | Kitchen Utensils | 65 | 88 |
| Knife-rests | Kitchen Utensils | 65 | 88 |
| Knife-sharpeners | Grinding and Polishing | 51 | 44 |
| Knitted goods | Knitting and Netting | | 89 |
| Knitting-burrs, Manufacture of | Hardware Manufactures | 53 | 89 |
| Knitting-burrs, Making | Needles and Pins | 86 | 89 |
| Knitting-machines | Knitting and Netting | 66 | 89 |
| Knitting-machines, Burrs for | Knitting and Netting | 66 | 89 |
| Knitting-machines, Needles for | Knitting and Netting | 66 | 89 |
| Knitting-machines, Setting-up devices for | Knitting and Netting | 66 | 89 |
| Knitting-machines, Sinkers for | Knitting and Netting | 66 | 89 |
| Knitting-machines, Stop-motions for | Knitting and Netting | 66 | 89 |
| Knitting-machines, Take-up devices for | Knitting and Netting | 66 | 89 |
| Knitting-machines, Tension-devices for | Knitting and Netting | 66 | 89 |
| Knitting-machines, Transferring devices for | Knitting and Netting | 66 | 89 |
| Knitting-needles, Making | Needles and Pins | 86 | 89 |
| Knives | Cutlery | 30 | 44 |
| Knives, Amputating | Cutlery | 30 | 156 |
| Knives and forks combined | Cutlery | 30 | 44 |
| Knives and spoons combined | Cutlery | 30 | 44 |
| Knives, Attaching handles to | Cutlery | 30 | 44 |

18

| SUBJECT. | Name of class—official classification. | No. of class—official classification. | No. of class—subscription classification. |
|---|---|---|---|
| Knives, Bread | Cutlery | 30 | 44 |
| Knives, Budding | Cutlery | 30 | 63 |
| Knives, Butcher | Cutlery | 30 | 44 |
| Knives, Cane | Cutlery | 30 | 44 |
| Knives, Carving | Cutlery | 30 | 44 |
| Knives, Corn | Cutlery | 30 | 44 |
| Knives, Curriers' | Cutlery | 30 | 44 |
| Knives, Dirk | Cutlery | 30 | 44 |
| Knives, Drawing | Wood-Working; 4 | 145 | 49 |
| Knives for cutting corn from the cob | Cutlery | 30 | 44 |
| Knives, Fruit | Cutlery | 30 | 44 |
| Knives, Grafting | Cutlery | 30 | 63 |
| Knives, Handles for | Cutlery | 30 | 44 |
| Knives, Hay | Cutlery | 30 | 49 |
| Knives, Hoof-paring | Cutlery | 30 | 60 |
| Knives, Machete | Cutlery | 30 | 44 |
| Knives, Manufacture of | Hardware Manufactures | 53A | 44 |
| Knives, Meat | Cutlery | 30 | 44 |
| Knives, Oyster | Cutlery | 30 | 44 |
| Knives, Pocket | Cutlery | 30 | 44 |
| Knives, Pruning | Cutlery | 30 | 63 |
| Knives, Shoe | Cutlery | 30 | 16 |
| Knob-fasteners | Locks and Latches | 70 | 93 |
| Knob-latches | Locks and Latches | 70 | 93 |
| Knobs, Door | Locks and Latches | 70 | 93 |
| Knobs for doors and bell-pulls, Manufacture of | Hardware Manufactures | 53B | 93 |
| Knobs to spindles, Attaching | Locks and Latches | 70 | 93 |
| Knuckle-shields for scrubbing | Brushes and Brooms | 15 | 22 |
| **L.** | | | |
| Label-fasteners | Stationery | 120 | 146 |
| Label-holders | Stationery | 120 | 146 |
| Label machines, Paper | Paper Manufactures | 93 | 117 |
| Labels | Stationery | 120 | 146 |
| Labels, Adhesive | Stationery | 120 | 146 |
| Ladling glass | Glass | 49 | 68 |
| Lace-machines | Knitting and Netting | 66 | 89 |
| Lace-paper machines | Paper Manufactures | 93 | 108 |
| Lacing-hooks | Clasps and Buckles | 24 | 23 |
| Lacing, Leather shoe | Boots and Shoes | 12 | 16 |
| Lactometers | Measuring-Instruments | 73 | 45 |
| Lacquers | Paint | 91 | 107 |
| Ladders | Carpentry | 20 | 30 |
| Ladders, Combination | Carpentry | 20 | 30 |
| Ladders, Extension | Carpentry | 20 | 30 |
| Ladders, Fire | Carpentry | 20 | 30 |
| Ladders, Garden | Carpentry | 20 | 30 |
| Ladders, Step | Carpentry | 20 | 30 |
| Ladies' figure-stands | Apparel | 2 | 38 |
| Ladles for molten metal | Casting | 22 | 33 |
| Lagging for drums | Journals and Bearings | 64 | 87 |
| Lambrequins | Furniture | 45 | 62 |
| Lampblack | Paint | 91 | 107 |
| Lamp-chimney adjusters | Lamps and Gas-Fittings | 67 | 90 |
| Lamp-chimney cleaners | Kitchen Utensils | 65 | 90 |
| Lamp-chimneys | Lamps and Gas-Fittings | 67 | 90 |
| Lamp-cones | Lamps and Gas-Fittings | 67 | 90 |
| Lamp-elevators | Lamps and Gas-Fittings | 67 | 90 |

| SUBJECT. | Name of class—official classification. | No. of class—official classification. | No. of class—subscription classification. |
|---|---|---|---|
| Lamp-fillers | Journals and Bearings | 64 | 90 |
| Lamp-posts | Lamps and Gas-Fittings | 67 | 90 |
| Lamp-reflectors | Lamps and Gas-Fittings | 67 | 90 |
| Lamps | Lamps and Gas-Fittings | 67 | 90 |
| Lamps, Electric | Electricity | 36 | 90 |
| Lamp-shade holders | Lamps and Gas-Fittings | 67 | 90 |
| Lamp-shades | Lamps and Gas-Fittings | 67 | 90 |
| Lamps, Miners' | Lamps and Gas-Fittings | 67 | 62 |
| Lamp-stands | Lamps and Gas-Fittings | 67 | 90 |
| Lamps, Submarine | Lamps and Gas-Fittings | 67 | 90 |
| Lamps, Vapor | Lamps and Gas-Fittings | 67 | 90 |
| Lamp-wick adjusters | Lamps and Gas-Fittings | 67 | 90 |
| Lamp-wick trimmers | Lamps and Gas-Fittings | 67 | 90 |
| Lamp-wicks | Lamps and Gas-Fittings | 67 | 90 |
| Lamp-wicks, Chemical treatment of | Chemical Miscellaneous | 23 | 35 |
| Lamp-wick scissors | Cutlery | 30 | 44 |
| Lances, Bomb | Projectiles | 102 | 105 |
| Lancets | Cutlery | 30 | 156 |
| Landing-platforms | Carpentry | 61 | 79 |
| Land-markers | Seeders and Planters | 111 | 133 |
| Land-paring machines | Plows | 97 | 114 |
| Lanterns | Lamps and Gas-Fittings | 67 | 90 |
| Lanterns, Magic | Optics | 88 | 157 |
| Lanterns, Signal | Signals | 116 | 120 |
| Lapidary tools | Grinding and Polishing | 51 | 70 |
| Lapidary wheels | Grinding and Polishing | 51 | 70 |
| Lapidary work | Grinding and Polishing | 51 | 70 |
| Lapping-machines | Carding | 19 | 28 |
| Lapping, Printers' | Printing | 101 | 117 |
| Lap-rings | Clasps and Buckles | 24 | 23 |
| Laps | Grinding and Polishing | 51 | 70 |
| Lap-shavers | Boots and Shoes | 12 | 16 |
| Lard-coolers | Oil, Fat, and Glue | 87 | 26 |
| Lard-oil presses | Presses | 100 | 26 |
| Lard-oil rendering-tanks | Oil, Fat, and Glue | 87 | 47 |
| Lard-presses | Presses | 100 | 26 |
| Lard-renderers | Oil, Fat, and Glue | 87 | 26 |
| Lard-squeezers | Presses | 100 | 26 |
| Lard-tanks for rendering | Stoves and Furnaces | 126 | 26 |
| Lard, Treatment of | Oil, Fat, and Glue | 87 | 26 |
| Lassoes | Harness | 54 | 72 |
| Last-holders | Boots and Shoes | 12 | 16 |
| Lasting-machines | Boots and Shoes | 12 | 16 |
| Lasts | Boots and Shoes | 12 | 16 |
| Lasts, Machines for cutting | Wood-Working; 1 | 142 | 172 |
| Lasts, Machines for making | Wood-Working; 1 | 142 | 172 |
| Lasts, Machines for polishing | Wood-Working; 1 | 142 | 70 |
| Lasts, Machines for turning | Wood-Working; 1 | 142 | 172 |
| Lasts, Toe-pieces for | Boots and Shoes | 12 | 16 |
| Latches | Locks and Latches | 70 | 93 |
| Latches, Manufacture of | Hardware Manufactures | 53B | 93 |
| Lathe-dogs for wood-lathes | Wood-Working; 1 | 142 | 173 |
| Lathes for milling | Metal-Working; 7 | 82 | 100 |
| Lathes for screw-cutting | Metal-Working; 7 | 82 | 100 |
| Lathes for turning | Metal-Working; 7 | 82 | 100 |
| Lathes for turning irregular forms | Wood-Working; 1 | 142 | 172 |
| Lathes for watchmakers | Metal-Working; 7 | 82 | 76 |
| Lathes, Mandrels for wood | Wood-Working; 1 | 142 | 172 |
| Lathes, Wood | Wood-Working; 1 | 142 | 172 |
| Lathing | Carpentry | 20 | 30 |

| SUBJECT. | Name of class—official classification. | No. of class—official classification. | No. of class—subscription classification. |
|---|---|---|---|
| Lathing-machines | Carpentry | 20 | 29 |
| Lathing, Metallic | Masonry | 72 | 4 |
| Lath, Machines for cutting | Wood-Working; 3 | 144 | 174 |
| Lath, Machines for splitting | Wood-Working; 3 | 144 | 174 |
| Latitude, Instruments for determining | Optics | 88 | 157 |
| Launching vessels | Hydraulic Engineering | 61 | 82 |
| Laundries | Laundry | 68 | 91 |
| Lawn-mowers | Harvesters | 56 | 75 |
| Layering | Garden and Orchard | 47 | 63 |
| Laying pavements | Paving | 94 | 110 |
| Laying rope | Cordage | 28 | 127 |
| Laying submarine cables | Hydraulic Engineering | 61 | 82 |
| Laying-top for cordage-machines | Cordage | 28 | 127 |
| Lazy-jacks for sails | Ships; 1 | 114 | 138 |
| Leaches | Water-Distribution | 137 | 167 |
| Leaches, Ash | Oil, Fat, and Glue | 87 | 26 |
| Lead-cutters for printers | Sheet-Metal | 113A | 117 |
| Leaders for spouting | Roofing | 108 | 126 |
| Lead pipe | Tubing and Wire | 134 | 163 |
| Lead-pipe lining | Tubing and Wire | 134 | 163 |
| Lead-pipe lining by electro-plating process | Plating | 96 | 163 |
| Lead pipe, Machines for making | Tubing and Wire | 134 | 163 |
| Leaf-turners for music-books | Music | 84 | 102 |
| Leak-signals | Signals | 116 | 137 |
| Leak-stoppers | Ships; 1 | 114 | 137 |
| Leather, Artificial | Paint | 91 | 92 |
| Leather-boarding | Tanning | 129 | 92 |
| Leather carpet | Leather | 69 | 56 |
| Leather cloth | Paint | 91 | 56 |
| Leather, Compositions for treating | Tanning | 129 | 92 |
| Leather, Compounds of waste | Tanning | 129 | 92 |
| Leather-cutting machines | Leather | 69 | 92 |
| Leather-dressing | Tanning | 129 | 92 |
| Leather-embossing machines | Leather | 69 | 92 |
| Leather, Enameling | Paint | 91 | 92 |
| Leather-finishing | Tanning | 129 | 92 |
| Leather-glueing machines | Leather | 69 | 92 |
| Leather-graining | Tanning | 129 | 92 |
| Leather-hammering machines | Leather | 69 | 92 |
| Leather hats | Felting and Hats | 38 | 76 |
| Leathering tacks | Nails | 85 | 103 |
| Leather, Japanning | Paint | 91 | 107 |
| Leather, Manufacture of | Tanning | 129 | 92 |
| Leather pegs for shoes | Boots and Shoes | 12 | 16 |
| Leather-piercing machines | Leather | 69 | 16 |
| Leather-pressing machines | Leather | 69 | 16 |
| Leather-punches | Clasps and Buckles | 24 | 92 |
| Leather-punching machines | Clasps and Buckles | 24 | 92 |
| Leather-quilting machines | Sewing-Machines | 112 | 134 |
| Leather rings, Tools for cutting | Leather | 69 | 92 |
| Leather-rolling machines | Leather | 69 | 92 |
| Leather-rounding machines | Leather | 69 | 92 |
| Leather-scalloping machines | Leather | 69 | 92 |
| Leather-shaving machines | Leather | 69 | 92 |
| Leather-skiving machines | Leather | 69 | 92 |
| Leather-sorting machines | Leather | 69 | 92 |
| Leather-splitting machines | Leather | 69 | 92 |
| Leather straps | Leather | 69 | 92 |

| SUBJECT. | Name of class—official classification. | No. of class—official classification. | No. of class—sub-scription classification. |
|---|---|---|---|
| Leather-stretching machines | Leather | 69 | 92 |
| Leather-stripping machines | Leather | 69 | 92 |
| Leather tubs, Machines for making | Leather | 69 | 92 |
| Leather washers and wads, Tools for cutting | Leather | 69 | 92 |
| Leather, Water-proofing | Tanning | 129 | 92 |
| Leather whips, Machines for making | Leather | 69 | 72 |
| Leather-working tools, Manufacture of | Metal-Working; 6 | 81B | 92 |
| Leeches, Mechanical | Surgery | 128 | 156 |
| Leers | Glass | 59 | 68 |
| Lee-way indicators | Measuring-Instruments | 73 | 137 |
| Leggings | Boots and Shoes | 12 | 16 |
| Legs, Artificial | Artificial Limbs | 3 | 5 |
| Lemon-squeezers | Kitchen Utensils | 65 | 88 |
| Lenses | Optics | 88 | 157 |
| Lenses, Grinding | Grinding and Polishing | 51 | 157 |
| Lenses, Polishing | Grinding and Polishing | 51 | 157 |
| Let-off motions for looms | Weaving | 139 | 95 |
| Letter-balances | Measuring-Instruments | 73 | 132 |
| Letter-boxes | Stationery | 120 | 146 |
| Letter-clips | Stationery | 120 | 146 |
| Letter-files | Stationery | 120 | 146 |
| Letter-openers | Stationery | 120 | 146 |
| Letter-receiving boxes | Stationery | 120 | 146 |
| Leucorrhea, Instruments for treating | Surgery | 128 | 156 |
| Levees | Hydraulic Engineering | 61 | 82 |
| Leveling millstones | Mills | 83 | 57 |
| Levels | Optics | 88 | 157 |
| Level-staffs | Optics | 88 | 157 |
| Lever-powers | Mechanical Powers | 74 | 79 |
| Levers, Differential | Mechanical Powers | 74 | 79 |
| Lewis for lifting stone | Stone, Lime, and Cement | 125 | 79 |
| Lid lifters, Stove | Stoves and Furnaces | 126 | 154 |
| Life-boats | Boats | 9 | 13 |
| Life-buckets | Boats | 9 | 138 |
| Life, Instruments for stimulating | Surgery | 128 | 156 |
| Life-preservers | Boats | 9 | 138 |
| Life-rafts | Boats | 9 | 13 |
| Lifters for kettles, ovens, &c | Stoves and Furnaces | 126 | 154 |
| Lifters for stove-lids | Stoves and Furnaces | 126 | 154 |
| Lifting-apparatus | Hoisting | 57 | 79 |
| Lifting-jacks | Hoisting | 57 | 79 |
| Lifting-screws | Hoisting | 57 | 79 |
| Ligatures | Surgery | 128 | 156 |
| Ligatures, Instruments for adjusting | Surgery | 128 | 156 |
| Lighters | Ships; 1 | 114 | 13 |
| Light-houses | Masonry | 72 | 4 |
| Lighting-devices | Lamps and Gas-Fittings | 67 | 90 |
| Lighting gas by electricity | Electricity | 36 | 64 |
| Lightning-arresters | Electricity | 36 | 50 |
| Lightning-protectors | Electricity | 36 | 50 |
| Lightning-rods | Electricity | 36 | 50 |
| Lightning-rods, Rolling | Metal-Working; 5 | 80 | 125 |
| Lights, Deck | Ships; 1 | 114 | 137 |
| Lights, Electric | Electricity | 36 | 50 |
| Light-ships | Ships; 1 | 114 | 137 |
| Lights, Side | Ships; 1 | 114 | 137 |
| Likenesses, Mounting | Fine Arts | 41 | 54 |
| Lime-furnaces | Stone, Lime, and Cement | 125 | 153 |
| Lime-kilns | Dryers and Kilns | 34 | 153 |

| SUBJECT. | Name of class—official classification. | No. of class—official classification. | No. of class—subscription classification. |
|---|---|---|---|
| Lime-spreaders | Seeders and Planters | 111 | 133 |
| Linchpins | Carriages and Wagons | 21 | 31 |
| Line-fasteners | Laundry | 68 | 91 |
| Line-reel, Gardeners' | Garden and Orchard | 47 | 63 |
| Line-reels | Laundry | 68 | 91 |
| Line-wires | Electricity | 36 | 50 |
| Line-wires, Covering | Electricity | 36 | 50 |
| Line-wires, Insulating | Electricity | 36 | 50 |
| Line-wires, Protecting | Electricity | 36 | 50 |
| Line-wires, Supporting | Electricity | 36 | 50 |
| Line-wires, Underground | Electricity | 36 | 50 |
| Liniments | Chemical Miscellaneous | 23 | 98 |
| Lining canals | Hydraulic Engineering | 61 | 82 |
| Lining cylinders | Tubing and Wire | 134 | 163 |
| Lining for safes | Safes | 109 | 128 |
| Lining for ships | Ships; 1 | 114 | 137 |
| Lining gun-barrels | Tubing and Wire | 134 | 163 |
| Lining metallic pipes | Tubing and Wire | 134 | 163 |
| Linings for stoves | Stoves and Furnaces | 126 | 154 |
| Lining tubes | Tubing and Wire | 134 | 154 |
| Link motion for steam-engines | Steam; 3 | 123 | 150 |
| Liquid-meters | Pumps | 103 | 118 |
| Liquor-pourers | Kitchen Utensils | 65 | 88 |
| Lithographic presses | Printing | 101 | 117 |
| Lithographic printing | Fine Arts | 41 | 54 |
| Lithography | Fine Arts | 41 | 54 |
| Lithotriptic instruments | Surgery | 128 | 156 |
| Lithotritors | Surgery | 128 | 156 |
| Litters | Carriages and Wagons | 21 | 31 |
| Loading machinery for wagons, vessels, and cars | Hoisting | 57 | 79 |
| Lock-bolt keepers | Locks and Latches | 70 | 93 |
| Lock-bolts | Locks and Latches | 70 | 93 |
| Locket-cases | Jewelry | 63 | 86 |
| Lockets | Jewelry | 63 | 86 |
| Lockets, Manufacture of | Hardware Manufactures | 53E | 86 |
| Locking apparatus for ferry-boats | Ships; 1 | 114 | 79 |
| Locking-nuts | Bolts, Nuts, and Rivets | 10 | 14 |
| Locks | Locks and Latches | 70 | 93 |
| Locks, Bag | Locks and Latches | 70 | 93 |
| Locks, Canal | Hydraulic Engineering | 61 | 89 |
| Locks, Car-door | Locks and Latches | 70 | 93 |
| Locks, Car-seat arm | Locks and Latches | 70 | 93 |
| Locks, Chest | Locks and Latches | 70 | 93 |
| Locks, Door | Locks and Latches | 70 | 93 |
| Locks, Drawer | Locks and Latches | 70 | 93 |
| Locks for fire-arms | Fire-Arms | 42 | 55 |
| Locks for ordnance | Ordnance | 89 | 105 |
| Locks, Hasp | Locks and Latches | 70 | 93 |
| Locks, Manufacture of | Hardware Manufactures | 53B | 93 |
| Locksmiths' tools | Metal-Working; 6 | 81A | 93 |
| Locks, Pad | Locks and Latches | 70 | 93 |
| Locks, Permutation | Locks and Latches | 70 | 93 |
| Locks, Piano | Locks and Latches | 70 | 93 |
| Locks, Safe | Locks and Latches | 70 | 93 |
| Locks, Sash | Locks and Latches | 70 | 93 |
| Locks, Seal | Locks and Latches | 70 | 93 |
| Locks, Shutter | Locks and Latches | 70 | 93 |
| Locks, Time | Locks and Latches | 70 | 93 |
| Locks, Trunk | Locks and Latches | 70 | 93 |

| SUBJECT. | Name of class—official classification. | No. of class—official classification. | No. of class—sub-scription classification. |
|---|---|---|---|
| Locks, Valise | Locks and Latches | 70 | 93 |
| Locks, Window | Builders' Hardware | 16 | 24 |
| Lock-trimmings | Locks and Latches | 70 | 93 |
| Lock-trimmings, Manufacture of | Hardware Manufactures | 53B | 93 |
| Lock-up safety-valves | Valves | 136 | 152 |
| Lock valves, Canal | Hydraulic Engineering | 61 | 82 |
| Locomotive-axles, Making | Railways; 4 | 107 | 94 |
| Locomotive-boilers | Steam; 3 | 123 | 94 |
| Locomotive-engines | Steam; 3 | 123 | 94 |
| Locomotive-frames, Making | Railways; 4 | 107 | 94 |
| Locomotive-furnaces | Steam; 3 | 123 | 94 |
| Locomotive head-lights | Lamps and Gas-Fittings | 67 | 90 |
| Locomotive-pumps | Steam; 3 | 123 | 118 |
| Locomotives | Steam; 3 | 123 | 94 |
| Locomotives, Fire-boxes for | Steam; 3 | 123 | 94 |
| Locomotives, Furnace-doors for | Steam; 3 | 123 | 94 |
| Locomotive smoke-stacks | Steam; 3 | 123 | 94 |
| Locomotives, Reversing-gear for | Steam; 3 | 123 | 94 |
| Locomotive valve-gear | Steam; 3 | 123 | 94 |
| Logs, Booming | Boats | 9 | 82 |
| Logs, Electrical | Electricity | 36 | 50 |
| Logs, Implements for loading | Hoisting | 57 | 79 |
| Logs, Implements for piling | Hoisting | 57 | 79 |
| Logs, Implements for rolling | Hoisting | 57 | 79 |
| Logs, Rafting | Boats | 9 | 13 |
| Logs, Ships' | Measuring-Instruments | 73 | 137 |
| Longitude, Instruments for determining | Optics | 88 | 157 |
| Looking-glasses | Furniture | 45 | 62 |
| Looking-glasses, Hanging | Furniture | 45 | 62 |
| Looms, Electrical | Electricity | 36 | 95 |
| Looms | Weaving | 139 | 95 |
| Looms for cross-weaving | Weaving | 139 | 95 |
| Looms for curvilinear weaving | Weaving | 139 | 95 |
| Looms for weaving bags | Weaving | 139 | 95 |
| Looms for weaving baskets | Weaving | 139 | 95 |
| Looms for weaving belting | Weaving | 139 | 95 |
| Looms for weaving blinds | Weaving | 139 | 95 |
| Looms for weaving cane | Weaving | 139 | 95 |
| Looms for weaving caps | Weaving | 139 | 95 |
| Looms for weaving carpets | Weaving | 139 | 95 |
| Looms for weaving cases for bottles | Weaving | 139 | 95 |
| Looms for weaving corded fabrics | Weaving | 139 | 95 |
| Looms for weaving counterpanes | Weaving | 139 | 95 |
| Looms for weaving crinoline | Weaving | 139 | 95 |
| Looms for weaving cut-piled fabrics | Weaving | 139 | 95 |
| Looms for weaving double-piled fabrics | Weaving | 139 | 95 |
| Looms for weaving elastic fabrics | Weaving | 139 | 95 |
| Looms for weaving figured fabrics | Weaving | 139 | 95 |
| Looms for weaving fringes | Weaving | 139 | 95 |
| Looms for weaving hats | Weaving | 139 | 95 |
| Looms for weaving horse-hair | Weaving | 139 | 95 |
| Looms for weaving hose | Weaving | 139 | 95 |
| Looms for weaving ingrain carpets | Weaving | 139 | 95 |
| Looms for weaving matting | Weaving | 139 | 95 |
| Looms for weaving metallic tissues | Weaving | 139 | 95 |
| Looms for weaving piled fabrics | Weaving | 139 | 95 |
| Looms for weaving ribbons | Weaving | 139 | 95 |
| Looms for weaving satin | Weaving | 139 | 95 |
| Looms for weaving scarfs | Weaving | 139 | 95 |

| SUBJECT. | Name of class—official classification. | No. of class—official classification. | No. of class—subscription classification. |
|---|---|---|---|
| Looms for weaving shaped fabrics | Weaving | 139 | 95 |
| Looms for weaving shawls | Weaving | 139 | 95 |
| Looms for weaving sieve-cloth | Weaving | 139 | 95 |
| Looms for weaving silk | Weaving | 139 | 140 |
| Looms for weaving slats | Weaving | 139 | 95 |
| Looms for weaving straw fabrics | Weaving | 139 | 95 |
| Looms for weaving terry fabrics | Weaving | 139 | 95 |
| Looms for weaving tubular fabrics | Weaving | 139 | 95 |
| Looms for weaving velvets | Weaving | 139 | 95 |
| Looms for weaving wire | Weaving | 139 | 170 |
| Looms for weaving woolen cloth | Weaving | 139 | 95 |
| Loop-banding machines | Cordage | 28 | 127 |
| Looped fabrics | Knitting and Netting | 66 | 89 |
| Loopers for carpet-rags | Weaving | 139 | 95 |
| Lopping-shears | Cutlery | 30 | 63 |
| Lorgnettes | Optics | 88 | 157 |
| Lowering boats | Boats | 9 | 13 |
| Low-water alarms for steam-boilers | Steam; 2 | 122 | 148 |
| Low-water detectors for steam-boilers | Steam; 2 | 122 | 148 |
| Low-water indicators for steam-boilers | Steam; 2 | 122 | 148 |
| Lozenge-making machines | Sugar | 127 | 155 |
| Lubricants | Journals and Bearings | 64 | 87 |
| Lubricating compounds | Journals and Bearings | 64 | 87 |
| Lubricators | Journals and Bearings | 64 | 87 |
| Lubricators for pistons | Journals and Bearings | 64 | 87 |
| Lubricators for projectiles | Projectiles | 102 | 105 |
| Lubricators for spindles | Journals and Bearings | 64 | 87 |
| Lubricators, Mechanical | Journals and Bearings | 64 | 87 |
| Lubricators, Steam-operated | Journals and Bearings | 64 | 87 |
| Lumber-dryers | Dryers and Kilns | 34 | 48 |
| Lumber, Implements for loading and unloading | Hoisting | 57 | 79 |
| Lumber, Implements for piling | Hoisting | 57 | 79 |
| Lumbermen's tools, Manufacture of | Metal-Working; 6 | 81B | 49 |
| Lumber-tallies | Measuring-Instruments | 73 | 42 |
| Lumber-wagons | Carriages and Wagons | 21 | 31 |
| Lunch-boxes | Kitchen Utensils | 65 | 88 |
| Lunch-heaters | Stoves and Furnaces | 126 | 154 |
| Lusters | Lamps and Gas-Fittings | 67 | 90 |
| Lye-extracts | Oil, Fat, and Glue | 87 | 26 |
| **M.** | | | |
| Machetes | Cutlery | 30 | 44 |
| Machine-guns | Ordnance | 89 | 105 |
| Machines for making welt-bindings | Weaving | 139 | 95 |
| Machinists' tools | Metal-Working; 6 | 81A | 100 |
| Magazine small-arms | Fire-Arms | 42 | 55 |
| Magic-lanterns | Optics | 88 | 157 |
| Magnesium-lights | Lamps and Gas-Fittings | 67 | 91 |
| Magnetic apparatus | Electricity | 36 | 50 |
| Magnetic needles | Electricity | 36 | 50 |
| Magneto-electric machines | Electricity | 36 | 50 |
| Magnets | Electricity | 36 | 50 |
| Magnets, Receiving | Electricity | 36 | 50 |
| Magnets, Sounding | Electricity | 36 | 50 |
| Magnifying-glasses | Optics | 88 | 157 |
| Mail-bag fasteners, (*except Locks, Class* 70) | Trunks | 133 | 162 |
| Mail-bag receiving and delivering apparatus | Railways; 2 | 105 | 121 |

| SUBJECT. | Name of class—official classification. | No. of class—official classification. | No. of class—sub-cription classification. |
|---|---|---|---|
| Mail-bags | Trunks | 133 | 162 |
| Mail-distributing boxes | Stationery | 120 | 146 |
| Mains, Gas | Water-Distribution | 137 | 64 |
| Mains, Water | Water-Distribution | 137 | 167 |
| Malaxators | Clay | 25 | 115 |
| Malaxators, (*see Class* 25) | Stone, Lime, and Cement | 125 | 115 |
| Malleable iron castings, Furnaces for | Metallurgy | 75 | 99 |
| Mallets | Wood-Working; 4 | 145 | 174 |
| Mallets, Croquet | Games and Toys | 46 | 143 |
| Mallets, Dental | Dental | 32 | 46 |
| Malting | Beer and Wine | 7 | 19 |
| Malt-kilns | Dryers and Kilns | 34 | 19 |
| Malt liquors, Cooling | Beer and Wine | 7 | 19 |
| Malt liquors, Preserving | Beer and Wine | 7 | 19 |
| Malt liquors | Beer and Wine | 7 | 19 |
| Mandrels | Metal-Working; 7 | 82 | 100 |
| Mangers | Stabling | 119 | 144 |
| Mangles | Laundry | 68 | 91 |
| Man-holes for boilers and tanks | Steam; 2 | 122 | 148 |
| Man-holes for sewers | Masonry | 72 | 4 |
| Manipulators | Electricity | 36 | 50 |
| Mantelets | Ordnance | 89 | 105 |
| Manufacture of aluminium | Metallurgy | 75 | 99 |
| Manufacture of cobalt | Metallurgy | 75 | 99 |
| Manufacture of copper | Metallurgy | 75 | 99 |
| Manufacture of iron | Metallurgy | 75 | 85 |
| Manufacture of lead | Metallurgy | 75 | 99 |
| Manufacture of magnesium | Metallurgy | 75 | 99 |
| Manufacture of nickel | Metallurgy | 75 | 99 |
| Manufacture of potassium | Metallurgy | 75 | 99 |
| Manufacture of sodium | Metallurgy | 75 | 99 |
| Manufacture of steel | Metallurgy | 75 | 85 |
| Manufacture of tin | Metallurgy | 75 | 99 |
| Manufacture of zinc | Metallurgy | 75 | 99 |
| Manure, Artificial | Manures | 71 | 96 |
| Manure-carts, Distributing | Seeders and Planters | 111 | 133 |
| Manure-drags | Harvesters | 56 | 77 |
| Manure-drills | Seeders and Planters | 111 | 133 |
| Manure-drills, Liquid | Seeders and Planters | 111 | 133 |
| Manure-forks | Garden and Orchard | 47 | 77 |
| Manure-hooks | Garden and Orchard | 47 | 77 |
| Manure-loading rakes | Harvesters | 56 | 77 |
| Manures, Composition, processes, machinery | Manures | 71 | 96 |
| Manure-screens | Seeders and Planters | 111 | 133 |
| Manure-spreaders | Seeders and Planters | 111 | 77 |
| Manure-sowers, Broadcast | Seeders and Planters | 111 | 77 |
| Manuring, Green | Seeders and Planters | 111 | 77 |
| Map-holders | Educational | 35 | 146 |
| Maple-sugar buckets | Garden and Orchard | 47 | 63 |
| Maple-sugar, Manufacture of | Sugar | 127 | 155 |
| Maple-sugar spiles | Garden and Orchard | 47 | 63 |
| Map projectors | Drafting | 33 | 97 |
| Maps | Educational | 35 | 146 |
| Maps, Dissected | Educational | 35 | 146 |
| Maps, Outline | Educational | 35 | 146 |
| Marble, Artificial | Stone, Lime, and Cement | 125 | 153 |
| Marble-crushing machines | Ore | 90 | 3 |
| Marble-grinding | Grinding and Polishing | 51 | 3 |

| SUBJECT. | Name of class—official classification. | No. of class—official classification. | No. of class—subscription classification. |
|---|---|---|---|
| Marble-grinding, Compositions for | Grinding and Polishing | 51 | 3 |
| Marble-polishing | Grinding and Polishing | 51 | 153 |
| Marble-working machines | Stone, Lime, and Cement | 125 | 153 |
| Marine railways | Hydraulic Engineering | 61 | 82 |
| Mariners' compasses | Electricity | 36 | 50 |
| Marine safes | Boats | 9 | 13 |
| Marine signals | Signals | 116 | 158 |
| Marine steam-boilers | Steam; 2 | 122 | 148 |
| Markers for seeding-machines | Seeders and Planters | 111 | 133 |
| Markers for sewing-machines | Sewing-Machines | 112 | 134 |
| Market-stalls | Furniture | 45 | 62 |
| Mark-holders for bales | Stationery | 120 | 146 |
| Marl, Treatment of | Manures | 71 | 96 |
| Marquetry | Fine Arts | 41 | 54 |
| Marshes, Machines for filling | Hydraulic Engineering | 61 | 82 |
| Marshes, Reclaiming | Hydraulic Engineering | 61 | 82 |
| Martingales | Harness | 54 | 72 |
| Martin-houses | Garden and Orchard | 47 | 143 |
| Mash-tubs | Beer and Wine | 7 | 19 |
| Masks | Games and Toys | 46 | 143 |
| Masonry | Masonry | 72 | 4 |
| Mast-hoops | Ships; 1 | 114 | 138 |
| Masting-sheers | Hoisting | 57 | 79 |
| Masts | Ships; 1 | 114 | 138 |
| Mast-scrapers | Wood-Working; 4 | 145 | 138 |
| Metallic blinds, doors, floors, gratings, and shutters | Masonry | 72 | 4 |
| Match-blocks, Machines for making | Wood-Working; 3 | 144 | 174 |
| Match-boxes | Lamps and Gas-Fittings | 67 | 90 |
| Matches, Compositions for | Gunpowder | 52 | 71 |
| Matches, Dipping frames for | Wood-Working; 3 | 144 | 71 |
| Matches in frames, Machines for putting | Wood-Working; 3 | 144 | 71 |
| Matches, Machines for pointing | Wood-Working; 3 | 144 | 71 |
| Matches, Machines for making | Wood-Working; 3 | 144 | 71 |
| Matches, Machines for rounding | Wood-Working; 3 | 144 | 174 |
| Matches, Machines for splitting | Wood-Working; 3 | 144 | 174 |
| Match-planes | Wood-Working; 4 | 145 | 29 |
| Match-plates | Casting | 22 | 33 |
| Match-safes | Lamps and Gas-Fittings | 67 | 90 |
| Match-splints, Framing | Wood-Working; 3 | 144 | 174 |
| Match-splints, Machines for making | Wood-Working; 3 | 144 | 174 |
| Mathematical instruments | Drafting | 33 | 97 |
| Mathematical scales | Drafting | 33 | 132 |
| Materials for bands, belts, and hose | Hose and Belting | 60 | 81 |
| Mats, Car | Brushes and Brooms | 15 | 22 |
| Mats, Door | Brushes and Brooms | 15 | 22 |
| Mats for cars, Wooden | Brushes and Brooms | 15 | 22 |
| Mats for oil-presses | Presses | 100 | 7 |
| Mats, Slat | Brushes and Brooms | 15 | 22 |
| Matting | Weaving | 139 | 56 |
| Mattocks | Garden and Orchard | 47 | 141 |
| Mattresses | Beds | 5 | 62 |
| Mattresses, Machines for making | Beds | 5 | 62 |
| Mattresses, Materials for | Beds | 5 | 62 |
| Meal-bins | Kitchen Utensils | 65 | 88 |
| Meal-chests | Kitchen Utensils | 65 | 88 |
| Measures, Board | Measuring-Instruments | 73 | 42 |
| Measures, Cloth | Measuring-Instruments | 73 | 37 |
| Measures, Dry | Measuring-Instruments | 73 | 136 |

| SUBJECT. | Name of class—official classification. | No. of class—official classification. | No. of class—subscription classification. |
|---|---|---|---|
| Measures, Liquid | Measuring-Instruments | 73 | 132 |
| Measures, Lumber | Measuring-Instruments | 73 | 42 |
| Measures of capacity | Measuring-Instruments | 73 | 132 |
| Measures, Tailors' | Measuring-Instruments | 73 | 97 |
| Measures, Tape | Measuring-Instruments | 73 | 97 |
| Measures, Tire | Measuring-Instruments | 73 | 97 |
| Measuring cloth, Machines for | Cloth | 26 | 37 |
| Measuring devices of faucets | Measuring-Instruments | 73 | 132 |
| Measuring devices of funnels | Measuring-Instruments | 73 | 132 |
| Measuring devices of pumps | Measuring-Instruments | 73 | 132 |
| Measuring-sticks | Measuring-Instruments | 73 | 97 |
| Measuring yarns, threads, &c | Cordage | 28 | 97 |
| Meat cans and jars | Preserving Food | 99 | 58 |
| Meat-curing | Preserving Food | 99 | 58 |
| Meat-cutters, (*for Mincers see Class* 17) | Kitchen Utensils | 65 | 88 |
| Meat-dryers | Dryers and Kilns | 34 | 58 |
| Meat-grinders | Butchering | 17 | 25 |
| Meat-holders | Kitchen Utensils | 65 | 88 |
| Meat-packing | Preserving Food | 99 | 58 |
| Meat-presses | Presses | 100 | 7 |
| Meat-safes | Kitchen Utensils | 65 | 88 |
| Meat-saws | Saws | 110 | 25 |
| Meats, Compositions for preserving | Preserving Food | 99 | 58 |
| Meat-spits | Kitchen Utensils | 65 | 88 |
| Meats, Processes for preserving | Preserving Food | 99 | 58 |
| Meat-tenderers | Kitchen Utensils | 65 | 88 |
| Mechanical leeches | Surgery | 128 | 156 |
| Mechanical memoranda | Stationery | 120 | 146 |
| Mechanical movements | Mechanical Powers | 74 | 67 |
| Mechanical typographers | Printing | 101 | 117 |
| Medals | Fine Arts | 41 | 54 |
| Medical compounds | Chemical Miscellaneous | 23 | 98 |
| Medical-electric apparatus | Electricity | 36 | 50 |
| Medicators | Surgery | 128 | 98 |
| Medicinal powders, Instruments for putting up | Surgery | 128 | 98 |
| Medicine-cases | Surgery | 128 | 156 |
| Medicine-glasses, Graduated | Measuring-Instruments | 73 | 98 |
| Medicines | Chemical Miscellaneous | 23 | 98 |
| Medicines, Instruments for administering | Surgery | 128 | 98 |
| Medicines, Instruments for compounding | Surgery | 128 | 98 |
| Medicine-spoons | Surgery | 128 | 98 |
| Meerschaums | Tobacco | 131 | 160 |
| Melodeons | Music | 84 | 106 |
| Melting glass | Glass | 49 | 68 |
| Memorandum-books | Stationery | 120 | 146 |
| Menstruation, Instruments for arresting | Surgery | 128 | 156 |
| Meridian-finders | Optics | 88 | 157 |
| Mess-chests | Kitchen Utensils | 65 | 88 |
| Mess-kits | Kitchen Utensils | 65 | 88 |
| Metallic blinds, doors, shutters, and gratings | Carpentry | 20 | 4 |
| Metallic boats | Boats | 9 | 13 |
| Metallic roofing | Roofing | 108 | 126 |
| Metallic surfaces, Compositions for coating | Paint | 91 | 107 |
| Metallurgic furnaces | Metallurgy | 75 | 99 |
| Metallurgic processes | Metallurgy | 75 | 99 |

| SUBJECT. | Name of class—official classification. | No. of class—official classification. | No. of class—subscription classification. |
|---|---|---|---|
| Metal-saws | Saws | 110 | 130 |
| Metals, Furnaces for reducing | Metallurgy | 75 | 99 |
| Metals, Manufacture of | Metallurgy | 75 | 99 |
| Metals, Purification of | Metallurgy | 75 | 99 |
| Metals with metals, Coating | Plating | 96 | 113 |
| Metal-working tools | Metal-Working; 6 | 81A | 100 |
| Meteorological instruments | Measuring-Instruments | 73 | 101 |
| Meteorometers, Telegraphic | Electricity | 36 | 50 |
| Meters | Measuring-Instruments | 73 | 132 |
| Meters, Air | Pneumatics | 98 | 64 |
| Meters, Current | Measuring-Instruments | 73 | 82 |
| Meters, Drop | Measuring-Instruments | 73 | 98 |
| Meters, Electrical | Electricity | 36 | 50 |
| Meters, fluid, Registering devices of | Measuring-Instruments | 73 | 167 |
| Meters, Gas | Pneumatics | 98 | 64 |
| Meters, gas, Registering devices of | Measuring-Instruments | 73 | 64 |
| Meters, Liquid | Pumps | 103 | 167 |
| Meters, liquid, Registering devices of | Measuring-Instruments | 73 | 167 |
| Meters, Registering devices for | Measuring-Instruments | 73 | 167 |
| Meters, Registering devices of water | Measuring-Instruments | 73 | 167 |
| Meters, Spirit | Pumps | 103 | 167 |
| Meters, spirit, Registering devices of | Measuring-Instruments | 73 | 167 |
| Meters, Water | Pumps | 103 | 167 |
| Metronomes | Horology | 58 | 36 |
| Mica lamp-chimneys | Lamps and Gas-Fittings | 67 | 90 |
| Micrometers | Optics | 88 | 157 |
| Microscopes | Optics | 88 | 157 |
| Military accouterments | Trunks | 133 | 162 |
| Milk, Air-tight packages for | Preserving Food | 99 | 58 |
| Milk-cans | Dairy | 31 | 45 |
| Milk-carriers | Dairy | 31 | 45 |
| Milk-closets | Dairy | 31 | 45 |
| Milk, Condensing | Preserving Food | 99 | 58 |
| Milk-coolers | Ice | 62 | 45 |
| Milkers, Cow | Dairy | 31 | 45 |
| Milk-houses | Dairy | 31 | 45 |
| Milking-shields | Surgery | 128 | 45 |
| Milking-stools | Dairy | 31 | 62 |
| Milk-pails | Dairy | 31 | 45 |
| Milk-pans | Dairy | 31 | 45 |
| Milk-safes | Dairy | 31 | 45 |
| Milk-shelves | Dairy | 31 | 45 |
| Milk-strainers | Dairy | 31 | 45 |
| Milk-testers, (*except Lactometers, Class* 73.) | Dairy | 31 | 45 |
| Mill-board | Paper-Making | 92 | 108 |
| Mill-dams | Hydraulic Engineering | 61 | 82 |
| Millers' alarms | Mills | 83 | 57 |
| Milling-cloth | Cloth | 26 | 37 |
| Milling-machines | Metal-Working; 7 | 82 | 100 |
| Mill-gearing | Mechanical Powers | 74 | 67 |
| Mills | Mills | 83 | 57 |
| Mills, Cane | Mills | 83 | 155 |
| Mills, Clay | Clay | 25 | 20 |
| Mills, Gig | Cloth | 26 | 37 |
| Mills, Pug | Clay | 25 | 20 |
| Mills, Sorghum | Mills | 83 | 155 |
| Mills, Sugar | Mills | 83 | 155 |
| Millstone-bridges | Mills | 83 | 57 |
| Millstone bush-hammers | Stone, Lime, and Cement | 125 | 57 |

| SUBJECT. | Name of class—official classification. | No. of class—official classification. | No. of class—subscription classification. |
|---|---|---|---|
| Millstone-curbs | Mills | 83 | 57 |
| Millstone-dress | Mills | 83 | 57 |
| Millstone-eyes | Mills | 83 | 57 |
| Millstone-hammers | Stone, Lime, and Cement | 125 | 57 |
| Millstone-picks | Stone, Lime, and Cement | 125 | 57 |
| Millstones | Stone, Lime, and Cement | 125 | 57 |
| Millstones | Mills | 83 | 57 |
| Millstones, Balancing | Mills | 83 | 57 |
| Millstone-scrapers | Stone, Lime, and Cement | 125 | 57 |
| Millstones, Dressing | Mills | 83 | 57 |
| Millstones, Elevating | Mills | 83 | 57 |
| Millstones, Hanging | Mills | 83 | 57 |
| Millstones, Leveling | Mills | 83 | 57 |
| Millstones, Regulating | Mills | 83 | 57 |
| Millstones, Supporting | Mills | 83 | 57 |
| Millstones, Tramming | Mills | 83 | 57 |
| Millstones, Ventilating | Mills | 83 | 57 |
| Mincing-machines | Butchering | 17 | 25 |
| Mineral-waters, Compositions for | Aëration and Bottling | 1 | 1 |
| Miners' tools, Manufacture of | Metal-Working; 6 | 81B | 124 |
| Mining | Stone, Lime, and Cement | 125 | 124 |
| Mirrors | Furniture | 45 | 62 |
| Mirrors for windows | Furniture | 45 | 62 |
| Miter-boxes | Wood-Working; 4 | 145 | 29 |
| Mitering printers' rules | Sheet-Metal | 113A | 117 |
| Mitrailleuses | Ordnance | 89 | 105 |
| Mittens | Apparel | 2 | 38 |
| Moccasins | Boots and Shoes | 12 | 16 |
| Moire metallique | Plating | 96 | 113 |
| Molasses-gates | Valves | 136 | 167 |
| Molasses, Treatment of | Sugar | 127 | 155 |
| Mold-boards for plows | Plows | 97 | 114 |
| Mold-boards, Materials for | Plows | 97 | 114 |
| Molders' flasks | Casting | 22 | 33 |
| Molders' tables | Casting | 22 | 33 |
| Molding glass | Glass | 49 | 68 |
| Molding-machines, (wood) | Wood-Working; 2 | 143 | 173 |
| Molding pipe, Machines for | Casting | 22 | 33 |
| Molding-planes | Wood-Working; 4 | 145 | 173 |
| Mold-facing | Casting | 22 | 33 |
| Molds, Composition for glass | Glass | 49 | 68 |
| Molds, Dental | Dental | 32 | 46 |
| Molds for bricks and ceramics | Clay | 25 | 20 |
| Molds for casting metals | Casting | 22 | 33 |
| Molds for cornice | Masonry | 72 | 4 |
| Molds for glass | Glass | 49 | 68 |
| Molds for rubber goods | Caoutchouc | 18 | 84 |
| Mole-plows | Plows | 97 | 114 |
| Monitors | Ships; 1 | 114 | 137 |
| Monitor-turrets | Ships; 1 | 114 | 137 |
| Monkey-wrenches | Metal-Working; 6 | 81A | 100 |
| Monuments | Fine Arts | 41 | 54 |
| Mop-handles | Brushes and Brooms | 15 | 22 |
| Mop-heads | Brushes and Brooms | 15 | 22 |
| Mops | Brushes and Brooms | 15 | 22 |
| Mop-wringers | Brushes and Brooms | 15 | 22 |
| Mortar | Stone, Lime, and Cement | 125 | 153 |
| Mortar, Applications of | Masonry | 72 | 153 |
| Mortar-machines | Clay | 25 | 153 |
| Mortar-mixers | Clay | 25 | 153 |

| SUBJECT. | Name of class—official classification. | No. of class—official classification. | No. of class—sub-scription classification. |
|---|---|---|---|
| Mortars | Ordnance | 89 | 105 |
| Mortars, Pharmaceutists' | Surgery | 128 | 98 |
| Mortising-machines | Wood-Working; 2 | 143 | 173 |
| Mosaic | Fine Arts | 41 | 54 |
| Mosaic tiles | Clay | 25 | 20 |
| Mosquito-netting, Frames for | Beds | 5 | 62 |
| Moth-destroyers | Bee-Hives | 6 | 9 |
| Moth-traps | Bee-Hives | 6 | 9 |
| Motion, Converting | Mechanical Powers | 74 | 67 |
| Motion, Multiplying | Mechanical Powers | 74 | 67 |
| Motion, Transmitting | Mechanical Powers | 74 | 67 |
| Motors, Mechanical | Mechanical Powers | 74 | 67 |
| Mottled yarns | Spinning | 118 | 17 |
| Mottled yarns, Machines for making | Spinning | 118 | 142 |
| Mounting ordnance | Ordnance | 89 | 105 |
| Mousing-hooks | Clasps and Buckles | 24 | 23 |
| Mouth-pieces for cigars and pipes | Tobacco | 131 | 160 |
| Movement-cure apparatus | Surgery | 128 | 156 |
| Mowing-machines | Harvesters | 56 | 74 |
| Mucilage-bottles | Stationery | 120 | 146 |
| Mucilage-holders | Stationery | 120 | 146 |
| Mud-boats | Ships; 1 | 114 | 14 |
| Mud-collectors for steam-boilers | Steam; 2 | 122 | 148 |
| Mud-machines | Hydraulic Engineering | 61 | 82 |
| Muffs | Apparel | 2 | 38 |
| Mules | Spinning | 118 | 142 |
| Muley-saws | Saws | 110 | 130 |
| Music | Music | 84 | 102 |
| Musical instruments | Music | 84 | 102 |
| Musical instruments, Electric attachments to | Electricity | 36 | 102 |
| Musical notation | Music | 84 | 102 |
| Music-boxes | Music | 84 | 102 |
| Music-stands | Furniture | 45 | 62 |
| Mustache-guard | Toilet | 132 | 161 |
| Muzzle-rings | Harness | 54 | 72 |
| Muzzles | Harness | 54 | 72 |
| N. | | | |
| Nail-brushes | Brushes and Brooms | 15 | 161 |
| Nail-distributing machines | Nails | 85 | 103 |
| Nail-extractors | Wood-Working; 4 | 145 | 29 |
| Nailing-machines, Boot | Boots and Shoes | 12 | 16 |
| Nail-plate feeders | Nails | 85 | 103 |
| Nails | Nails | 85 | 103 |
| Nails, Cut | Nails | 85 | 103 |
| Nails, Machines for making | Nails | 85 | 103 |
| Nails, Wrought | Nails | 85 | 103 |
| Napkin-holders | Kitchen Utensils | 65 | 88 |
| Napkins | Kitchen Utensils | 65 | 88 |
| Napping-machines for cloth | Cloth | 26 | 37 |
| Napping-machines, Hat | Felting and Hats | 38 | 76 |
| Nasal douches | Surgery | 128 | 156 |
| Nasal tubes | Surgery | 128 | 156 |
| Nautical alarms | Signals | 116 | 137 |
| Nautical blocks, Machines for making | Wood-Working; 3 | 144 | 174 |
| Naval architecture | Ships; 1 | 114 | 137 |
| Neck-tie clasps | Clasps and Buckles | 24 | 38 |
| Neck-tie retainers | Apparel | 2 | 38 |

| SUBJECT. | Name of class—official classification. | No. of class—official classification. | No. of class—subscription classification. |
|---|---|---|---|
| Neck-ties | Apparel | 2 | 38 |
| Neck-yokes | Carriages | 21 | 31 |
| Needle-boxes | Apparel | 2 | 38 |
| Needle-cases | Apparel | 2 | 38 |
| Needle-guards | Apparel | 2 | 38 |
| Needle-guns | Fire-Arms | 42 | 18 |
| Needles | Apparel | 2 | 104 |
| Needle-setters for sewing-machines | Sewing-Machines | 112 | 134 |
| Needles for knitting-machines | Knitting and Netting | 66 | 89 |
| Needles for sewing-machines | Sewing-Machines | 112 | 104 |
| Needles, Machines for making | Needles and Pins | 86 | 104 |
| Needles, Machines for papering | Needles and Pins | 86 | 104 |
| Needles, Surgeons' | Surgery | 128 | 156 |
| Needle-threaders | Apparel | 2 | 38 |
| Needle-threaders for sewing-machines | Sewing-Machines | 112 | 134 |
| Needle-wrappers | Apparel | 2 | 104 |
| Nests for fowls | Stabling | 119 | 116 |
| Nests, Bird | Games and Toys | 46 | 143 |
| Nets, Fishing | Fishing | 43 | 143 |
| Nets, Ladies' head | Felting and Hats | 38 | 160 |
| Netting-machines | Knitting and Netting | 66 | 89 |
| Newspaper-clamps | Stationery | 120 | 146 |
| Newspaper-covers | Stationery | 120 | 146 |
| Newspaper-folders | Paper Manufactures | 93 | 117 |
| Nicking horses, Apparatus for | Stabling | 119 | 144 |
| Night-carts | Carriages and Wagons | 21 | 31 |
| Night-latches | Locks and Latches | 70 | 93 |
| Night-soil, Treatment of | Manures | 71 | 96 |
| Nipple-guards for fire-arms | Fire-Arms | 42 | 55 |
| Nipples, Artificial | Surgery | 128 | 156 |
| Nipple-shields | Surgery | 128 | 156 |
| Nipples of fire-arms | Fire-Arms | 42 | 55 |
| Nitro-glycerine | Gunpowder | 52 | 71 |
| Nitrous oxide, Apparatus for making | Gas | 48 | 46 |
| Nitrous oxide gas, Instruments for administering | Surgery | 128 | 46 |
| Nose-bags | Harness | 54 | 72 |
| Nose-rings | Harness | 54 | 144 |
| Nostril-pinchers | Harness | 54 | 144 |
| Nozzles for fire-engines | Water-Distribution | 137 | 167 |
| Nozzles for hose | Water-Distribution | 137 | 167 |
| Nozzles for oil-cups | Water-Distribution | 137 | 87 |
| Nozzles, Manufacture of | Hardware Manufactures | 53J | 136 |
| Nozzles, Variable exhaust | Steam ; 3 | 123 | 94 |
| Nubias | Felting and Hats | 38 | 38 |
| Numbering machines | Printing | 101 | 15 |
| Numeral-frames | Educational | 35 | 146 |
| Nursery-bottles | Surgery | 128 | 156 |
| Nursery-gates | Furniture | 45 | 62 |
| Nursing couches | Beds | 5 | 62 |
| Nut-blanks, Making | Bolts, Nuts, and Rivets | 10 | 14 |
| Nut-crackers | Kitchen Utensils | 65 | 88 |
| Nut-locks | Bolts, Nuts, and Rivets | 10 | 14 |
| Nut-making-machines | Bolts, Nuts, and Rivets | 10 | 14 |
| Nutmeg-graters | Kitchen Utensils | 65 | 88 |
| Nut-pickers | Kitchen Utensils | 65 | 88 |
| Nuts | Bolts, Nuts, and Rivets | 10 | 14 |
| Nut-threading machines | Bolts, Nuts, and Rivets | 10 | 14 |

| SUBJECT. | Name of class—official classification. | No. of class—official classification. | No. of class—subscription classification. |
|---|---|---|---|
| O. | | | |
| Oakum-pickers | Brakes and Gins | 13 | 78 |
| Oar-locks | Boats | 9 | 13 |
| Oars | Boats | 9 | 13 |
| Oars, Feathering | Boats | 9 | 13 |
| Oars, Machines for dressing | Wood-Working; 3 | 144 | 13 |
| Oars, Machines for planing | Wood-Working; 3 | 144 | 13 |
| Oars, Machines for riving | Wood-Working; 3 | 144 | 13 |
| Oars, Machines for sawing | Saws | 110 | 13 |
| Oars, Machines for turning | Wood-Working; 1 | 142 | 13 |
| Oasts | Dryers and Kilns | 34 | 19 |
| Object-teaching, Devices for | Educational | 35 | 146 |
| Obstetric apparatus | Surgery | 128 | 156 |
| Obstructing rivers and harbors | Hydraulic Engineering | 61 | 82 |
| Octants | Optics | 88 | 157 |
| Odometers | Measuring-Instruments | 73 | 42 |
| Offal, Deodorizing | Manures | 71 | 96 |
| Offal-dryers | Dryers and Kilns | 34 | 48 |
| Offal for fertilizers, Treatment of | Manures | 71 | 96 |
| Offal, Treating | Oil, Fat, and Glue | 87 | 26 |
| Office time-indicators, (*non-automatic*) | Stationery | 120 | 146 |
| Oil-cake trimmers | Presses | 100 | 26 |
| Oil-cans, Air-tight | Preserving Food | 99 | 58 |
| Oil-cloth | Paint | 91 | 56 |
| Oil-cloth cutters | Furniture | 45 | 44 |
| Oil-cloth printing | Printing | 101 | 56 |
| Oil-cups | Journals and Bearings | 64 | 87 |
| Oilers | Journals and Bearings | 64 | 87 |
| Oil extracting, Chemical modes of | Oil, Fat, and Glue | 87 | 26 |
| Oil-filters | Oil, Fat, and Glue | 87 | 26 |
| Oil-mills | Presses | 100 | 26 |
| Oil-presses | Presses | 100 | 26 |
| Oil-refining | Stills | 124 | 47 |
| Oils and fats from fiber and meal, Removing | Oil, Fat, and Glue | 87 | 26 |
| Oils, Apparatus for boiling | Stoves and Furnaces | 126 | 26 |
| Oils, fats, and glue | Oil, Fat, and Glue | 87 | 26 |
| Oil-stills | Stills | 124 | 47 |
| Oil-stones | Grinding and Polishing | 51 | 70 |
| Oil-tanks, Sealed | Preserving Food | 99 | 58 |
| Oil-wells, Blasting | Projectiles | 102 | 124 |
| Oil-wells, Torpedoes for | Projectiles | 102 | 124 |
| Ointments | Chemical Miscellaneous | 23 | 98 |
| Oleometers | Measuring-Instruments | 73 | 47 |
| Olivers | Metal-Working; 3 | 78 | 60 |
| Omnibuses | Carriages and Wagons | 21 | 31 |
| Open links | Clasps and Buckles | 24 | 23 |
| Open rings | Clasps and Buckles | 24 | 23 |
| Opera-glasses | Optics | 88 | 157 |
| Operating ordnance | Ordnance | 89 | 105 |
| Ophthalmoscopes | Surgery | 128 | 157 |
| Optical instruments | Optics | 88 | 157 |
| Ordnance | Ordnance | 89 | 105 |
| Ordnance, Attaching trunnions to | Ordnance | 89 | 105 |
| Ordnance, Instruments for disabling | Ordnance | 89 | 105 |
| Ordnance-manufacturing | Metal-Working; 3 | 78 | 105 |
| Ordnance, Mounting | Ordnance | 89 | 105 |
| rdnance, Operating | Ordnance | 89 | 105 |
| Ordnance-projectiles | Projectiles | 102 | 105 |

| SUBJECT. | Name of class—official classification. | No. of class—official classification. | No. of class—subscription classification. |
|---|---|---|---|
| Ore-crushers | Ore | 90 | 3 |
| Ore-disintegrators | Ore | 90 | 3 |
| Ore-dressers | Ore | 90 | 3 |
| Ore-mills | Ore | 90 | 3 |
| Ores, Calcining | Metallurgy | 75 | 99 |
| Ores, Desulphurizing | Metallurgy | 75 | 99 |
| Ore-separators | Ore | 90 | 3 |
| Ores, Furnaces for reducing | Metallurgy | 75 | 99 |
| Ores, Oxidizing | Metallurgy | 75 | 99 |
| Ores, Roasting | Metallurgy | 75 | 99 |
| Ores, Smelting | Metallurgy | 75 | 99 |
| Ore-stamps | Ore | 90 | 3 |
| Ore-washers | Ore | 90 | 3 |
| Organ-actions | Music | 84 | 106 |
| Organ-bellows | Music | 84 | 106 |
| Organ-couplers | Music | 84 | 106 |
| Organ key-boards | Music | 84 | 106 |
| Organ-pipes | Music | 84 | 106 |
| Organs | Music | 84 | 106 |
| Organ sound-boards | Music | 84 | 106 |
| Organ-stops | Music | 84 | 106 |
| Organs, Tremolo attachments to | Music | 84 | 106 |
| Organ-swells | Music | 84 | 106 |
| Organ-tuning | Music | 84 | 106 |
| Organ wind-chests | Music | 84 | 106 |
| Ornamental yarns | Cordage | 28 | 17 |
| Ornamentation | Fine Arts | 41 | 54 |
| Ornamenting glass | Glass | 49 | 54 |
| Ornaments, Compositions for plaster | Stone, Lime, and Cement | 125 | 153 |
| Ornaments for sheet-metal ware, Die | Sheet-Metal | 113A | 136 |
| Ornaments, Small metallic | Clasps and Buckles | 24 | 23 |
| Ornaments, (small metallic,) Manufacture of | Hardware Manufactures | 53E | 23 |
| Orreries | Educational | 35 | 146 |
| Orthopedic apparatus | Artificial Limbs | 3 | 156 |
| Oscillating steam-engines | Steam; 1 | 121 | 150 |
| Oscillating-valves | Valves | 136 | 152 |
| Osier-cutting | Garden and Orchard | 47 | 63 |
| Osier-peelers | Wood-Working; 3 | 144 | 169 |
| Otoscopes | Surgery | 128 | 156 |
| Ottomans | Furniture | 45 | 62 |
| Outriggers | Boats | 9 | 13 |
| Oval frames, Machines for turning | Wood-Working; 1 | 142 | 175 |
| Ovens, Bakers' | Stoves and Furnaces | 126 | 6 |
| Ovens, Coke | Dryers and Kilns | 34 | 48 |
| Ovens for biscuit, (Clay) | Dryers and Kilns | 34 | 6 |
| Ovens for enameling | Dryers and Kilns | 34 | 115 |
| Ovens for pottery | Dryers and Kilns | 34 | 115 |
| Ovens, Portable | Stoves and Furnaces | 126 | 154 |
| Ovens, Reel | Stoves and Furnaces | 126 | 154 |
| Ovens, Rotary | Stoves and Furnaces | 126 | 154 |
| Ovens, Stove | Stoves and Furnaces | 126 | 154 |
| Overshoes | Boots and Shoes | 12 | 16 |
| Oxshoes | Horseshoes | 59 | 60 |
| Oxygen, Processes and apparatus for obtaining | Gas | 48 | 64 |
| Oxhydrogen lights | Lamps and Gas-Fittings | 67 | 90 |
| Ox-yokes, bows, and pins | Carriages and Wagons | 21 | 31 |
| Oyster-cans | Preserving Food | 99 | 58 |
| Oyster-culture | Fishing | 43 | 143 |

| SUBJECT. | Name of class—official classification. | No. of class—official classification. | No. of class—subscription classification. |
|---|---|---|---|
| Oyster-dredging | Fishing | 43 | 143 |
| Oyster-openers | Kitchen Utensils | 65 | 88 |
| Oyster-rakes | Fishing | 43 | 143 |
| Oyster-tongs | Fishing | 43 | 143 |
| Ozone, Apparatus for obtaining | Gas | 48 | 64 |
| P. | | | |
| Packaging fabrics | Cloth | 26 | 37 |
| Packaging yarns, threads, &c., (*except Bundling Presses, Class* 100) | Cordage | 28 | 17 |
| Packers for flour, fruit, sugar, salt, &c. | Presses | 100 | 57 |
| Packing-bands | Presses | 100 | 7 |
| Packing for pistons | Steam; 1 | 121 | 151 |
| Packing for steam-joints | Steam; 1 | 121 | 151 |
| Packing for well-tubes | Water-Distribution | 137 | 167 |
| Packing furniture | Furniture | 45 | 62 |
| Packing glass | Glass | 49 | 68 |
| Packing houses, Meat | Preserving Food | 99 | 58 |
| Packing projectiles | Projectiles | 102 | 105 |
| Pack-saddles | Harness | 54 | 72 |
| Pad-crimps | Harness | 54 | 72 |
| Paddle-box bridges | Ships; 1 | 114 | 139 |
| Paddle-wheels | Ships; 2 | 115 | 139 |
| Pad-locks | Locks and Latches | 70 | 93 |
| Pads | Harness | 54 | 72 |
| Pads for trusses | Surgery | 128 | 156 |
| Pad-trees | Harness | 54 | 72 |
| Paging-machines | Printing | 101 | 15 |
| Paint-brushes | Brushes and Brooms | 15 | 22 |
| Paint-burners | Stoves and Furnaces | 126 | 154 |
| Paint-cans, Air-tight | Preserving Food | 99 | 107 |
| Painted fabrics | Paint | 91 | 56 |
| Painters' tools, (*except Brushes, Class* 15) | Paint | 91 | 107 |
| Painting, Decorative | Fine Arts | 41 | 54 |
| Painting on glass | Fine Arts | 41 | 54 |
| Painting-machines | Paint | 91 | 107 |
| Paint-mills | Mills | 83 | 107 |
| Paint-mullers | Mills | 83 | 107 |
| Paint-pot hangers | Paint | 91 | 107 |
| Paints | Paint | 91 | 107 |
| Paints, Apparatus for compounding | Paint | 91 | 107 |
| Paints, Apparatus for mixing | Paint | 91 | 107 |
| Palates, Artificial | Artificial Limbs | 3 | 5 |
| Palates, Artificial | Dental | 32 | 5 |
| Palliasses | Beds | 5 | 62 |
| Palm-leaf hats | Felting and Hats | 38 | 76 |
| Palm-leaves, Machines for splitting | Wood-Working; 3 | 144 | 165 |
| Panels of wire | Wire-Working | 140 | 170 |
| Panels, Rabbeting door | Wood-Working; 2 | 143 | 173 |
| Panniers | Crinoline and Corsets | 29 | 38 |
| Pans for cooking *in vacuo* | Sugar | 127 | 155 |
| Pans for making vegetable extracts | Sugar | 127 | 155 |
| Pans for manufacturing glue | Oil, Fat, and Glue | 87 | 26 |
| Pans, Kitchen and pantry | Kitchen Utensils | 65 | 136 |
| Pans, Sugar | Sugar | 127 | 155 |
| Pans, Vacuum | Sugar | 127 | 155 |
| Pantalets | Apparel | 2 | 38 |
| Pantaloon-protectors | Apparel | 2 | 38 |
| Pantaloons | Apparel | 2 | 38 |

| SUBJECT. | Name of class—official classification. | No. of class—official classification. | No. of class—sub-scription classification. |
|---|---|---|---|
| Pantaloon-straps | Apparel | 2 | 38 |
| Pantaloon-stretchers | Apparel | 2 | 38 |
| Pantographs | Drafting | 33 | 97 |
| Paper | Paper-Making | 92 | 108 |
| Paper apparel | Paper Manufactures | 93 | 109 |
| Paper-bag machines | Paper Manufactures | 93 | 109 |
| Paper bags | Paper Manufactures | 93 | 109 |
| Paper-box machines | Paper Manufactures | 93 | 109 |
| Paper boxes | Paper Manufactures | 93 | 109 |
| Paper, Calendering | Paper-Making | 92 | 108 |
| Paper-clamps | Stationery | 120 | 146 |
| Paper-clasps, Manufacture of | Hardware Manufactures | 53 I | 146 |
| Paper-clips | Stationery | 120 | 146 |
| Paper collars | Paper Manufactures | 93 | 109 |
| Paper cuffs | Paper Manufactures | 93 | 109 |
| Paper-cutters | Paper Manufactures | 93 | 109 |
| Paper-cutting machines | Paper Manufactures | 93 | 109 |
| Paper-dampers | Printing | 101 | 117 |
| Paper, Drying | Paper-Making | 92 | 108 |
| Paper, Enameling | Paper-Making | 92 | 108 |
| Paper-envelope machines | Paper Manufactures | 93 | 109 |
| Paper-fasteners | Clasps and Buckles | 24 | 146 |
| Paper-feeders | Paper Manufactures | 93 | 117 |
| Paper-files | Stationery | 120 | 146 |
| Paper, Flocking | Paper-Making | 92 | 108 |
| Paper floor-cloth | Paper Manufactures | 93 | 109 |
| Paper-folders | Paper Manufactures | 93 | 146 |
| Paper for roofs, Machines for preparing | Roofing | 108 | 109 |
| Paper-hanging appliances | Fine Arts | 41 | 107 |
| Paper hats | Paper Manufactures | 93 | 76 |
| Paper labels, Manufacture of | Paper Manufactures | 93 | 109 |
| Paper, Lace | Paper Manufactures | 93 | 109 |
| Paper-making machines | Paper-Making | 92 | 108 |
| Paper manufactures | Paper Manufactures | 93 | 108 |
| Paper, Materials for | Paper-Making | 92 | 108 |
| Paper-perforating machines | Paper Manufactures | 93 | 108 |
| Paper, Processes of making | Paper-Making | 92 | 108 |
| Paper-pulp | Paper-Making | 92 | 108 |
| Paper-pulping machines | Paper-Making | 92 | 108 |
| Paper-punches | Clasps and Buckles | 24 | 146 |
| Paper-sizing | Paper-Making | 92 | 108 |
| Paper tags, Manufacture of | Paper Manufactures | 93 | 109 |
| Paper twine | Paper Manufactures | 93 | 109 |
| Paper-twine machines | Cordage | 28 | 109 |
| Paper water-proofing | Paper-Making | 92 | 109 |
| Paper-weights | Stationery | 120 | 146 |
| Papier-maché | Stationery | 92 | 108 |
| Paraffine | Oil, Fat, and Glue | 87 | 26 |
| Parallel rulers | Drafting | 33 | 97 |
| Parasols | Umbrellas and Fans | 135 | 165 |
| Parceling-rope | Cordage | 28 | 127 |
| Parchment | Tanning | 129 | 92 |
| Parquetry | Fine Arts | 41 | 54 |
| Parrels | Ships; 1 | 114 | 138 |
| Partitions of buildings, Wooden | Carpentry | 20 | 30 |
| Paste-board | Paper-Making | 92 | 109 |
| Paste, Compositions for | Bleaching and Dyeing, | 8 | 11 |
| Paste-heating pots | Stoves and Furnaces | 126 | 154 |
| Paste-making | Bleaching and Dyeing | 8 | 11 |
| Pastilles | Chemical Miscellaneous | 23 | 98 |

| SUBJECT. | Name of class—official classification. | No. of class—official classification. | No. of class—sub-scription classification. |
|---|---|---|---|
| Pasting cops | Spinning | 118 | 142 |
| Pattern-cards | Weaving | 139 | 95 |
| Pattern-chains | Weaving | 139 | 95 |
| Pattern-cylinders | Weaving | 139 | 95 |
| Pattern-lifters for molders | Casting | 22 | 33 |
| Patterns for garments | Apparel | 2 | 38 |
| Pavements, Asphalt | Paving | 94 | 110 |
| Pavements, Composition | Paving | 94 | 110 |
| Pavements, Concrete | Paving | 94 | 110 |
| Pavements, Iron | Paving | 94 | 110 |
| Pavements, Stone | Paving | 94 | 110 |
| Pavements, Wooden | Paving | 94 | 110 |
| Paving-blocks, Machines for making | Saws | 110 | 174 |
| Paving compositions | Paving | 94 | 110 |
| Paving | Paving | 94 | 110 |
| Paving-stone, Preparation of | Stone, Lime, and Cement | 125 | 153 |
| Paving-tiles | Clay | 25 | 20 |
| Paviors' tools, machines, and processes | Paving | 94 | 110 |
| Pawls and ratchets | Mechanical Powers | 74 | 67 |
| Paying seams | Ships; 1 | 114 | 138 |
| Peach-parers | Kitchen Utensils | 65 | 88 |
| Peach-stoners | Kitchen Utensils | 65 | 88 |
| Pea-nut cleaners | Mills | 83 | 58 |
| Pea-nut roasters | Stoves and Furnaces | 126 | 154 |
| Pea-shellers | Kitchen Utensils | 65 | 88 |
| Peat-dryers | Dryers and Kilns | 34 | 48 |
| Peat for fuel, Treatment of | Fuel | 44 | 61 |
| Peat-machines | Fuel | 44 | 61 |
| Peat-presses | Presses | 100 | 7 |
| Pedometers | Measuring-Instruments | 73 | 42 |
| Peeling almonds | Kitchen Utensils | 65 | 88 |
| Peg-cutters | Boots and Shoes | 12 | 16 |
| Pegging-jacks | Boots and Shoes | 12 | 16 |
| Pegging-machines | Boots and Shoes | 12 | 16 |
| Pegging soles | Boots and Shoes | 12 | 16 |
| Pegs, Leather | Boots and Shoes | 12 | 16 |
| Pegs, Machines for arranging | Boots and Shoes | 12 | 16 |
| Pegs, Machines for driving | Boots and Shoes | 12 | 16 |
| Pegs, Machines for making | Wood-Working; 3 | 144 | 175 |
| Pegs, Machines for making shoe | Wood-Working; 3 | 144 | 16 |
| Pegs, Machines for pointing | Wood-Working; 3 | 144 | 175 |
| Pegs, Machines for splitting | Wood-Working; 3 | 144 | 175 |
| Pegs, Machines for turning | Wood-Working; 1 | 142 | 175 |
| Pegs, Shoe | Boots and Shoes | 12 | 16 |
| Pen-cases | Stationery | 120 | 146 |
| Pencil-cases | Stationery | 120 | 146 |
| Pencil-cleaners | Stationery | 120 | 146 |
| Pencil-mark erasers | Stationery | 120 | 146 |
| Pencils | Stationery | 120 | 146 |
| Pencils, Compositions for indelible | Stationery | 120 | 146 |
| Pencil-sharpeners | Stationery | 120 | 146 |
| Pencils, Machines for making | Wood-Working; 3 | 144 | 146 |
| Pendulum-powers | Mechanical Powers | 74 | 36 |
| Pendulums | Horology | 58 | 36 |
| Pendulums, Electrical | Electricity | 36 | 36 |
| Pen-holders | Stationery | 120 | 146 |
| Pen-holders, Manufacture of | Hardware Manufactures | 53 I | 146 |
| Pen-makers, Quill | Stationery | 120 | 146 |
| Penmanship, Appliances for teaching | Educational | 35 | 146 |
| Pen-racks | Stationery | 120 | 146 |

| SUBJECT. | Name of class—official classification. | No. of class—official classification. | No. of class—sub-scription classification. |
|---|---|---|---|
| Pens | Stationery | 120 | 146 |
| Pens, Cattle and hog | Stabling | 119 | 144 |
| Pens, Fountain | Stationery | 120 | 146 |
| Pens, Galvanic | Electricity | 36 | 146 |
| Pens, Manufacture of | Hardware Manufactures | 53 1 | 136 |
| Pens, Rubber | Stationery | 120 | 146 |
| Pens, Ruling | Drafting | 33 | 15 |
| Penstocks | Water-Wheels | 138 | 167 |
| Pen-trays | Stationery | 120 | 146 |
| Pepper-boxes | Kitchen Utensils | 65 | 88 |
| Perambulators | Carriages and Wagons | 21 | 31 |
| Perches, Wagon | Carriages | 21 | 31 |
| Perch-plates, Manufacture of | Hardware Manufactures | 53C | 31 |
| Percolators for medicinal solutions | Surgery | 128 | 98 |
| Percussion-cap-filling machines | Projectiles | 102 | 55 |
| Percussion-cap holders | Projectiles | 102 | 55 |
| Percussion-cap-lining machines | Projectiles | 102 | 55 |
| Percussion-caps | Projectiles | 102 | 55 |
| Percussion-caps, Manufacture of | Hardware Manufactures | 53A | 55 |
| Percussion-cartridges | Projectiles | 102 | 32 |
| Percussion-shells | Projectiles | 102 | 105 |
| Perfuming-apparatus | Chemical Miscellaneous | 23 | 86 |
| Permutation-locks | Locks and Latches | 70 | 93 |
| Perspective apparatus | Drafting | 33 | 97 |
| Pessaries | Surgery | 128 | 156 |
| Petroleum, Apparatus for treating | Stills | 124 | 47 |
| Petroleum-burners for stoves | Stoves and Furnaces | 126 | 154 |
| Petroleum-cars | Railways; 2 | 105 | 121 |
| Petroleum-stills | Stills | 124 | 47 |
| Pew-fittings | Furniture | 45 | 62 |
| Pewter-ware, Casting processes for | Casting | 22 | 33 |
| Pharmaceutical apparatus | Surgery | 128 | 156 |
| Philosophical toys | Games and Toys | 46 | 143 |
| Phosphates for fertilizers, Treatment of | Manures | 71 | 96 |
| Photogalvanography | Photography | 95 | 111 |
| Photographic albums | Stationery | 120 | 146 |
| Photographic apparatus | Photography | 95 | 111 |
| Photographic cameras | Photography | 95 | 111 |
| Photographic card-mounts | Fine Arts | 41 | 54 |
| Photographic paper | Photography | 95 | 111 |
| Photographic-plate holders | Photography | 95 | 111 |
| Photographic printing | Photography | 95 | 111 |
| Photographic prints, Coloring | Photography | 95 | 111 |
| Photographic processes | Photography | 95 | 111 |
| Photographic toning | Photography | 95 | 111 |
| Photography | Photography | 95 | 111 |
| Photolithography | Photography | 95 | 54 |
| Photometers | Measuring-Instruments | 73 | 157 |
| Photo-sculpture | Photography | 95 | 54 |
| Piano-forte actions | Music | 84 | 112 |
| Piano-forte attachments | Music | 84 | 112 |
| Piano-forte cases | Music | 84 | 112 |
| Piano-forte frames | Music | 84 | 112 |
| Piano-forte keys | Music | 84 | 112 |
| Piano-fortes | Music | 84 | 112 |
| Piano-fortes combined with reed instruments | Music | 84 | 112 |
| Piano-forte sounding-boards | Music | 84 | 112 |
| Piano-fortes, Scales for | Music | 84 | 112 |
| Piano-fortes, Stringing | Music | 84 | 112 |

| SUBJECT. | Name of class—official classification. | No. of class—official classification. | No. of class—sub-scription classification. |
|---|---|---|---|
| Piano-fortes, Swell-attachment to | Music | 84 | 112 |
| Piano-fortes, Tuning | Music | 84 | 112 |
| Piano-locks | Locks and Latches | 70 | 93 |
| Piano-movers | Furniture | 45 | 62 |
| Piano-stools | Furniture | 45 | 62 |
| Picker-checks | Weaving | 139 | 95 |
| Picker-motions | Weaving | 139 | 95 |
| Pickers | Weaving | 139 | 95 |
| Picker-staves | Weaving | 139 | 95 |
| Pickers, Wool or Cotton, (Factory) | Carding | 19 | 28 |
| Picket-fences | Fences | 39 | 52 |
| Picket-pointers | Wood-Working; 3 | 144 | 174 |
| Pickets, Machines for making | Wood-Working; 3 | 144 | 174 |
| Picking fur | Felting and Hats | 38 | 76 |
| Picking rope | Cordage | 28 | 127 |
| Picks | Garden and Orchard | 47 | 49 |
| Picks, Millstone | Stone, Lime, and Cement | 125 | 57 |
| Picture-cases | Fine Arts | 41 | 54 |
| Picture-frames | Fine Arts | 41 | 54 |
| Picture-hangers | Furniture | 45 | 62 |
| Picture-holders, Photographic | Photography | 95 | 111 |
| Picture-knobs | Nails | 85 | 103 |
| Picture-mounting | Fine Arts | 41 | 54 |
| Picture-nails | Nails | 85 | 103 |
| Pie-crimpers | Kitchen Utensils | 65 | 88 |
| Pier-glasses | Furniture | 45 | 62 |
| Piers, Bridge and abutment | Bridges | 14 | 21 |
| Piers, Landing | Hydraulic Engineering | 61 | 82 |
| Pie-trimmers | Kitchen Utensils | 65 | 88 |
| Pigeon-holes, Desk | Stationery | 120 | 146 |
| Pigments | Paint | 91 | 107 |
| Pigments, Manufacture of | Paint | 91 | 107 |
| Pikes | Cutlery | 30 | 44 |
| Pile-cutters | Hydraulic Engineering | 61 | 82 |
| Pile-drivers | Hydraulic Engineering | 61 | 82 |
| Piles | Hydraulic Engineering | 61 | 82 |
| Piles and fagots for rolling | Metal-Working; 5 | 80 | 125 |
| Piles, Instruments for treating | Surgery | 128 | 156 |
| Piles, Protection of | Hydraulic Engineering | 61 | 82 |
| Piles, Screw | Hydraulic Engineering | 61 | 82 |
| Pile-wire actuating-mechanisms | Weaving | 139 | 96 |
| Pile-wires | Weaving | 139 | 96 |
| Pills, Machines for making | Surgery | 128 | 98 |
| Pillows | Beds | 5 | 62 |
| Pilot-stands | Ships; 1 | 114 | 138 |
| Pin-boxes | Apparel | 2 | 38 |
| Pinchers | Metal-Working; 6 | 81 | 100 |
| Pinchers, Lasting | Boots and Shoes | 12 | 16 |
| Pinchers, Shoemakers' | Boots and Shoes | 12 | 16 |
| Pin-cushions | Apparel | 2 | 38 |
| Pinking-machines | Apparel | 2 | 38 |
| Pinking-punches | Clasps and Buckles | 24 | 38 |
| Pins, Breast | Jewelry | 63 | 86 |
| Pins, Diaper | Jewelry | 63 | 104 |
| Pins, Dress | Jewelry | 63 | 104 |
| Pins, Machines for cutting clothes | Wood-Working; 3 | 144 | 175 |
| Pins, Machines for making | Needles and Pins | 86 | 104 |
| Pins, Machines for papering | Needles and Pins | 86 | 104 |
| Pins, Machines for slotting clothes | Wood-Working; 3 | 144 | 175 |

| SUBJECT. | Name of class—official classification. | No. of class—official classification. | No. of class—subscription classification. |
|---|---|---|---|
| Pins, Machines for turning bedstead and clothes | Wood-Working; 1 | 142 | 175 |
| Pins, Scarf | Jewelry | 63 | 86 |
| Pins, Shawl | Jewelry | 63 | 86 |
| Pipe, Cement | Clay | 25 | 20 |
| Pipe-couplings | Water-Distribution | 137 | 167 |
| Pipe-cutters | Metal-Working; 6 | 81A | 100 |
| Pipe-protectors | Water-Distribution | 137 | 167 |
| Pipes, Chain-locker | Ships; 1 | 114 | 138 |
| Pipes, Clay | Clay | 25 | 20 |
| Pipes, Compositions for making | Clay | 25 | 20 |
| Pipes, Draining | Clay | 25 | 20 |
| Pipes, Non-conducting covering for steam | Steam; 2 | 122 | 148 |
| Pipes, Tobacco | Tobacco | 131 | 160 |
| Pipes, Water and gas | Water-Distribution | 137 | 167 |
| Pipes, Wooden | Wood-Working; 1 | 142 | 173 |
| Pipe-tongs | Metal-Working; 6 | 81A | 100 |
| Pipettes | Measuring-Instruments | 73 | 156 |
| Pipe-wrenches | Metal-Working; 6 | 81A | 100 |
| Pisciculture | Fishing | 43 | 143 |
| Pisé-building, Implements and modes for | Masonry | 72 | 4 |
| Pistols | Fire-Arms | 42 | 55 |
| Pistol-swords | Fire-Arms | 42 | 55 |
| Piston-blowers | Pneumatics | 98 | 12 |
| Piston-packing | Steam; 1 | 121 | 151 |
| Piston-rod packing | Steam; 1 | 121 | 151 |
| Piston-rods | Steam; 1 | 121 | 151 |
| Pistons | Steam; 1 | 121 | 151 |
| Piston-valves | Valves | 139 | 152 |
| Pitchforks | Garden and Orchard | 47 | 141 |
| Pitch-heaters | Stoves and Furnaces | 126 | 126 |
| Pitchers, Beer | Kitchen Utensils | 65 | 88 |
| Pitchers, Ice | Kitchen Utensils | 65 | 88 |
| Pitchers, Molasses | Kitchen Utensils | 65 | 88 |
| Pitman connections | Mechanical Powers | 74 | 67 |
| Pitmen | Mechanical Powers | 74 | 67 |
| Pitmen for harvesters | Mechanical Powers | 74 | 74 |
| Pitmen for sewing-machines | Mechanical Powers | 74 | 134 |
| Plaiters for sewing-machines | Sewing-Machines | 112 | 134 |
| Plaiting-machines | Apparel | 2 | 38 |
| Planchets from bars and plates, Cutting | Metal-Working; 4 | 79 | 119 |
| Plane-bits | Wood-Working; 4 | 145 | 29 |
| Plane-making | Wood-Working; 4 | 145 | 29 |
| Planers | Metal-Working; 7 | 82 | 100 |
| Planes, Carpenters' | Wood-Working; 4 | 145 | 29 |
| Planes, Coopers' | Wood-Working; 4 | 145 | 40 |
| Planes, Joiners' | Wood-Working; 4 | 145 | 29 |
| Planes, Splint | Wood-Working; 2 | 143 | 29 |
| Plane-stocks | Wood-Working; 2 | 145 | 29 |
| Plane-stocks, Machines for making | Wood-Working; 3 | 144 | 29 |
| Plane-tables | Optics | 88 | 157 |
| Planetariums | Educational | 35 | 157 |
| Planimeters | Measuring-Instruments | 73 | 97 |
| Planing machines, Metal | Metal-Working; 7 | 82 | 100 |
| Planing-machines, (wood) | Wood-Working; 2 | 143 | 173 |
| Planing stone | Stone, Lime, and Cement | 125 | 153 |
| Planispheres | Educational | 35 | 157 |
| Planking-clamps | Wood-Working; 4 | 145 | 29 |

| SUBJECT. | Name of class—official classification. | No. of class—official classification. | No. of class—subscription classification. |
|---|---|---|---|
| Planking hats | Felting and Hats | 38 | 76 |
| Planning-instruments | Drafting | 33 | 97 |
| Plant-boxes | Garden and Orchard | 47 | 63 |
| Planters, Corn | Seeders and Planters | 111 | 133 |
| Planters, Cotton-seed | Seeders and Planters | 111 | 133 |
| Planters, Hand | Seeders and Planters | 111 | 133 |
| Planters, Seed | Seeders and Planters | 111 | 133 |
| Planters, Tobacco-seed | Seeders and Planters | 111 | 133 |
| Plant-frames | Garden and Orchard | 47 | 63 |
| Planting-tools | Garden and Orchard | 47 | 63 |
| Plant-labels | Stationery | 120 | 63 |
| Plastering-machines | Masonry | 72 | 4 |
| Plastering-tools | Masonry | 72 | 4 |
| Plaster-mills | Ore | 90 | 3 |
| Plasters, Adhesive | Chemical Miscellaneous | 23 | 98 |
| Plasters, Machines for spreading | Surgery | 128 | 98 |
| Plate-engraving | Fine Arts | 41 | 54 |
| Plate-glass | Glass | 49 | 68 |
| Platening glass | Glass | 49 | 68 |
| Plates | Kitchen Utensils | 65 | 88 |
| Plates, Bending metallic | Metal-Working; 1 | 76 | 125 |
| Plates, Corrugating metallic | Metal-Working; 1 | 76 | 125 |
| Plates for artificial teeth | Dental | 32 | 46 |
| Plates, Straightening metallic | Metal-Working; 1 | 76 | 125 |
| Platform-bridges for railway-cars | Railways; 3 | 106 | 121 |
| Platform-cars | Railways; 2 | 105 | 121 |
| Platforms for railway-cars | Railways; 3 | 106 | 121 |
| Platforms, Railway | Railways; 1 | 104 | 120 |
| Platinizing | Plating | 96 | 113 |
| Playing-cards | Games and Toys | 46 | 143 |
| Pliers | Wire-Working | 140 | 170 |
| Pliers, Wire-cutting | Wire-Working | 140 | 170 |
| Plotting-instruments | Drafting | 33 | 97 |
| Plow-beams | Plows | 97 | 114 |
| Plow-cleaners | Plows | 97 | 114 |
| Plow-clevises | Plows | 97 | 114 |
| Plow-colters | Plows | 97 | 114 |
| Plow-handles, Machines for bending | Wood-Working; 3 | 144 | 176 |
| Plow-handles, Machines for dressing | Wood-Working; 2 | 143 | 173 |
| Plow-handles, Machines for turning | Wood-Working; 1 | 142 | 172 |
| Plow-holders | Plows | 97 | 114 |
| Plow-irons | Plows | 97 | 114 |
| Plow-landsides | Plows | 97 | 114 |
| Plow mold-boards | Plows | 97 | 114 |
| Plow-points | Plows | 97 | 114 |
| Plow-regulators | Plows | 97 | 114 |
| Plows, Book-binders' | Book-Binding | 11 | 15 |
| Plows, Ditching | Plows | 97 | 114 |
| Plows, Double mold-board | Plows | 97 | 114 |
| Plows, Double-shovel | Plows | 97 | 114 |
| Plows, Draining | Plows | 97 | 114 |
| Plows, Drain-tile | Plows | 97 | 114 |
| Plows, Expanding | Plows | 97 | 114 |
| Plows, Gang | Plows | 97 | 114 |
| Plows, Garden | Plows | 97 | 114 |
| Plows, Hand | Plows | 97 | 114 |
| Plowshares | Plows | 97 | 114 |
| Plows, Hill-side | Plows | 97 | 114 |
| Plows, Marking | Plows | 97 | 114 |
| Plows, Mole | Plows | 97 | 114 |

| SUBJECT. | Name of class—official classification. | No. of class—official classification. | No. of class—subscription classification. |
|---|---|---|---|
| Plow-soles | Plows | 97 | 114 |
| Plows, Paring | Plows | 97 | 114 |
| Plows, Prairie | Plows | 97 | 114 |
| Plows, Rotary | Plows | 97 | 114 |
| Plows, Seeding | Plows | 97 | 114 |
| Plows, Single-shovel | Plows | 97 | 114 |
| Plows, Snow | Railways; 3 | 106 | 122 |
| Plow-standards | Plows | 97 | 114 |
| Plows, Steam | Plows | 97 | 114 |
| Plows, Subsoil | Plows | 97 | 114 |
| Plows, Sulky | Plows | 97 | 114 |
| Plows, Tile-laying | Plows | 97 | 114 |
| Plows, Wheel | Plows | 97 | 114 |
| Plow-wheel colters | Plows | 97 | 114 |
| Pluggers | Dental | 32 | 46 |
| Plugs, Machines for cutting wooden | Wood-Working; 1 | 142 | 174 |
| Plugs, Machines for turning | Wood-Working; 1 | 142 | 172 |
| Plumbers' hooks, manufacture of | Hardware Manufactures | 53J | 136 |
| Plumbers' tools | Metal-Working; 6 | 81A | 136 |
| Plumbs | Measuring-Instruments | 73 | 157 |
| Pneumatic car-brakes | Steam; 1 | 121 | 122 |
| Pneumatic dispatch-tubes | Railways; 1 | 104 | 120 |
| Pneumatic hammers | Metal-Working; 3 | 78 | 60 |
| Pneumatic railways | Railways; 1 | 104 | 120 |
| Pocket-book fastenings | Stationery | 120 | 146 |
| Pocket-books | Stationery | 120 | 146 |
| Pockets | Apparel | 2 | 38 |
| Pockets, Safety | Apparel | 2 | 38 |
| Pointing rails, pickets, matches, &c | Wood-Working; 3 | 144 | 174 |
| Pointing wire | Wire-Working | 140 | 170 |
| Pokers, Fire | Stoves and Furnaces | 126 | 154 |
| Pokes for animals | Harness | 54 | 144 |
| Polariscopes | Optics | 88 | 157 |
| Polarizing-apparatus | Optics | 88 | 157 |
| Poles, Wagon | Carriages and Wagons | 21 | 31 |
| Police batons | Games and Toys | 46 | 143 |
| Polishing-compounds | Grinding and Polishing | 51 | 70 |
| Polishing glass | Grinding and Polishing | 51 | 68 |
| Polishing-machines | Grinding and Polishing | 51 | 70 |
| Polishing-machines, (wood) | Wood-Working; 2 | 143 | 173 |
| Polishing marble | Grinding and Polishing | 51 | 153 |
| Polishing-materials | Grinding and Polishing | 51 | 70 |
| Polishing metals | Grinding and Polishing | 51 | 70 |
| Polishing slate | Grinding and Polishing | 51 | 153 |
| Polishing stone | Grinding and Polishing | 51 | 153 |
| Polishing yarns, threads, &c | Cordage | 28 | 17 |
| Pomatum-cases | Toilet | 132 | 161 |
| Pontils | Glass | 49 | 68 |
| Ponton-bridges | Boats | 9 | 13 |
| Pontons | Boats | 9 | 13 |
| Porcelain materials and manufacture | Clay | 25 | 115 |
| Porcelain, Printing | Printing | 101 | 54 |
| Portable engines | Steam; 3 | 123 | 94 |
| Portable forges | Metal-Working; 3 | 78 | 60 |
| Portable houses | Carpentry | 20 | 30 |
| Portable houses, (metallic) | Masonry | 72 | 4 |
| Portable presses | Presses | 100 | 7 |
| Portable railways | Railways; 1 | 104 | 120 |
| Portable safes | Safes | 109 | 128 |
| Porte-caustic | Surgery | 128 | 156 |

| SUBJECT. | Name of class—official classification. | No. of class—official classification. | No. of class—sub-scription classification. |
|---|---|---|---|
| Porte-crayons | Stationery | 120 | 146 |
| Portemonnaies | Stationery | 120 | 146 |
| Portfolios | Stationery | 120 | 146 |
| Portfolios, Making | Book-Binding | 11 | 146 |
| Portfolio-stands | Furniture | 45 | 62 |
| Port-hole closers | Ships; 1 | 114 | 137 |
| Port-holes, Submarine | Ships; 1 | 114 | 137 |
| Portmanteaus | Trunks | 133 | 162 |
| Postage-stamp holders | Stationery | 120 | 146 |
| Postage-stamps | Fine Arts | 41 | 54 |
| Postal cards | Stationery | 120 | 146 |
| Post-borers | Wood-Working; 3 | 144 | 52 |
| Post-drivers | Fences | 39 | 52 |
| Post-hole borers | Excavators | 37 | 52 |
| Posting bills and placards | Fine Arts | 41 | 54 |
| Post-office boxes | Stationery | 120 | 146 |
| Post-office letter-stamping apparatus | Printing | 101 | 146 |
| Post-drivers | Hydraulic Engineering | 61 | 52 |
| Post-pullers | Fences | 39 | 52 |
| Potato-digging machines | Plows | 97 | 114 |
| Potato-forks | Garden and Orchard | 47 | 141 |
| Potato-hooks | Garden and Orchard | 47 | 141 |
| Potato-mashers | Kitchen Utensils | 65 | 88 |
| Potato-peelers | Kitchen Utensils | 65 | 88 |
| Potato-planters | Garden and Orchard | 47 | 133 |
| Potato-scoops | Garden and Orchard | 47 | 133 |
| Pot-covers | Stoves and Furnaces | 126 | 154 |
| Pots, Glass | Glass | 49 | 68 |
| Pottery materials and manufacture | Clay | 25 | 115 |
| Pouches, Tobacco | Tobacco | 131 | 160 |
| Poudrette, Treatment of | Manures | 71 | 96 |
| Poultices, Compositions for | Chemical Miscellaneous | 23 | 98 |
| Poultry-coops | Stabling | 119 | 116 |
| Poultry-feeder and houses | Stabling | 119 | 116 |
| Poultry-houses | Stabling | 119 | 116 |
| Poultry-nests | Stabling | 119 | 116 |
| Pouncing hats | Felting and Hats | 38 | 76 |
| Powder-blowers | Garden and Orchard | 47 | 63 |
| Powder boxes, Cosmetic | Toilet | 132 | 161 |
| Powder-cans, Air-tight | Preserving Food | 99 | 58 |
| Powder-flask chargers | Projectiles | 102 | 55 |
| Powder-flasks | Projectiles | 102 | 55 |
| Powder-injectors | Surgery | 128 | 156 |
| Powder-kegs | Gunpowder | 52 | 58 |
| Powder-mills | Mills | 83 | 71 |
| Powder-mixers | Kitchen Utensils | 65 | 88 |
| Powders, Appliances for putting up | Surgery | 128 | 98 |
| Power-hammers | Metal-Working; 3 | 78 | 60 |
| Power-looms | Weaving | 139 | 95 |
| Precious stones, Setting | Jewelry | 63 | 86 |
| Preserving-cans | Preserving Food | 99 | 58 |
| Preserving-cars | Preserving Food | 99 | 121 |
| Preserving fermented liquors | Beer and Wine | 7 | 19 |
| Preserving food | Preserving Food | 99 | 58 |
| Preserving-houses | Preserving Food | 99 | 58 |
| Preserving-processes for food | Preserving Food | 99 | 58 |
| Preserving timber | Chemical Miscellaneous | 23 | 35 |
| Press and strainer combined | Presses | 100 | 7 |
| Presser-feet for sewing-machines | Sewing-Machines | 112 | 134 |
| Presser-flyers | Spinning | 118 | 142 |

| SUBJECT. | Name of class—official classification. | No. of class—official classification. | No. of class—sub-scription classification. |
|---|---|---|---|
| Presses, Brick | Clay | 25 | 20 |
| Presses, Canceling | Printing | 101 | 146 |
| Presses, Copper-plate | Printing | 101 | 117 |
| Presses, Copying | Printing | 101 | 146 |
| Presses, Embossing | Fine Arts | 41 | 54 |
| Presses for obtaining relief-printing surfaces | Fine Arts | 41 | 54 |
| Presses for oils, stearine, fats, &c | Presses | 100 | 26 |
| Presses, Hand-printing | Printing | 101 | 117 |
| Presses, Hydraulic | Pumps | 103 | 118 |
| Presses, Lithographic | Printing | 101 | 117 |
| Presses, Printing | Printing | 101 | 117 |
| Presses, Seal | Printing | 101 | 146 |
| Pressing garments | Apparel | 2 | 38 |
| Pressing glass | Glass | 49 | 68 |
| Pressing machines, Leather | Leather | 69 | 92 |
| Pressing metals, scrap, &c | Metal-Working; 3 | 78 | 60 |
| Pressing silk | Silk | 117 | 140 |
| Pressure-indicators, Registering devices of steam | Measuring-Instruments | 73 | 148 |
| Primers | Projectiles | 102 | 105 |
| Primers, Friction | Projectiles | 102 | 105 |
| Priming for paint | Paint | 91 | 107 |
| Printers' blankets | Printing | 101 | 117 |
| Printers' blankets, Washing | Printing | 101 | 117 |
| Printers' boxes | Printing | 101 | 117 |
| Printers' cases | Printing | 101 | 117 |
| Printers' furniture | Printing | 101 | 117 |
| Printers' lapping | Printing | 101 | 117 |
| Printers' quoins | Printing | 101 | 117 |
| Printers' rollers | Printing | 101 | 117 |
| Printers' rules | Printing | 101 | 117 |
| Printers' rules, Cutting and mitering | Sheet-Metal | 113A | 117 |
| Printing cloth | Printing | 101 | 11 |
| Printing fabrics | Printing | 101 | 11 |
| Printing floor-cloth | Printing | 101 | 56 |
| Printing-frames, Photographic | Photography | 95 | 111 |
| Printing oil-cloth | Printing | 101 | 56 |
| Printing-presses | Printing | 101 | 117 |
| Printing-processes | Printing | 101 | 117 |
| Printing surfaces | Printing | 101 | 117 |
| Printing surfaces, Presses for obtaining relief | Printing | 101 | 54 |
| Printing wall-paper | Printing | 101 | 109 |
| Printing-wheels | Printing | 101 | 117 |
| Prisms | Optics | 88 | 157 |
| Privies | Masonry | 72 | 51 |
| Privies, Cleaning | Manures | 71 | 96 |
| Probangs | Surgery | 128 | 156 |
| Probes | Surgery | 128 | 156 |
| Profiling apparatus | Optics | 88 | 157 |
| Projectiles, Casting | Casting | 22 | 105 |
| Projectiles, Compositions for coating and packing | Projectiles | 102 | 105 |
| Projectiles for fire-arms | Projectiles | 102 | 32 |
| Projectiles for ordnance | Projectiles | 102 | 105 |
| Prolapsus, Instruments for treating | Surgery | 128 | 156 |
| Propagating-boxes | Garden and Orchard | 47 | 63 |
| Propagating by cuttings | Garden and Orchard | 47 | 63 |
| Propellers | Ships; 2 | 115 | 139 |

| SUBJECT. | Name of class—official classification. | No. of class—official classification. | No. of class—subscription classification. |
|---|---|---|---|
| Propellers, Screw | Ships; 2 | 115 | 139 |
| Propelling in shoal water | Ships; 2 | 115 | 139 |
| Propulsion of vessels | Ships; 2 | 115 | 139 |
| Protecting bulwarks | Ships; 1 | 114 | 137 |
| Protecting tiller-ropes | Ships; 1 | 114 | 127 |
| Protectors, Plant and shrub | Garden and Orchard | 47 | 63 |
| Protectors, Tree | Garden and Orchard | 47 | 63 |
| Protractors | Drafting | 33 | 97 |
| Provision cans, jars, &c., Air-tight | Preserving Food | 99 | 58 |
| Provision-cars | Preserving Food | 99 | 58 |
| Provision-safes | Kitchen Utensils | 65 | 58 |
| Pruning-knives | Cutlery | 30 | 63 |
| Pruning-saws | Saws | 110 | 63 |
| Pruning-shears | Cutlery | 30 | 63 |
| Puddlers' balls, Rolling and squeezing | Metal-Working; 5 | 80 | 85 |
| Puddling | Metallurgy | 75 | 85 |
| Puddling-furnaces | Metallurgy | 75 | 85 |
| Puddling-processes | Metallurgy | 75 | 85 |
| Puffing-irons | Toilet | 132 | 161 |
| Puffs | Toilet | 132 | 161 |
| Pug-mills | Clay | 25 | 20 |
| Pulley-blocks | Hoisting | 57 | 79 |
| Pulleys, Belt | Journals and Bearings | 64 | 67 |
| Pulleys, Calisthenic | Games and Toys | 46 | 143 |
| Pulleys, Fast and loose | Journals and Bearings | 64 | 67 |
| Pulleys, Sash | Builders' Hardware | 16 | 24 |
| Pulls, (bell,) Manufacture of | Hardware Manufactures | 53B | 24 |
| Pulmometers | Mechanical Powers | 73 | 67 |
| Pulping-machines | Paper-Making | 92 | 108 |
| Pulp, Reducing materials to | Paper-Making | 92 | 108 |
| Pump-brakes | Pumps | 103 | 118 |
| Pump-motions | Pumps | 103 | 118 |
| Pump-piston rods | Pumps | 103 | 118 |
| Pump-pistons | Pumps | 103 | 118 |
| Pumps | Pumps | 103 | 118 |
| Pumps, Air | Pumps | 98 | 12 |
| Pumps, Cattle | Pumps | 103 | 118 |
| Pumps, Centrifugal | Pumps | 103 | 118 |
| Pumps, Chain | Pumps | 103 | 118 |
| Pumps, Circular brake | Pumps | 103 | 118 |
| Pumps, Double-acting | Pumps | 103 | 118 |
| Pumps, Feed | Pumps | 103 | 118 |
| Pumps, Force | Pumps | 103 | 118 |
| Pumps, Hydraulic-press | Pumps | 103 | 118 |
| Pumps, Lift | Pumps | 103 | 118 |
| Pumps, Oil | Pumps | 103 | 118 |
| Pumps, Oscillating | Pumps | 103 | 118 |
| Pumps, Packing for | Pumps | 103 | 151 |
| Pumps, Rotary | Pumps | 103 | 118 |
| Pumps, Ships' | Pumps | 103 | 118 |
| Pumps, (steam,) Engines of | Steam; 1 | 121 | 118 |
| Pumps, Stomach | Surgery | 128 | 156 |
| Pumps, Submerged | Pumps | 103 | 118 |
| Pumps, Water | Pumps | 103 | 118 |
| Pumps, Wind-propelled | Pumps | 103 | 118 |
| Pump-valves | Valves | 136 | 118 |
| Punches, Leather, paper, and ticket | Clasps and Buckles | 24 | 72 |
| Punching-machines for bar or plate | Metal-Working; 4 | 79 | 119 |
| Punching machines, Leather | Leather | 69 | 72 |
| Punching sheet-metal | Sheet-Metal | 113A | 136 |

| SUBJECT. | Name of class—official classification. | No. of class—official classification. | No. of class—subscription classification. |
|---|---|---|---|
| Puppet-valves | Valves | 136 | 152 |
| Purifiers, Gas | Gas | 48 | 66 |
| Puttying-machines for sash | Glass | 49 | 68 |
| Puttying-machines for vessel-seams | Ships; 1 | 114 | 137 |
| Puzzles | Games and Toys | 46 | 143 |
| Pyrometers | Measuring-Instruments | 73 | 101 |
| Pyrotechnic compounds | Gunpowder | 52 | 71 |
| Pyrotechnics | Gunpowder | 52 | 71 |
| Pyroxyline | Chemical Miscellaneous | 23 | 71 |
| Q. | | | |
| Quadrants | Optics | 88 | 157 |
| Quarrying | Stone, Lime, and Cement | 153 | 124 |
| Quarrying-machines | Stone, Lime, and Cement | 153 | 124 |
| Quartz-mills | Ore | 90 | 3 |
| Quàys | Hydraulic Engineering | 61 | 82 |
| Quilling yarn | Cordage | 28 | 17 |
| Quill-pen makers | Stationery | 120 | 146 |
| Quills, Yarn and silk | Cordage | 28 | 17 |
| Quill thread fastener | Cordage | 28 | 17 |
| Quilting-frames | Apparel | 2 | 38 |
| Quoins for ordnance | Ordnance | 89 | 105 |
| Quoins, Printers' | Printing | 101 | 117 |
| R. | | | |
| Rabbeting-machines | Wood-Working; 2 | 143 | 173 |
| Racking wines, &c | Beer and Wine | 7 | 19 |
| Racks, Boot, hat, and brush | Furniture | 45 | 62 |
| Racks, Hay-feeding | Stabling | 119 | 144 |
| Racks, Hay-wagon | Carriages and Wagons | 21 | 31 |
| Racks, Sheep | Stabling | 119 | 135 |
| Radiators | Stoves and Furnaces | 126 | 154 |
| Rafter-hooks | Hoisting | 57 | 79 |
| Rafting-dogs | Boats | 9 | 13 |
| Rafting logs | Boats | 9 | 13 |
| Rafts | Boats | 9 | 13 |
| Rag carpet, Cutting cloth for | Cloth | 26 | 95 |
| Rag-engines | Paper-Making | 92 | 108 |
| Rail-jacks | Hoisting | 57 | 79 |
| Railroad earth-works | Excavators | 37 | 51 |
| Railroad excavating | Excavators | 37 | 51 |
| Railroad-rail pieces, Piling | Metal-Working; 5 | 80 | 125 |
| Railroad-rails, Capping | Metal-Working; 5 | 80 | 125 |
| Railroad-rails, Repairing | Metal-Working; 5 | 80 | 125 |
| Railroad-rails, Re-rolling | Metal-Working; 5 | 80 | 125 |
| Railroad-rails, Rolling | Metal-Working; 5 | 80 | 125 |
| Railroad-signals | Signals | 116 | 120 |
| Railroad-spikes | Nails | 85 | 120 |
| Rails, Bending or straightening | Metal-Working; 1 | 76 | 125 |
| Rails, Railway | Railways; 1 | 104 | 120 |
| Railway bumping-posts | Railways; 1 | 104 | 120 |
| Railway-car heaters | Stoves and Furnaces | 126 | 154 |
| Railway-car irons, Making | Railways; 4 | 107 | 121 |
| Railway-car spittoons | Furniture | 45 | 121 |
| Railway cars, Refrigerating | Ice | 62 | 121 |
| Railway cattle-guards | Railways; 1 | 104 | 120 |
| Railway-chairs | Railways; 1 | 104 | 120 |
| Railway-chairs, Making | Railways; 4 | 107 | 120 |

| SUBJECT. | Name of class—official classification. | No. of class—official classification. | No. of class—subscription classification. |
|---|---|---|---|
| Railway-crossings | Railways; 1 | 104 | 120 |
| Railway-crossing, Making | Railways; 4 | 107 | 120 |
| Railway fish-bars | Railways; 1 | 104 | 120 |
| Railway-frogs | Railways; 1 | 104 | 120 |
| Railway-frogs, Making | Railways; 4 | 107 | 120 |
| Railway-gates | Railways; 1 | 104 | 120 |
| Railway-joints | Railways; 1 | 104 | 120 |
| Railway metallic mountings, Making | Railways; 4 | 107 | 121 |
| Railway-rails, Making | Railways; 4 | 107 | 120 |
| Railway-rails | Railways; 1 | 104 | 120 |
| Railways | Railways; 1 | 104 | 120 |
| Railway safety-guards | Railways; 1 | 104 | 120 |
| Railways, Elevated | Railways; 1 | 104 | 120 |
| Railway-signals, Electrical | Electricity | 36 | 120 |
| Railway-sleepers | Railways; 1 | 104 | 120 |
| Railways, Marine | Hydraulic Engineering | 61 | 82 |
| Railways, Pneumatic | Railways; 1 | 104 | 120 |
| Railway-splice pieces | Railways; 1 | 104 | 120 |
| Railway-splice pieces, Making | Railways; 4 | 107 | 120 |
| Railways, Portable | Railways; 1 | 104 | 120 |
| Railways, Street | Railways; 1 | 104 | 120 |
| Railways, Track-laying | Railways; 1 | 104 | 120 |
| Railway-switches | Railways; 1 | 104 | 120 |
| Railway-switches, Electrical | Electricity | 36 | 120 |
| Railway-telegraphs, Electrical | Electricity | 36 | 120 |
| Railway-ticket holders | Stationery | 120 | 121 |
| Railway-ties | Railways; 1 | 104 | 120 |
| Railway-ties, Machines for dressing | Wood-Working; 2 | 143 | 173 |
| Railway-track layers | Railways; 1 | 104 | 120 |
| Railway-track platforms | Railways; 1 | 104 | 120 |
| Railway turn-outs | Railways; 1 | 104 | 120 |
| Railway turn-tables | Railways; 1 | 104 | 120 |
| Railway-wheels, Making | Railways; 4 | 107 | 122 |
| Rain-water cut-offs | Water-Distribution | 137 | 126 |
| Raising sunken vessels | Ships; 1 | 114 | 82 |
| Raisin-seeders | Kitchen Utensils | 65 | 88 |
| Rakes for harvesters | Harvesters | 56 | 74 |
| Rakes, Garden | Garden and Orchard | 47 | 141 |
| Rakes, Hand hay | Garden and Orchard | 47 | 141 |
| Rakes, Horse hay | Harvesters | 56 | 74 |
| Rakes, Oyster | Fishing | 43 | 143 |
| Rake-teeth, Machines for making | Wood-Working; 1 | 142 | 175 |
| Rammers for ordnance | Ordnance | 89 | 105 |
| Rammers for small-arms | Fire-Arms | 42 | 55 |
| Rammers, Paviors' | Paving | 94 | 110 |
| Ramrods | Fire-Arms | 42 | 55 |
| Rams, Naval | Ships; 1 | 114 | 137 |
| Rams, Water | Pumps | 103 | 118 |
| Rand-machines | Boots and Shoes | 12 | 16 |
| Rands | Boots and Shoes | 12 | 16 |
| Ranges | Stoves and Furnaces | 126 | 154 |
| Rasps | Files | 40 | 53 |
| Rasps, Machines for making | Files | 40 | 53 |
| Rasps, Wood | Wood-Working; 4 | 145 | 53 |
| Ratchet-drills | Metal-Working; 2 | 77 | 100 |
| Ratchet-wrenches | Metal-Working; 6 | 81A | 100 |
| Ration-packages for animals | Stabling | 119 | 59 |
| Rats | Toilet | 132 | 161 |
| Rattan, Machines for cutting | Wood-Working; 3 | 144 | 169 |
| Rattan, Machines for dressing | Wood-Working; 3 | 144 | 169 |

| SUBJECT. | Name of class—official classification. | No. of class—official classification. | No. of class—subscription classification. |
|---|---|---|---|
| Rattan, Machines for polishing | Wood-Working; 3 | 144 | 169 |
| Rattan, Machines for reducing | Wood-Working; 3 | 144 | 169 |
| Rattan, Machines for slitting | Wood-Working; 3 | 144 | 169 |
| Rattan, Machines for splitting | Wood-Working; 3 | 144 | 169 |
| Rattan, Machines for stripping | Wood-Working; 3 | 144 | 169 |
| Razors | Cutlery | 30 | 44 |
| Razor-strops | Grinding and Polishing | 51 | 70 |
| Razor-strops, Compositions for | Grinding and Polishing | 51 | 70 |
| Reading-cards | Games and Toys | 46 | 143 |
| Reading-desks | Furniture | 45 | 62 |
| Reamers for metal | Metal-Working; 2 | 77 | 100 |
| Reaming-bits, (Wood) | Wood-Working; 4 | 145 | 29 |
| Reaper-knife sharpeners | Grinding and Polishing | 51 | 70 |
| Reaping-hooks | Harvesters | 56 | 74 |
| Reaping-machines | Harvesters | 56 | 74 |
| Rebaling-presses | Presses | 100 | 7 |
| Reciprocating steam-engines | Steam; 1 | 121 | 150 |
| Reckoning-instruments | Measuring-Instruments | 73 | 42 |
| Record-books | Stationery | 120 | 146 |
| Recording-instruments | Measuring-Instruments | 73 | 42 |
| Rectification of spirits | Stills | 124 | 47 |
| Rectifying liquors | Stills | 124 | 47 |
| Reducing furnaces, Metal | Metallurgy | 75 | 99 |
| Reducing processes, Metal | Metallurgy | 75 | 99 |
| Reed instruments | Music | 84 | 106 |
| Reed-setter | Weaving | 139 | 95 |
| Reeds for looms | Weaving | 139 | 95 |
| Reeds for looms, Machines for making | Weaving | 139 | 95 |
| Reefing-apparatus | Ships; 1 | 114 | 138 |
| Reefing sails | Ships; 1 | 114 | 138 |
| Reeling silk | Silk | 117 | 140 |
| Reels, Fishing-line | Fishing | 43 | 143 |
| Reels for clothes-lines | Laundry | 68 | 91 |
| Reels for harvesters | Harvesters | 56 | 74 |
| Reels for spinning-machines | Spinning | 118 | 142 |
| Reels for winding yarns and thread | Cordage | 28 | 17 |
| Reels, Telegraph | Electricity | 36 | 50 |
| Refining-furnaces | Metallurgy | 75 | 99 |
| Refining liquors | Stills | 124 | 47 |
| Refining oils | Stills | 124 | 47 |
| Reflecting-circles | Optics | 88 | 157 |
| Reflectors for lamps | Lamps and Gas-Fittings | 67 | 90 |
| Reflectors for windows | Lamps and Gas-Fittings | 67 | 90 |
| Refrigerating-cars | Ice | 62 | 58 |
| Refrigerating-cellars | Ice | 62 | 58 |
| Refrigerating processes | Ice | 62 | 58 |
| Refrigerators | Ice | 62 | 58 |
| Register-books | Stationery | 120 | 146 |
| Register books, Hotel | Stationery | 120 | 146 |
| Registering-devices for meters | Measuring-Instruments | 73 | 66 |
| Registering-devices of air, steam, and water gages | Measuring-Instruments | 73 | 167 |
| Registering devices of fluid-meters | Measuring-Instruments | 73 | 167 |
| Register-locks | Locks and Latches | 70 | 93 |
| Register-points for printing-presses | Printing | 101 | 117 |
| Registers | Measuring-Instruments | 73 | 42 |
| Registers, Billiard | Measuring-Instruments | 73 | 143 |
| Registers, Counting | Measuring-Instruments | 73 | 42 |
| Registers, Grain | Measuring-Instruments | 73 | 132 |
| Registers, Hot-air | Stoves and Furnaces | 126 | 154 |

| SUBJECT. | Name of class—official classification. | No. of class—official classification. | No. of class—sub-scription classification. |
|---|---|---|---|
| Registers, Measuring | Measuring-Instruments | 73 | 132 |
| Registers, Passenger | Measuring-Instruments | 73 | 42 |
| Registers, Printing-press | Measuring-Instruments | 73 | 117 |
| Registers, Steam-engine | Measuring-Instruments | 73 | 150 |
| Registers, Street-car | Measuring-Instruments | 73 | 42 |
| Registers, Telegraph | Electricity | 36 | 50 |
| Registers, Time | Horology | 58 | 36 |
| Registers, Weighing | Measuring-Instruments | 73 | 132 |
| Regulators, Draft | Stoves and Furnaces | 126 | 154 |
| Regulators, Exhaust | Steam; 3 | 123 | 150 |
| Regulators for millstones | Mills | 83 | 57 |
| Regulators, Heat | Stoves and Furnaces | 126 | 154 |
| Regulators, Steam | Steam; 2 | 122 | 148 |
| Rein-holders | Carriages and Wagons | 21 | 31 |
| Rein-pulls | Harness | 54 | 72 |
| Reins, Safety | Harness | 54 | 72 |
| Reins, Slides for | Harness | 54 | 72 |
| Relays | Electricity | 36 | 50 |
| Relief valves | Valves | 136 | 152 |
| Rendering lard and tallow | Oil, Fat, and Glue | 87 | 26 |
| Rendering-tanks | Stoves and Furnaces | 126 | 154 |
| Repeaters | Electricity | 36 | 50 |
| Repeating fire-arms | Fire-Arms | 42 | 55 |
| Reservoirs, Canal | Hydraulic Engineering | 61 | 82 |
| Resistance-boxes | Electricity | 36 | 50 |
| Resistance-coils | Electricity | 36 | 50 |
| Resist printing | Bleaching and Dyeing | 8 | 11 |
| Respirators | Surgery | 128 | 156 |
| Respiring-apparatus | Surgery | 128 | 156 |
| Rests for metal-lathes | Metal-Working; 7 | 82 | 100 |
| Rests for wood-lathes | Wood-Working; 1 | 142 | 172 |
| Retorts, (*except Coal, Class* 48) | Chemical Miscellaneous | 23 | 35 |
| Retorts for gas-making | Gas | 48 | 66 |
| Retorts for glass | Glass | 49 | 68 |
| Retort-stands | Chemical Miscellaneous | 23 | 35 |
| Revenue guards | Stills | 124 | 47 |
| Revenue-stamp cancelers | Printing | 101 | 146 |
| Revenue-stamps | Fine Arts | 41 | 146 |
| Reversing-gear for steam-engines | Steam; 1 | 121 | 67 |
| Reversing motion, Apparatus for | Mechanical Powers | 74 | 67 |
| Revolving-breech cannon | Ordnance | 89 | 105 |
| Revolving cars | Railways; 2 | 105 | 121 |
| Revolving fire-arms | Fire-Arms | 42 | 123 |
| Rheometers | Electricity | 36 | 50 |
| Rheotomes | Electricity | 36 | 50 |
| Rheostats | Electricity | 36 | 50 |
| Rheotropes | Electricity | 36 | 50 |
| Ribbons for hand-stamps, Preparing | Stationery | 120 | 117 |
| Ribs, Iron | Ships; 1 | 114 | 137 |
| Rice-cleaning machines | Mills | 83 | 57 |
| Rice-hullers | Mills | 83 | 57 |
| Rice-polishers | Mills | 83 | 57 |
| Riding-bitts | Ships; 1 | 114 | 138 |
| Rifled ordnance | Ordnance | 89 | 105 |
| Rifles | Fire-Arms | 42 | 55 |
| Rifles for whetting | Grinding and Polishing | 51 | 70 |
| Rifling-machines | Metal-Working; 7 | 82 | 100 |
| Rifling small-arms | Fire-Arms | 42 | 55 |
| Rigging | Ships; 1 | 114 | 138 |
| Rigging-clasps | Ships; 1 | 114 | 138 |

| SUBJECT. | Name of class—official classification. | No. of class—official classification. | No. of class—subscription classification. |
|---|---|---|---|
| Rigging, Running | Ships; 1 | 114 | 138 |
| Rigging, Setting up | Ships; 1 | 114 | 138 |
| Rigging, Standing | Ships; 1 | 114 | 138 |
| Rigging-stoppers | Ships; 1 | 114 | 138 |
| Rings | Clasps and Buckles | 24 | 23 |
| Rings for spinning | Spinning | 118 | 142 |
| Rings, Machines for cutting wooden | Wood-Working; 1 | 142 | 175 |
| Rings, Tools for cutting leather | Leather | 69 | 72 |
| Ring-travelers | Spinning | 118 | 142 |
| Ring-travelers, Manufacture of | Hardware Manufactures | 53H | 142 |
| Rinks | Carpentry | 20 | 30 |
| Ripping-tool | Apparel | 2 | 38 |
| Ripples for flax, broom-corn, &c | Brakes and Gins | 13 | 78 |
| River-bars, Removing | Hydraulic Engineering | 61 | 82 |
| Rivers, Improving channels of | Hydraulic Engineering | 61 | 82 |
| Rivers, Protection against inundations of | Hydraulic Engineering | 61 | 82 |
| Riveting | Metal-Working; 3 | 78 | 100 |
| Riveting-machines | Metal-Working; 3 | 78 | 100 |
| Rivet-making | Bolts, Nuts, and Rivets | 10 | 14 |
| Rivets | Bolts, Nuts, and Rivets | 10 | 14 |
| Rivets, (harness,) Manufacture of | Hardware Manufacture | 53F | 14 |
| Riving splints, Tools for | Wood-Working; 2 | 143 | 174 |
| Road-making | Excavators | 37 | 51 |
| Road-paving | Paving | 94 | 110 |
| Road-rollers | Paving | 94 | 110 |
| Road-scrapers | Excavators | 37 | 51 |
| Roasters | Stoves and Furnaces | 126 | 154 |
| Roasting ores | Metallurgy | 75 | 99 |
| Rock-blasting | Projectiles | 102 | 124 |
| Rock-drills | Stone, Lime, and Cement | 125 | 124 |
| Rock-drills, Air-engines for | Pneumatics | 98 | 2 |
| Rock-drills, Steam-engines for | Steam; 1 | 121 | 150 |
| Rocket-harpoons | Projectiles | 102 | 143 |
| Rockets | Gunpowder | 52 | 71 |
| Rockets, War | Projectiles | 102 | 71 |
| Rocking-horses | Games and Toys | 46 | 143 |
| Rod-cutters | Metal-Working; 6 | 81A | 100 |
| Rods, Bending or straightening | Metal-Working; 1 | 76 | 119 |
| Rods for gathering glass | Glass | 49 | 68 |
| Roll-blotters | Stationery | 120 | 146 |
| Roller and harrow, Combined | Harrows | 55 | 73 |
| Roller-heaters | Stoves and Furnaces | 126 | 154 |
| Rollers, Barn-door | Builders' Hardware | 16 | 24 |
| Rollers, Covering | Spinning | 118 | 142 |
| Rollers, Flanged | Harrows | 55 | 73 |
| Rollers for delivering silk | Silk | 117 | 140 |
| Roller-sheaves for doors | Builders' Hardware | 16 | 24 |
| Rollers, Land | Harrows | 55 | 73 |
| Rollers, Printers' | Printing | 101 | 117 |
| Rollers, Sash | Builders' Hardware | 16 | 24 |
| Rollers, Sliding-door | Builders' Hardware | 16 | 24 |
| Rolling glass | Glass | 49 | 68 |
| Rolling machines, Leather | Leather | 69 | 92 |
| Rolling metals | Metal-Working; 5 | 80 | 100 |
| Rolling pins | Kitchen Utensils | 65 | 88 |
| Rolling wire | Tubing and Wire | 134 | 163 |
| Roofing-compositions | Roofing | 108 | 126 |
| Roofing-paper | Paper-Making | 92 | 126 |
| Roofing-processes | Roofing | 108 | 126 |

| SUBJECT. | Name of class—official classification. | No. of class—official classification. | No. of class—subscription classification. |
|---|---|---|---|
| Roofing-tiles | Roofing | 108 | 20 |
| Roofing-tiles, Manufacture of | Clay | 25 | 20 |
| Roofs, Compositions and means for preserving | Roofing | 108 | 126 |
| Roof-seaming machines | Sheet-Metal | 113A | 126 |
| Roofs, (*except Trusses, Class* 14) | Roofing | 108 | 126 |
| Roofs | Roofing | 108 | 126 |
| Roosts for fowls | Stabling | 119 | 116 |
| Root-cutters | Stabling | 119 | 59 |
| Root-diggers | Plows | 97 | 114 |
| Root-peelers | Kitchen Utensils | 65 | 88 |
| Root-washers | Stabling | 119 | 59 |
| Rope-clutches | Mechanical Powers | 74 | 138 |
| Rope coiling | Cordage | 28 | 127 |
| Rope-making machines | Cordage | 28 | 127 |
| Rope-ways | Railways; 1 | 104 | 120 |
| Rope, Wire | Wire-Working | 140 | 170 |
| Roping bales | Presses | 100 | 7 |
| Rose-engines | Metal-Working; 7 | 82 | 100 |
| Rose-heads | Water-Distribution | 137 | 167 |
| Roses for door-knobs | Locks and Latches | 70 | 93 |
| Rosettes for horses | Harness | 54 | 23 |
| Rossing bark | Wood-Working; 3 | 144 | 130 |
| Rossing dye-stuffs | Wood-Working; 3 | 144 | 176 |
| Rossing-machines | Wood-Working; 3 | 144 | 130 |
| Rotary blowers | Pneumatics | 98 | 12 |
| Rotary brushes | Brushes and Brooms | 15 | 22 |
| Rotary harrows | Harrows | 55 | 73 |
| Rotary spaders | Plows | 97 | 114 |
| Rotary steam-engines | Steam; 1 | 121 | 150 |
| Rotary valves | Valves | 136 | 152 |
| Rotting flax and hemp | Brakes and Gins | 13 | 78 |
| Roughening glass | Grinding and Polishing | 51 | 68 |
| Roughening pavements | Paving | 94 | 110 |
| Roulettes | Drafting | 33 | 97 |
| Rounding machines, Rein | Harness | 54 | 72 |
| Roving-frames | Spinning | 118 | 142 |
| Row-locks | Boats | 9 | 13 |
| Rubber boots | Boots and Shoes | 12 | 16 |
| Rubber erasers | Stationery | 120 | 146 |
| Rubber, Hard | Caoutchouc | 18 | 84 |
| Rubber hose | Hose and Belting | 60 | 81 |
| Rubber, India | Caoutchouc | 18 | 84 |
| Rubber rolls for wringers | Laundry | 68 | 91 |
| Rubber shoes | Boots and Shoes | 12 | 16 |
| Rubber soles | Boots and Shoes | 12 | 16 |
| Rubber type | Printing | 101 | 117 |
| Rubber, Vulcanizing | Caoutchouc | 18 | 84 |
| Rubber, White | Caoutchouc | 18 | 84 |
| Rubbing-apparatus for medical treatment | Surgery | 128 | 156 |
| Rudders | Ships; 1 | 114 | 137 |
| Rudders, Attachments for | Ships; 1 | 114 | 137 |
| Rudders, Hanging | Ships; 1 | 114 | 137 |
| Rudders, Protecting | Ships; 1 | 114 | 137 |
| Rufflers for sewing-machines | Sewing-Machines | 112 | 134 |
| Ruffles | Apparel | 2 | 38 |
| Ruffles, Machines for making | Apparel | 2 | 38 |
| Rule-cutters for printers | Sheet-Metal | 113A | 117 |
| Rulers, Desk | Stationery | 120 | 146 |

| SUBJECT. | Name of class—official classification. | No. of class—official classification. | No. of class—subscription classification. |
|---|---|---|---|
| Rulers, Parallel | Drafting | 33 | 97 |
| Rules, Measuring | Measuring-Instruments | 73 | 97 |
| Rules, Printing | Printing | 101 | 117 |
| Ruling-apparatus, Electrical | Electricity | 36 | 50 |
| Ruling-machines | Book-Binding | 11 | 15 |
| Ruling-machines, Pens for | Book-Binding | 11 | 15 |
| Ruling-pens | Drafting | 33 | 97 |
| Running-gear of vehicles | Carriages | 21 | 31 |
| "Russian" sheet-iron, Imitations of | Metal-Working; 5 | 80 | 136 |
| S. | | | |
| Saber-bayonets | Fire-Arms | 42 | 55 |
| Sabers, Manufacture of | Hardware Manufactures | 53A | 44 |
| Sabots for projectiles | Projectiles | 102 | 105 |
| Saccharometers | Measuring-Instruments | 73 | 155 |
| Sack-fasteners | Mills | 83 | 57 |
| Saddle-bags | Trunks | 133 | 162 |
| Saddle-blankets | Harness | 54 | 72 |
| Saddle-cloths | Harness | 54 | 72 |
| Saddle-girths | Harness | 54 | 72 |
| Saddle-pad machines | Harness | 54 | 72 |
| Saddles | Harness | 54 | 72 |
| Saddles, Cart | Harness | 54 | 72 |
| Saddles for suspension-bridges | Bridges | 14 | 21 |
| Saddles, Harness | Harness | 54 | 72 |
| Saddles, Pack | Harness | 54 | 72 |
| Saddles, Pad | Harness | 54 | 72 |
| Saddles, Riding | Harness | 54 | 72 |
| Saddles, Side | Harness | 54 | 72 |
| Saddle-trees, Manufacture of | Hardware Manufacture | 53F | 72 |
| Saddle-trees | Harness | 54 | 72 |
| Saddle trees, Side | Harness | 54 | 72 |
| Sad-iron heaters | Stoves and Furnaces | 126 | 91 |
| Sad-iron holders | Laundry | 68 | 91 |
| Sad-irons | Laundry | 68 | 91 |
| Safe-locks | Locks and Latches | 70 | 93 |
| Safes | Safes | 109 | 128 |
| Safes, Lining for | Safes | 109 | 128 |
| Safes, Match | Lamps and Gas-Fittings | 67 | 90 |
| Safes, Meat | Kitchen Utensils | 65 | 58 |
| Safes, Wire-work for kitchen | Wire-Working | 140 | 88 |
| Safety-cars | Railways; 2 | 105 | 121 |
| Safety-fuses | Projectiles | 102 | 105 |
| Safety-guards for car-platforms | Railways; 3 | 106 | 121 |
| Safety-guards for gun-hammers | Fire-Arms | 42 | 55 |
| Safety-guards for hatchways | Carpentry | 20 | 29 |
| Safety-guards for railway-tracks | Railways; 1 | 104 | 120 |
| Safety-guards for watches and pocket-books | Apparel | 2 | 38 |
| Safety-guards for windows | Carpentry | 20 | 29 |
| Safety-guards in paper | Paper-Making | 92 | 108 |
| Safety-guards on cars | Railways; 3 | 106 | 121 |
| Safety-guards to locks | Locks and Latches | 70 | 93 |
| Safety-plugs for steam-boilers | Steam; 2 | 122 | 148 |
| Safety-straps of vehicles | Carriages and Wagons | 21 | 31 |
| Safety-tubes for steam-boilers | Steam; 2 | 122 | 148 |
| Safety-valves | Valves | 136 | 152 |
| Sail-clutches | Ships; 1 | 114 | 138 |
| Sail-hoops | Ships; 1 | 114 | 138 |

| SUBJECT. | Name of class—official classification. | No. of class—official classification. | No. of class—sub-scription classification. |
|---|---|---|---|
| Sailing-carriages | Carriages and Wagons | 21 | 31 |
| Sails | Ships; 1 | 114 | 138 |
| Sails, Furling | Ships; 1 | 114 | 138 |
| Sails, Making | Ships; 1 | 114 | 138 |
| Sails, Reefing | Ships; 1 | 114 | 138 |
| Salinometers | Measuring-Instruments | 73 | 129 |
| Saliva-removers | Dental | 32 | 46 |
| Salt-furnaces | Sugar | 127 | 129 |
| Salt-packers | Presses | 100 | 129 |
| Salves | Chemical Miscellaneous | 23 | 98 |
| Sample-frames | Furniture | 45 | 62 |
| Sampling-tubes | Water-Distribution | 137 | 47 |
| Sandals | Boots and Shoes | 12 | 16 |
| Sand-blowers | Pneumatics | 98 | 12 |
| Sand-cleaning machines | Stone, Lime, and Cement | 125 | 153 |
| Sand-paper | Grinding and Polishing | 51 | 70 |
| Sand-paper holders | Wood-Working; 2 | 143 | 172 |
| Sand-papering machines | Wood-Working; 2 | 143 | 70 |
| Sand-papering machines | Wood-Working; 3 | 144 | 70 |
| Sand-pumps | Excavators | 37 | 51 |
| Sand-screens | Wire-Working | 140 | 153 |
| Sand-washer | Ore | 90 | 153 |
| Sap-bucket hooks | Garden and Orchard | 47 | 63 |
| Sap-conductors | Garden and Orchard | 47 | 63 |
| Sap-spiles | Garden and Orchard | 47 | 63 |
| Sap-troughs | Garden and Orchard | 47 | 63 |
| Sarcophagi, Clay | Clay | 25 | 153 |
| Sash | Carpentry | 20 | 30 |
| Sash-bolts | Builders' Hardware | 16 | 24 |
| Sash-centers | Builders' Hardware | 16 | 24 |
| Sashes, Fasteners for meeting rails of | Builders' Hardware | 16 | 24 |
| Sash-fasteners | Builders' Hardware | 16 | 24 |
| Sash-frames | Carpentry | 20 | 30 |
| Sash-hangings | Builders' Hardware | 16 | 24 |
| Sash-holders, Manufacture of | Hardware Manufactures | 53B | 24 |
| Sash-locks | Builders' Hardware | 16 | 24 |
| Sash, Machines for making | Wood-Working; 3 | 144 | 173 |
| Sash, Machines for mortising | Wood-Working; 3 | 144 | 173 |
| Sash, Machines for planing | Wood-Working; 3 | 144 | 173 |
| Sash-pulleys | Builders' Hardware | 16 | 24 |
| Sash-pulleys, Manufacture of | Hardware Manufactures | 53B | 24 |
| Sash-rollers | Builders' Hardware | 16 | 24 |
| Sash-springs | Builders' Hardware | 16 | 24 |
| Sash-stops | Builders' Hardware | 16 | 24 |
| Sash-weights | Builders' Hardware | 16 | 24 |
| Sash-weights, Manufacture of | Hardware Manufactures | 53B | 24 |
| Satchel-fastenings | Trunks | 133 | 162 |
| Satchels | Trunks | 133 | 162 |
| Sauce-pans | Kitchen Utensils | 65 | 136 |
| Sausage-cutters | Butchering | 17 | 25 |
| Sausage-stuffers | Butchering | 17 | 25 |
| Saw-bore | Wood-Working; 4 | 145 | 131 |
| Saw-bucks | Wood-Working; 4 | 145 | 131 |
| Saw-dust carriers | Wood-Working; 4 | 110 | 130 |
| Saw-filing | Grinding and Polishing | 51 | 70 |
| Saw-filing machines | Grinding and Polishing | 51 | 131 |
| Saw-frames | Grinding and Polishing | 110 | 131 |
| Saw-gages | Grinding and Polishing | 110 | 131 |
| Saw-gates | Grinding and Polishing | 110 | 131 |
| Saw-grinding | Grinding and Polishing | 51 | 70 |

| SUBJECT. | Name of class—official classification. | No. of class—official classification. | No. of class—subscription classification. |
|---|---|---|---|
| Saw-grinding machines | Grinding and Polishing | 51 | 131 |
| Saw-gummers | Metal-Working; 4 | 79 | 131 |
| Saw-horses | Saws | 110 | 131 |
| Saw-jacks | Wood-Working; 4 | 145 | 131 |
| Sawing-machines | Wood-Working; 4 | 145 | 131 |
| Sawing-machines for chair-backs | Wood-Working; 3 | 144 | 130 |
| Sawing-machines for clapboards | Wood-Working; 3 | 144 | 130 |
| Sawing-machines for combs | Wood-Working; 3 | 144 | 161 |
| Sawing-machines for hoops | Wood-Working; 3 | 110 | 40 |
| Sawing-machines for laths | Wood-Working; 3 | 144 | 174 |
| Sawing-machines for match-splints | Wood-Working; 3 | 144 | 174 |
| Sawing-machines for oars | Saws | 110 | 13 |
| Sawing-machines for shingles | Saws | 110 | 174 |
| Sawing-machines for splints | Wood-Working; 2 | 143 | 174 |
| Sawing-machines for staves | Wood-Working; 3 | 144 | 40 |
| Sawing-machines for veneers | Wood-Working; 3 | 144 | 176 |
| Sawing stone | Stone, Lime, and Cement | 125 | 153 |
| Saw-jacks | Saws | 110 | 131 |
| Saw-making | Metal-Working; 4 | 79 | 131 |
| Saw-mills | Saws | 110 | 130 |
| Saw mills, Circular | Saws | 110 | 130 |
| Saw-mills, Dogs for | Saws | 110 | 130 |
| Saw-mills, Feed for | Saws | 110 | 130 |
| Saw-mills, Gates for | Saws | 110 | 130 |
| Saw-mills, Head-blocks for | Saws | 110 | 130 |
| Saws | Saws | 110 | 131 |
| Saws, Amputation | Saws | 110 | 156 |
| Saws, Annular | Saws | 110 | 40 |
| Saws, Band | Saws | 110 | 130 |
| Saws, Bow | Saws | 110 | 131 |
| Saws, Circular | Saws | 110 | 130 |
| Saws, Cross-cut | Saws | 110 | 131 |
| Saws, Cylindrical | Saws | 110 | 40 |
| Saws, Diamond | Stone, Lime, and Cement | 125 | 153 |
| Saws, Drag | Saws | 110 | 130 |
| Saw-sets | Metal-Working; 4 | 79 | 131 |
| Saws, Gate | Saws | 110 | 130 |
| Saws, Hand | Saws | 110 | 131 |
| Saws, Hanging | Saws | 110 | 130 |
| Saws, Ice | Saws | 110 | 131 |
| Saws, Key-hole | Saws | 110 | 131 |
| Saws, Meat | Saws | 110 | 131 |
| Saws, Metal | Saws | 110 | 131 |
| Saws, Pruning | Saws | 110 | 131 |
| Saws, Scroll | Saws | 110 | 130 |
| Saws, Stone | Stone, Lime, and Cement | 125 | 153 |
| Saws, Straining | Saws | 110 | 131 |
| Saws, Tempering | Metallurgy | 75 | 131 |
| Saws, Tenon | Saws | 110 | 131 |
| Saw-swages | Metal-Working; 4 | 79 | 131 |
| Saws, Web | Saws | 110 | 131 |
| Saws, Whip | Saws | 110 | 131 |
| Saw-teeth | Saws | 110 | 131 |
| Scaffolding | Carpentry | 20 | 29 |
| Scagliola | Stone, Lime, and Cement | 125 | 153 |
| Scalding-tubs and appliances | Butchering | 17 | 25 |
| Scale-board, Working | Wood-Working; 1 | 142 | 173 |
| Scales, Drafting | Drafting | 33 | 97 |
| Scales, Sliding | Measuring-Instruments | 73 | 97 |

| SUBJECT. | Name of class—official classification. | No. of class—official classification. | No. of class—sub-scription classification. |
|---|---|---|---|
| Scales, Tailors' | Measuring-Instruments | 73 | 38 |
| Scales, Weighing | Measuring-Instruments | 73 | 132 |
| Scaling boilers, Tools for | Steam; 2 | 122 | 150 |
| Scaling fish | Butchering | 17 | 25 |
| Scalloping glass | Glass | 49 | 68 |
| Scalpels | Cutlery | 30 | 156 |
| Scarfing and other timber-joints | Carpentry | 20 | 173 |
| Scarf-pins | Jewelry | 63 | 86 |
| Scarificators | Surgery | 128 | 156 |
| Scarifiers | Harrows | 55 | 73 |
| School-charts and maps | Educational | 35 | 146 |
| School-furniture | Furniture | 45 | 62 |
| Scissors | Cutlery | 30 | 44 |
| Scissors, Manufacture of | Hardware Manufactures | 53A | 44 |
| Scissors-sharpeners | Grinding and Polishing | 51 | 70 |
| Scoops, Flour | Kitchen Utensils | 65 | 88 |
| Scouring cloth | Cloth | 26 | 37 |
| Scouring glass | Grinding and Polishing | 51 | 70 |
| Scouring-machines | Mills | 83 | 57 |
| Scouring stone | Grinding and Polishing | 51 | 70 |
| Scrap-books | Stationery | 120 | 146 |
| Scrap, Bundling and rolling | Metal-Working; 5 | 80 | 125 |
| Scrapers, Box | Wood-Working; 4 | 145 | 29 |
| Scrapers, Cane | Plows | 97 | 43 |
| Scrapers, Cotton | Plows | 97 | 43 |
| Scrapers, Deck | Wood-Working; 4 | 145 | 137 |
| Scrapers, Earth | Excavators | 37 | 51 |
| Scrapers for fire-arms | Fire-Arms | 42 | 55 |
| Scrapers, Grading | Excavators | 37 | 51 |
| Scrapers, Mast | Wood-Working; 4 | 145 | 138 |
| Scrapers, Millstone | Stone, Lime, and Cement | 125 | 153 |
| Scrapers, Tree | Garden and Orchard | 47 | 63 |
| Scraping and dressing cord | Cordage | 28 | 127 |
| Screens, Fire | Furniture | 45 | 62 |
| Screens for window | Carpentry | 20 | 62 |
| Screens of wire | Wire-Working | 140 | 170 |
| Screw-blanks, Cutting | Wood-Screws | 141 | 175 |
| Screw-blanks, Feeding | Wood-Screws | 141 | 175 |
| Screw-blanks, Making | Wood-Screws | 141 | 175 |
| Screw-blanks, Threading | Wood-Screws | 141 | 175 |
| Screw-bolts | Bolts, Nuts, and Rivets | 10 | 14 |
| Screw-caps, Making sheet-metal | Sheet-Metal | 113B | 136 |
| Screw-cutting lathes | Metal-Working; 7 | 82 | 100 |
| Screw-drivers | Wood-Working; 4 | 145 | 29 |
| Screw-forming, Sheet-metal | Sheet-Metal | 113A | 136 |
| Screw-nails, Manufacture of | Hardware Manufactures | 53G | 103 |
| Screw-nuts | Bolts, Nuts, and Rivets | 10 | 14 |
| Screw-piles | Hydraulic Engineering | 61 | 82 |
| Screw-presses | Presses | 100 | 7 |
| Screw-propellers | Ships; 2 | 115 | 139 |
| Screw-rings, Making sheet-metal | Sheet-Metal | 113B | 136 |
| Screws, Swaging | Wood-Screws | 141 | 175 |
| Screw-taps | Bolts, Nuts. and Rivets | 10 | 14 |
| Scribing-gages | Wood-Working; 4 | 145 | 29 |
| Scroll-saws | Saws | 110 | 130 |
| Scrubbing-brushes | Brushes and Brooms | 15 | 22 |
| Scrubbing-machines | Brushes and Brooms | 15 | 22 |
| Scrubbing-pails | Brushes and Brooms | 15 | 40 |
| Scuffle-hoes | Garden and Orchard | 47 | 63 |
| Scufflers | Harrows | 55 | 73 |

| SUBJECT. | Name of class—official classification. | No. of class—official classification. | No. of class—subscription classification. |
|---|---|---|---|
| Scuppers | Ships; 1 | 114 | 137 |
| Scutcheons for key-holes | Locks and Latches | 70 | 93 |
| Scutching-machines | Brakes and Gins | 13 | 78 |
| Scythe-fastenings | Harvesters | 56 | 49 |
| Scythe-nibs | Harvesters | 56 | 49 |
| Scythe-rifles | Grinding and Polishing | 51 | 70 |
| Scythes | Harvesters | 56 | 49 |
| Scythe-snaths | Harvesters | 56 | 49 |
| Scythe-stones | Grinding and Polishing | 51 | 70 |
| Sealing cans and jars | Preserving Food | 99 | 58 |
| Seal-locks | Locks and Latches | 70 | 93 |
| Seals | Printing | 101 | 117 |
| Seaming-machines, Sheet-metal | Sheet-Metal | 113A | 136 |
| Seats, Carriage | Carriages | 21 | 31 |
| Seats for harvesters | Harvesters | 56 | 74 |
| Seats for railway-cars | Railways; 2 | 105 | 121 |
| Seats, Garden | Furniture | 45 | 62 |
| Seats, School | Furniture | 45 | 62 |
| Seats, Spring-wagon | Carriages | 21 | 31 |
| Sectional boats | Boats | 9 | 13 |
| Securing cables and hawsers | Ships; 1 | 114 | 137 |
| Securing hatches | Ships; 1 | 114 | 137 |
| Sediment-collectors for steam-boilers | Steam; 2 | 122 | 148 |
| Seed-drills | Seeders and Planters | 111 | 133 |
| Seed-droppers | Seeders and Planters | 111 | 133 |
| Seed for planting and sowing, Preparing | Seeders and Planters | 111 | 133 |
| Seed-planters | Seeders and Planters | 111 | 133 |
| Seeding-bags and pouches | Seeders and Planters | 111 | 133 |
| Self-fastening buttons | Clasps and Buckles | 24 | 23 |
| Self-loading carts | Carriages and Wagons | 21 | 31 |
| Self-loading fire-arms | Fire-Arms | 42 | 55 |
| Semaphores, Electrical | Electricity | 36 | 50 |
| Separating fibers of plants for use in textile fabrics | Brakes and Gins | 13 | 78 |
| Serving rope | Cordage | 28 | 127 |
| Settees | Furniture | 45 | 62 |
| Setting gems | Jewelry | 63 | 86 |
| Setting glaziers' diamonds | Glass | 49 | 68 |
| Sewage for fertilizers, Treatment of | Manures | 71 | 96 |
| Sewer-cleaners | Manures | 71 | 96 |
| Sewers | Masonry | 72 | 51 |
| Sewing-birds | Apparel | 2 | 38 |
| Sewing-boxes | Apparel | 2 | 38 |
| Sewing-guards | Apparel | 2 | 38 |
| Sewing-machines | Sewing-Machines | 112 | 134 |
| Sewing-machines for boots and shoes | Sewing-Machines | 112 | 16 |
| Sewing-machines for embroidery | Sewing-Machines | 112 | 134 |
| Sewing-machines for making button-holes | Sewing-Machines | 112 | 134 |
| Sewing-machines for sewing straw goods | Sewing-Machines | 112 | 134 |
| Sewing-machine needles, Making | Needles and Pins | 86 | 104 |
| Sewing-pins | Apparel | 2 | 38 |
| Sewing-presses | Book-Binding | 11 | 15 |
| ewing-stands | Apparel | 2 | 38 |
| Sewing-work holders | Apparel | 2 | 38 |
| Sextants | Optics | 88 | 157 |
| Shackles | Harness | 54 | 72 |
| Shades, Lamp | Lamps and Gas-Fittings | 67 | 90 |
| haft-couplings, Manufacture of | Hardware Manufactures | 53C | 87 |

| SUBJECT. | Name of class—official classification. | No. of class—official classification. | No. of class—subscription classification. |
|---|---|---|---|
| Shaft-couplings | Journals and Bearings | 64 | 87 |
| Shaft-hangers | Journals and Bearings | 64 | 87 |
| Shafting | Journals and Bearings | 64 | 87 |
| Shafts, Bending or straightening | Metal-Working; 1 | 76 | 119 |
| Shaft-tugs | Harness | 54 | 72 |
| Shampooing appliances | Toilet | 132 | 161 |
| Shank-cutters | Boots and Shoes | 12 | 16 |
| Shank-lasters | Boots and Shoes | 12 | 16 |
| Shank-markers | Boots and Shoes | 12 | 16 |
| Shanks, Boot | Boots and Shoes | 12 | 16 |
| Shanks for door-knobs | Locks and Latches | 70 | 93 |
| Shank-stiffeners | Boots and Shoes | 12 | 16 |
| Shaping machines, Metal | Metal-Working | 7 | 100 |
| Shaping ships' timbers | Ships; 1 | 114 | 137 |
| Sharpeners, Picket and pole | Wood-Working; 3 | 144 | 174 |
| Shaving-cases | Toilet | 132 | 161 |
| Shaving-horses | Wood-Working; 3 | 144 | 173 |
| Shawl-pins | Jewelry | 63 | 86 |
| Shawls | Apparel | 2 | 38 |
| Shawl-straps | Trunks | 133 | 162 |
| Shearing cloth | Cloth | 26 | 37 |
| Shearing-machines for bar or plate | Metal-Working; 4 | 79 | 119 |
| Shears | Cutlery | 30 | 44 |
| Shears, Bench | Cutlery | 30 | 44 |
| Shears, Button-hole | Cutlery | 30 | 44 |
| Shears, Cattle-marking | Cutlery | 30 | 135 |
| Shears, Clipping | Cutlery | 30 | 135 |
| Shears, Edging | Cutlery | 30 | 63 |
| Shears, Garden | Cutlery | 30 | 63 |
| Shears, Grass-edging | Cutlery | 30 | 63 |
| Shears, Hair-clipping | Cutlery | 30 | 161 |
| Shears, Hair-cutting | Cutlery | 30 | 44 |
| Shears, Hand | Cutlery | 30 | 44 |
| Shears, Hedging | Cutlery | 30 | 63 |
| Shears, Lopping | Cutlery | 30 | 63 |
| Shears, Manufacture of | Hardware Manufactures | 53A | 44 |
| Shears, Pruning | Cutlery | 30 | 63 |
| Shears, Sheep | Cutlery | 30 | 135 |
| Shears, Sheet-metal | Cutlery | 30 | 136 |
| Shears, Tailors' | Cutlery | 30 | 38 |
| Shears, Tinmen's | Cutlery | 30 | 136 |
| Sheathing-nails | Nails | 85 | 103 |
| Sheathing vessels | Ships; 1 | 114 | 137 |
| Sheaves, Manufacture of | Hardware Manufactures | 53B | 24 |
| Shedding-apparatus | Weaving | 139 | 95 |
| Sheds and stables, Portable | Carpentry | 20 | 144 |
| Sheep-holders | Stabling | 119 | 135 |
| Sheep-racks | Stabling | 119 | 135 |
| Sheep-shearing machines | Stabling | 119 | 135 |
| Sheep-shearing tables | Stabling | 119 | 135 |
| Sheep-shears | Cutlery | 30 | 135 |
| Sheep-shears | Cutlery | 30 | 135 |
| Sheep-sheds | Stabling | 119 | 135 |
| Sheep-taggers | Stabling | 119 | 135 |
| Sheep-troughs | Stabling | 119 | 135 |
| Sheers, Masting | Hoisting | 57 | 138 |
| Sheet-glass | Glass | 49 | 68 |
| Sheet-iron, Chemical processes in manufacture of | Metal-Working; 5 | 80 | 136 |
| Sheet-iron, Furnaces for | Metallurgy | 75 | 136 |

| SUBJECT. | Name of class—official classification. | No. of class—official classification. | No. of class—subscription classification. |
|---|---|---|---|
| Sheet-iron, Rolling | Metal-Working; 5 | 80 | 125 |
| Sheet-iron, Scraping and cleansing | Metal-Working; 5 | 80 | 136 |
| Sheet-lead, Rolling | Metal-Working; 5 | 80 | 125 |
| Sheet-metal barrels and casks, Making | Sheet-Metal | 113B | 136 |
| Sheet-metal boilers and cisterns, Making | Sheet-Metal | 113B | 136 |
| Sheet-metal boxes and cans, Making | Sheet-Metal | 113B | 136 |
| Sheet-metal buckets and pans, Making | Sheet-Metal | 113B | 136 |
| Sheet-metal capsules and rings, Making | Sheet-Metal | 113B | 136 |
| Sheet-metal gutters and spouting, Making | Sheet-Metal | 113B | 136 |
| Sheet-metal kitchen and table ware, Making | Sheet-Metal | 113B | 136 |
| Sheet-metal, Machines for working | Sheet-Metal | 113B | 136 |
| Sheet-metal making | Metal-Working; 5 | 80 | 136 |
| Sheet-metal planishing | Metal-Working; 5 | 80 | 136 |
| Sheet-metal, Processes for working | Sheet-Metal | 113A | 136 |
| Sheet-metal screw caps and rings, Making | Sheet-Metal | 113A | 136 |
| Sheet-metal table-ware, Making | Sheet-Metal | 113A | 136 |
| Sheet-metal, Tools for working | Sheet-Metal | 113A | 136 |
| Sheet-metal ware, Making | Sheet-Metal | 113A | 136 |
| Sheet-metal workers' tools | Metal-Working; 6 | 81A | 136 |
| Shelf-brackets | Furniture | 45 | 62 |
| Shellers, Corn | Mills | 83 | 59 |
| Shell-fuses | Projectiles | 102 | 105 |
| Shells | Projectiles | 102 | 105 |
| Shells, Incendiary | Projectiles | 102 | 105 |
| Shelves, Folding | Furniture | 45 | 62 |
| Shelving for stores, &c | Furniture | 45 | 62 |
| Shields for horses | Harness | 54 | 72 |
| Shields for tunneling | Hydraulic Engineering | 61 | 82 |
| Shingle-binders | Wood-Working; 3 | 144 | 174 |
| Shingles, Compositions and means for preserving | Roofing | 108 | 126 |
| Shingles, Machines for jointing | Wood-Working; 3 | 144 | 174 |
| Shingles, Machine for planing | Wood-Working; 3 | 144 | 174 |
| Shingles, Machines for riving | Wood-Working; 3 | 144 | 174 |
| Shingling-brackets | Carpentry | 20 | 29 |
| Ships' armor | Ships; 1 | 114 | 137 |
| Ships' berths | Ships; 1 | 114 | 137 |
| Ships' capstans | Hoisting | 57 | 79 |
| Ships' knees | Ships; 1 | 114 | 137 |
| Ships' logs | Measuring-Instruments | 73 | 138 |
| Ships, Propulsion of | Ships; 1 | 114 | 139 |
| Ships, Propulsion of | Ships; 2 | 115 | 139 |
| Shipwrights' clamps | Wood-Working; 4 | 145 | 137 |
| Shipwrights' tools | Wood-Working; 4 | 145 | 137 |
| Shipwrights' tools, Manufacture of | Metal-Working; 6 | 81B | 137 |
| Shirt and collar fastenings, Manufacture of | Hardware Manufactures | 53E | 23 |
| Shirt-bosoms | Apparel | 2 | 38 |
| Shirts | Apparel | 2 | 38 |
| Shoal-alarms | Signals | 116 | 138 |
| Shock-tyers | Harvesters | 56 | 77 |
| Shoe-binding | Boots and Shoes | 12 | 16 |
| Shoe-brushes | Brushes and Brooms | 15 | 16 |
| Shoe-buttoners | Clasps and Buckles | 24 | 23 |
| Shoe-buttons, Manufacture of | Hardware Manufacture | 53E | 23 |
| Shoe-clasps | Boots and Shoes | 12 | 23 |

| SUBJECT. | Name of class—official classification. | No. of class—official classification. | No. of class—subscription classification. |
|---|---|---|---|
| Shoe-distenders | Boots and Shoes | 12 | 16 |
| Shoe-fasteners | Boots and Shoes | 12 | 16 |
| Shoe-hammers | Boots and Shoes | 12 | 16 |
| Shoe-holders | Boots and Shoes | 12 | 16 |
| Shoe-hooks | Boots and Shoes | 12 | 16 |
| Shoe-horns | Boots and Shoes | 12 | 16 |
| Shoeing-tools, Blacksmiths' | Metal-Working; 6 | 81A | 60 |
| Shoe-lacing | Boots and Shoes | 12 | 16 |
| Shoe-lacing, Machines for tagging | Boots and Shoes | 12 | 16 |
| Shoe-lacing hooks | Boots and Shoes | 12 | 23 |
| Shoe-linings | Boots and Shoes | 12 | 16 |
| Shoemakers' benches | Boots and Shoes | 12 | 16 |
| Shoemakers' tools | Boots and Shoes | 12 | 16 |
| Shoe-nails | Nails | 85 | 103 |
| Shoe-patterns | Boots and Shoes | 12 | 16 |
| Shoe-pegs, Machines for making | Wood-Working; 3 | 144 | 16 |
| Shoes | Boots and Shoes | 12 | 16 |
| Shoes for car-brakes | Railways; 3 | 106 | 122 |
| Shoe-soles | Boots and Shoes | 12 | 16 |
| Shoe-soles, Machines for making wooden | Wood-Working; 3 | 144 | 16 |
| Shoe-stays | Boots and Shoes | 12 | 16 |
| Shoe-straps | Boots and Shoes | 12 | 16 |
| Shoe-tips, Manufacture of | Hardware Manufactures | 53G | 16 |
| Short-hand reporting | Educational | 35 | 54 |
| Shot | Projectiles | 102 | 35 |
| Shot-cartridges | Projectiles | 102 | 32 |
| Shot-chargers | Projectiles | 102 | 55 |
| Shot-hole stoppers | Ships; 1 | 114 | 137 |
| Shot making, Cast | Casting | 22 | 32 |
| Shot, Making, (*Pressed*) | Hardware Manufactures | 53A | 32 |
| Shot, Manufacture of | Hardware Manufactures | 53A | 32 |
| Shot-pouches | Projectiles | 102 | 55 |
| Shot-sorting machines | Hardware Manufactures | 53A | 32 |
| Shotting-machines | Casting | 22 | 32 |
| Shoulder-braces | Apparel | 2 | 38 |
| Shoulder-straps | Apparel | 2 | 38 |
| Shoulder-supporters | Apparel | 2 | 38 |
| Shovel-handles, Machines for making | Wood-Working; 3 | 144 | 141 |
| Shovel-plows | Plows | 97 | 114 |
| Shovels | Garden and Orchard | 47 | 141 |
| Shovels and tongs | Stoves and Furnaces | 126 | 154 |
| Show-cards | Stationery | 120 | 146 |
| Show-cases | Furniture | 45 | 54 |
| Show-stands | Furniture | 45 | 54 |
| Shutter-bars | Builders' Hardware | 16 | 24 |
| Shutter-fasteners | Builders' Hardware | 16 | 24 |
| Shutter-fasteners, Manufacture of | Hardware Manufactures | 53B | 24 |
| Shutter-hinges | Builders' Hardware | 16 | 24 |
| Shutter-lifts | Builders' Hardware | 16 | 24 |
| Shntter-locks | Builders' Hardware | 16 | 24 |
| Shutter-operators | Builders' Hardware | 16 | 24 |
| Shutters | Carpentry | 20 | 30 |
| Shutters, Fire-proof | Masonry | 72 | 4 |
| Shutter-workers, Manufacture of | Hardware Manufacture | 53B | 24 |
| Shuttle-boxes | Weaving | 139 | 95 |
| Shuttle-boxes, Actuating | Weaving | 139 | 95 |
| Shuttle-checks | Weaving | 139 | 95 |
| Shuttle-drivers for sewing-machines | Sewing-Machines | 112 | 134 |
| Shuttle-frames, Manufacture of | Hardware Manufactures | 53H | 95 |
| Shuttles for sewing-machines | Sewing-Machines | 112 | 134 |

| SUBJECT. | Name of class—official classification. | No. of class—official classification. | No. of class—subscription classification. |
|---|---|---|---|
| Shuttles | Weaving | 139 | 95 |
| Shuttle-spindles | Weaving | 139 | 95 |
| Shuttles, Tatting | Apparel | 2 | 38 |
| Shuttle-tips, Manufacture of | Hardware Manufactures | 53H | 95 |
| Sickles | Harvesters | 56 | 49 |
| Sidewalks | Paving | 94 | 110 |
| Siding of houses | Carpentry | 20 | 30 |
| Sieve-cloth, Weaving | Weaving | 139 | 95 |
| Sieves of wire | Wire-Working | 140 | 170 |
| Sifters | Wire-Working | 140 | 170 |
| Sifting-machines | Wire-Working | 140 | 170 |
| Sights for ordnance | Ordnance | 89 | 105 |
| Sights for small-arms | Fire-Arms | 42 | 105 |
| Signal-bells | Signals | 116 | 24 |
| Signal-boxes for electric telegraphs | Electricity | 36 | 50 |
| Signal-codes | Signals | 116 | 50 |
| Signaling-apparatus | Signals | 116 | 158 |
| Signal-lanterns | Signals | 116 | 158 |
| Signals, Electrical | Electricity | 36 | 50 |
| Signals, Fog | Signals | 116 | 138 |
| Signals, marine | Signals | 116 | 138 |
| Signals, Pyrotechnic | Signals | 116 | 158 |
| Signals, Railroad | Signals | 116 | 120 |
| Signals, Semaphoric | Signals | 116 | 158 |
| Signal-towers | Signals | 116 | 158 |
| Signs | Stationery | 120 | 146 |
| Signs, Illuminating | Stationery | 120 | 54 |
| Silk filature | Silk | 117 | 140 |
| Silk manufacture | Silk | 117 | 140 |
| Silk-worm culture | Stabling | 119 | 140 |
| Silk-worm frames | Silk | 117 | 140 |
| Silvering | Plating | 96 | 113 |
| Silver-plating | Plating | 96 | 113 |
| Singeing-apparatus | Butchering | 17 | 25 |
| Singeing cloth | Cloth | 26 | 37 |
| Singeing-tools | Stabling | 119 | 144 |
| Single-shovel plows | Plows | 97 | 114 |
| Single-trees | Carriages and Wagons | 21 | 31 |
| Sinkers for knitting-machines | Knitting and Netting | 66 | 89 |
| Sinks | Baths and Closets | 4 | 8 |
| Sinks, Kitchen | Baths and Closets | 4 | 8 |
| Siphon-bottles | Aëration and Bottling | 1 | 1 |
| Siphons | Water-Distribution | 137 | 1 |
| Sirup-fountains | Aëration and Bottling | 1 | 1 |
| Sirup-gages | Aëration and Bottling | 1 | 1 |
| Sirup, Treatment of | Sugar | 127 | 155 |
| Size-markers for hats | Felting and Hats | 38 | 76 |
| Size, Preparation of | Bleaching and Dyeing | 8 | 11 |
| Sizing fabrics | Bleaching and Dyeing | 8 | 11 |
| Sizing silk | Silk | 117 | 140 |
| Sizing yarn, thread, &c | Cordage | 28 | 17 |
| Skate-fasteners | Games and Toys | 46 | 143 |
| Skates | Games and Toys | 46 | 143 |
| Skate-sharpeners | Grinding and Polishing | 51 | 70 |
| Skates, Parlor | Games and Toys | 46 | 143 |
| Skates, Roller | Games and Toys | 46 | 143 |
| Skating-rinks | Carpentry | 20 | 30 |
| Skeins of axles | Carriages and Wagons | 21 | 31 |
| Skelp and blank furnaces | Metal-Working; 3 | 78 | 125 |
| Skewers, Machines for making | Wood-Working; 3 | 144 | 175 |

| SUBJECT. | Name of class—official classification. | No. of class—official classification. | No. of class—subscription classification. |
|---|---|---|---|
| Skids | Hoisting | 57 | 79 |
| Skillets | Stoves and Furnaces | 126 | 154 |
| Skimmers, Culinary | Kitchen Utensils | 65 | 88 |
| Skimmers for steam-boilers | Steam; 2 | 122 | 148 |
| Skimmers for sugar-pans | Sugar | 127 | 155 |
| Skinning appliances | Butchering | 17 | 25 |
| Skins, Treatment of | Tanning | 129 | 92 |
| Skips | Hoisting | 57 | 79 |
| Skirt-covers | Apparel | 2 | 38 |
| Skirt-elevators | Apparel | 2 | 38 |
| Skirt-hoops, Covering wire with wire for | Wire-Working | 140 | 170 |
| Skirts and corsets combined | Crinoline and Corsets | 29 | 38 |
| Skirts, Binding for | Apparel | 2 | 38 |
| Skirts, Clasps for | Clasps and Buckles | 24 | 23 |
| Skirts, Covered wire for hoop | Crinoline and Corsets | 29 | 80 |
| Skirts, Hoop | Crinoline and Corsets | 29 | 80 |
| Skirts, (*except Crinoline, Class* 29) | Apparel | 2 | 38 |
| Skirts, Printing | Printing | 101 | 117 |
| Skirt-supporters | Apparel | 2 | 38 |
| Skiving machines, Leather | Leather | 69 | 92 |
| Skylights | Carpentry | 20 | 30 |
| Skylights, Operating | Carpentry | 20 | 29 |
| Slashers | Cordage | 28 | 127 |
| Slate, Artificial | Stone, Lime, and Cement | 125 | 153 |
| Slate-cleaners | Educational | 35 | 146 |
| Slate-frames | Educational | 35 | 146 |
| Slate-frames, Machines for making | Wood-Working; 3 | 144 | 146 |
| Slate-pencils | Educational | 35 | 146 |
| Slates | Educational | 35 | 146 |
| Slates, Artificial | Educational | 35 | 146 |
| Slate-working machines | Stone, Lime, and Cement | 125 | 153 |
| Slating | Roofing | 108 | 126 |
| Slat-mats | Brushes and Brooms | 15 | 56 |
| Slatted floor-covering | Brushes and Brooms | 15 | 56 |
| Slaughtering-apparatus | Butchering | 17 | 25 |
| Sled-brakes | Carriages and Wagons | 21 | 31 |
| Sled-runners | Carriages and Wagons | 21 | 31 |
| Sleds | Carriages and Wagons | 21 | 31 |
| Sleds for furrowing ground | Plows | 97 | 114 |
| Sled-soles | Carriages and Wagons | 21 | 31 |
| Sleepers, Railway | Railways; 1 | 104 | 120 |
| Sleeping-cars | Railways; 2 | 105 | 121 |
| Sleeve-buttons | Clasps and Buckles | 24 | 23 |
| Sleeves | Apparel | 2 | 38 |
| Sleigh-bell attachments | Harness | 54 | 72 |
| Sleigh-bells | Harness | 54 | 72 |
| Sleigh-bells, Manufacture of | Hardware Manufactures | 53B | 10 |
| Sleighs | Carriages and Wagons | 21 | 31 |
| Slide-valves | Valves | 136 | 152 |
| Sliding-door rollers | Builders' Hardware | 16 | 24 |
| Sliding hills | Carpentry | 20 | 29 |
| Sliding-gages | Measuring-Instruments | 73 | 97 |
| Slinging-accouterments | Trunks | 133 | 162 |
| Slippers | Boots and Shoes | 12 | 16 |
| Slips, Marine | Hydraulic Engineering | 61 | 82 |
| Slitting timber, Machines for | Wood-Working; 3 | 144 | 173 |
| Slivering-machines | Wood-Working; 2 | 143 | 174 |
| Slivers, Cans for | Carding | 19 | 28 |
| Slop-buckets | Kitchen Utensils | 65 | 88 |

| SUBJECT. | Name of class—official classification. | No. of class—official classification. | No. of class—subscription classification. |
|---|---|---|---|
| Slop-jars | Kitchen Utensils | 65 | 88 |
| Slotting machines, Metal | Metal-Working; 7 | 82 | 100 |
| Slubbing-frames | Spinning | 118 | 142 |
| Sluice-gates | Hydraulic Engineering | 61 | 82 |
| Small-arms | Fire-Arms | 42 | 55 |
| Smelting-furnaces | Metallurgy | 75 | 88 |
| Smelting glass | Glass | 49 | 68 |
| Smoke-bells | Lamps and Gas-Fittings | 67 | 90 |
| Smoke-burners | Stoves and Furnaces | 126 | 154 |
| Smoke-houses | Dryers and Kilns | 34 | 48 |
| Smoke-jacks | Stoves and Furnaces | 126 | 154 |
| Smoke-pipes for boiler-furnaces | Steam; 2 | 122 | 149 |
| Smoking meat and fish | Dryers and Kilns | 34 | 58 |
| Smoking-pipes | Tobacco | 131 | 160 |
| Smoothing-irons | Laundry | 68 | 91 |
| Smut-machines | Mills | 83 | 159 |
| Smut-mills | Mills | 83 | 159 |
| Snag-boats | Hydraulic Engineering | 61 | 82 |
| Snags, Removing | Hydraulic Engineering | 61 | 82 |
| Snap-hooks | Clasps and Buckles | 24 | 23 |
| Snap-hooks, Manufacture of | Hardware Manufactures | 53F | 23 |
| Snap-rings | Clasps and Buckles | 24 | 23 |
| Snaths, Machines for bending | Wood-Working; 3 | 144 | 175 |
| Snaths, Machines for finishing | Wood-Working; 3 | 144 | 175 |
| Snaths, Machines for rounding | Wood-Working; 1 | 142 | 175 |
| Snaths, Machines for turning | Wood-Working; 1 | 142 | 175 |
| Snout-rings | Harness | 54 | 144 |
| Snow-plows | Railways; 3 | 106 | 122 |
| Snubbers | Ships; 1 | 114 | 137 |
| Snuff-boxes | Tobacco | 131 | 160 |
| Snuffers | Lamps and Gas-Fittings | 67 | 90 |
| Snuff-sifters | Tobacco | 131 | 160 |
| Soap | Oil, Fat, and Glue | 87 | 26 |
| Soap-cutting machines | Oil, Fat, and Glue | 87 | 26 |
| Soap-holders | Furniture | 45 | 26 |
| Soap-making machines | Oil, Fat, and Glue | 87 | 26 |
| Soap-making processes | Oil, Fat, and Glue | 87 | 26 |
| Soda-fountains | Aëration and Bottling | 1 | 1 |
| Soda-fountains, Apparatus for charging | Aëration and Bottling | 1 | 1 |
| Soda-water apparatus | Aëration and Bottling | 1 | 1 |
| Sod-cutters | Plows | 97 | 114 |
| Sofas | Furniture | 45 | 62 |
| Solar cameras | Optics | 88 | 111 |
| Solder | Sheet-Metal | 113A | 136 |
| Soldering-furnaces | Sheet-Metal | 113A | 136 |
| Soldering-implements | Sheet-Metal | 113A | 136 |
| Soldering-processes | Sheet-Metal | 113A | 136 |
| Soldering-tools | Sheet-Metal | 113A | 136 |
| Soles for boots and shoes | Boots and Shoes | 12 | 16 |
| Sounds | Surgery | 128 | 156 |
| Sorghum-evaporators | Sugar | 127 | 155 |
| Sorghum manufacture | Sugar | 127 | 155 |
| Sorghum-mills | Mills | 83 | 155 |
| Sorghum-strippers | Garden and Orchard | 47 | 155 |
| Sorting-apparatus, Electrical | Electricity | 36 | 50 |
| Sorting machines, Leather | Leather | 69 | 92 |
| Sorting silk | Silk | 117 | 140 |
| Sounders | Electricity | 36 | 50 |
| Sounding-apparatus | Measuring-Instruments | 73 | 137 |

| SUBJECT. | Name of class—official classification. | No. of class—official classification. | No. of class—subscription classification. |
|---|---|---|---|
| Sounding-boards for musical instruments | Music | 84 | 112 |
| Sounding, Instruments for marine | Measuring-Instruments | 73 | 137 |
| Sowers, Broad-cast | Seeders and Planters | 111 | 133 |
| Sowers, Fertilizer | Seeders and Planters | 111 | 133 |
| Sowers, Seed | Seeders and Planters | 111 | 133 |
| Spaces for printers | Printing | 101 | 117 |
| Spade-bayonets | Fire-Arms | 42 | 55 |
| Spades | Garden and Orchard | 47 | 141 |
| Spading-machines | Plows | 97 | 114 |
| Spangles to skirts, Attaching | Crinoline and Corsets | 29 | 38 |
| Spanners for hose-couplings | Hose and Belting | 60 | 81 |
| Spark-arresters | Steam; 2 | 122 | 94 |
| Spark-consumers | Steam; 2 | 122 | 94 |
| Sparring steamboats | Ships; 1 | 114 | 138 |
| Spars, Handling and hoisting | Ships; 1 | 114 | 138 |
| Spaying device | Surgery | 128 | 156 |
| Speaking-trumpets | Signals | 116 | 158 |
| Speaking-tubes | Signals | 116 | 158 |
| Specific-gravity apparatus | Measuring-Instruments | 73 | 132 |
| Spectacles | Optics | 88 | 157 |
| Speculums, Anal | Surgery | 128 | 156 |
| Speculums, Vaginal | Surgery | 128 | 156 |
| Speeder-flyers, Manufacture of | Hardware Manufactures | 53H | 142 |
| Speeders | Spinning | 118 | 142 |
| Speeders for capstans | Hoisting | 57 | 79 |
| Speed-indicators | Measuring-Instruments | 73 | 42 |
| Spermatorrhea, Instruments for treating | Surgery | 128 | 156 |
| Spherical-case shot | Projectiles | 102 | 105 |
| Sphygmographs | Measuring-Instruments | 73 | 156 |
| Spice-boxes | Kitchen Utensils | 65 | 88 |
| Spice-mills | Mills | 83 | 88 |
| Spiders | Stoves and Ranges | 126 | 154 |
| Spike-bar rolling | Metal-Working; 5 | 80 | 125 |
| Spike-extractors | Wood-Working; 4 | 145 | 29 |
| Spikes | Nails | 85 | 103 |
| Spikes, Machines for making | Nails | 85 | 103 |
| Spikes, Railroad | Railways; 4 | 107 | 120 |
| Spikes, Screw | Railways; 4 | 107 | 120 |
| Spikes, Self-clenching | Nails | 85 | 103 |
| Spinal curvatures, Instruments for treating | Artificial Limbs | 3 | 156 |
| Spindle-heads for flyers, Manufacture of | Hardware Manufactures | 53H | 142 |
| Spindles, Banding | Spinning | 118 | 142 |
| Spindles for door-knobs | Locks and Latches | 70 | 93 |
| Spindles for spinning-machines | Spinning | 118 | 142 |
| Spindles, Lubricating | Journals and Bearings | 64 | 87 |
| Spindles, Mill | Mills | 83 | 57 |
| Spindle-steps | Spinning | 118 | 142 |
| Spinning glass | Glass | 49 | 68 |
| Spinning-machinery | Spinning | 118 | 142 |
| Spinning sheet-metal ware | Sheet-Metal | 113A | 136 |
| Spinning silk | Silk | 117 | 140 |
| Spinning-wheels | Spinning | 118 | 142 |
| Spiral stock-drills | Metal-Working; 2 | 77 | 100 |
| Spirals, Instruments for drawing | Drafting | 33 | 97 |
| Spirit-meters | Pumps | 103 | 167 |
| Spirits, Distillation of | Stills | 124 | 47 |
| Spirits, Rectification of | Stills | 124 | 47 |

| SUBJECT. | Name of class—official classification. | No. of class—official classification. | No. of class—sub-scription classification. |
|---|---|---|---|
| Spirometers | Measuring-Instruments | 73 | 156 |
| Spits for roasting meat | Stoves and Ranges | 126 | 154 |
| Spittoon-envelopes | Furniture | 45 | 62 |
| Spittoons, Railway-car | Furniture | 45 | 121 |
| Spittoons | Furniture | 45 | 62 |
| Splice pieces, Railway | Railways; 1 | 104 | 120 |
| Splices for railways, Making | Railways; 4 | 107 | 120 |
| Splicing timbers | Carpentry | 20 | 30 |
| Splint-brooms | Brushes and Brooms | 15 | 22 |
| Splint-planes | Wood-Working; 2 | 143 | 29 |
| Splints | Artificial Limbs | 3 | 156 |
| Splints for baskets, blinds, brooms, and matches | Wood-Working; 2 | 143 | 174 |
| Split-rings | Clasps and Buckles | 24 | 23 |
| Split-rings, Manufacture of | Hardware Manufactures | 53E | 23 |
| Splitting fire-wood, Machines for | Wood-Working; 3 | 144 | 174 |
| Splitting machines, Leather | Leather | 69 | 92 |
| Splitting slate | Stone, Lime, and Cement | 125 | 153 |
| Spoke-shaves | Wood-Working; 4 | 145 | 31 |
| Spokes, Machines for driving | Wood-Working; 3 | 144 | 31 |
| Spokes, Machines for planing | Wood-Working; 3 | 144 | 31 |
| Spokes, Machines for polishing | Wood-Working; 3 | 144 | 31 |
| Spokes, Machines for tenoning | Wood-Working; 3 | 144 | 31 |
| Spokes, Machines for turning | Wood-Working; 1 | 142 | 31 |
| Spokes | Carriages and Wagons | 21 | 31 |
| Spoke-sockets | Carriages and Wagons | 21 | 31 |
| Sponge-cups | Stationery | 120 | 146 |
| Sponge for stuffing, Preparing | Beds | 5 | 62 |
| Sponges for ordnance | Ordnance | 89 | 105 |
| Sponging cloth | Cloth | 26 | 37 |
| Spool-cases | Apparel | 2 | 38 |
| Spool-holders | Apparel | 2 | 38 |
| Spooling-machines | Cordage | 28 | 17 |
| Spool-printing | Printing | 101 | 117 |
| Spools | Cordage | 28 | 17 |
| Spools, Machines for boring | Wood-Working; 1 | 142 | 173 |
| Spools, Machines for turning | Wood-Working; 1 | 142 | 172 |
| Spool-stands | Apparel | 2 | 38 |
| Spoons | Kitchen Utensils | 65 | 88 |
| Spoons, Making | Sheet-Metal | 113B | 136 |
| Spoons, Medicinal | Surgery | 128 | 156 |
| Spouting | Water-Distribution | 137 | 167 |
| Spouting for eaves | Roofing | 108 | 126 |
| Spouting, Making | Sheet-Metal | 113B | 136 |
| Spring-bolts | Builders' Hardware | 16 | 24 |
| Spring-brim hats | Felting and Hats | 38 | 76 |
| Spring-heads, Manufacture of | Hardware Manufactures | 53C | 24 |
| Spring-hinges | Builders' Hardware | 16 | 24 |
| Spring-punches | Clasps and Buckles | 24 | 72 |
| Springs, Car | Railways; 3 | 106 | 122 |
| Springs, Carriage | Carriages | 21 | 31 |
| Springs, Door | Builders' Hardware | 16 | 24 |
| Springs for artificial teeth | Dental | 32 | 46 |
| Springs for beds | Beds | 5 | 62 |
| Springs for sofas, chairs, &c | Furniture | 45 | 62 |
| Springs, Watch and clock | Horology | 58 | 36 |
| Spring-testing presses | Presses | 100 | 7 |
| Sprinklers for clothes | Laundry | 68 | 91 |
| Sprouting-beds for plants | Garden and Orchard | 47 | 63 |
| Sprue-patterns | Castings | 22 | 33 |

| SUBJECT. | Name of class—official classification. | No. of class—official classification. | No. of class—subscription classification. |
|---|---|---|---|
| Sprues | Castings | 22 | 33 |
| Spuds | Garden and Orchard | 47 | 63 |
| Spurs | Harness | 54 | 72 |
| Spurs for saw-mills | Saws | 110 | 130 |
| Squares | Measuring-Instruments | 73 | 97 |
| Squeezers for puddlers' balls | Metal-Working; 5 | 80 | 85 |
| Stable-cleaners | Stabling | 119 | 144 |
| Stables | Stabling | 119 | 144 |
| Stackers | Thrashing | 130 | 159 |
| Stage-properties | Fine Arts | 41 | 143 |
| Staging for boats | Hydraulic Engineering | 61 | 82 |
| Staging for loading and unloading vessels | Hydraulic Engineering | 61 | 82 |
| Staining | Paint | 91 | 107 |
| Staining-compositions for marble and stone | Stone, Lime, and Cement | 125 | 153 |
| Staining glass, wood, &c | Fine Arts | 41 | 54 |
| Stains, Composition for removing | Chemical Miscellaneous | 23 | 35 |
| Stair-covers | Furniture | 45 | 62 |
| Stair-pads | Furniture | 45 | 62 |
| Stair-plates | Furniture | 45 | 62 |
| Stair-rod fastenings | Builders' Hardware | 16 | 24 |
| Stair-rods | Builders' Hardware | 16 | 24 |
| Stair-rods | Furniture | 45 | 62 |
| Stair-rods, Manufacture of | Hardware Manufactures | 53D | 24 |
| Stairs | Carpentry | 20 | 30 |
| Staiths | Hydraulic Engineering | 61 | 82 |
| Stalk-choppers | Harrows | 55 | 73 |
| Stalk-pullers | Harrows | 55 | 73 |
| Stalls for horses and cattle | Stabling | 119 | 144 |
| Stamp-cancelers | Printing | 101 | 146 |
| Stamp-holders | Stationery | 120 | 146 |
| Stamping fabrics | Apparel | 2 | 37 |
| Stamping-mills | Ore | 90 | 3 |
| Stamps, Address | Printing | 101 | 117 |
| Stamps, Branding | Printing | 101 | 117 |
| Stamps for sheet-metal working | Sheet-Metal | 113A | 136 |
| Stamps, Hand | Printing | 101 | 146 |
| Stamps, Postage | Fine Arts | 41 | 146 |
| Stamps, Post-office | Printing | 101 | 146 |
| Stamps, Revenue | Fine Arts | 41 | 146 |
| Stamps, Seal | Printing | 101 | 146 |
| Stanchions | Stabling | 119 | 144 |
| Stands, Hat and cloak | Furniture | 45 | 62 |
| Stands, Towel | Furniture | 45 | 62 |
| Stands, Umbrella | Furniture | 45 | 62 |
| Staples | Nails | 85 | 103 |
| Staples, Machines for making | Nails | 85 | 103 |
| Starch, Manufacture of | Bleaching and Dyeing | 8 | 11 |
| Starting engines off their centers | Mechanical Powers | 74 | 67 |
| Stationary steam-engines | Steam; 1 | 121 | 150 |
| Stationery, Articles of | Stationery | 120 | 146 |
| Stationery, (metallic articles of,) Manufacture of | Hardware Manufactures | 53I | 124 |
| Station-indicators, (*non-automatic*) | Stationery | 120 | 146 |
| Staves, Machines for bilging | Wood-Working; 3 | 144 | 40 |
| Staves, Machines for crozing | Wood-Working; 3 | 144 | 40 |
| Staves, Machines for dressing | Wood-Working; 3 | 144 | 40 |
| Staves, Machines for howeling | Wood-Working; 3 | 144 | 40 |
| Staves, Machines for jointing | Wood-Working; 3 | 144 | 40 |

| SUBJECT. | Name of class—official classification. | No. of class—official classification. | No. of class—subscription classification. |
|---|---|---|---|
| Staves, Machines for riving | Wood-Working; 3 | 144 | 40 |
| Staving metals | Metal-Working; 3 | 78 | 100 |
| Steam-actuated car-brakes | Steam; 1 | 121 | 122 |
| Steam-boiler covering | Steam; 2 | 122 | 149 |
| Steam-boiler furnaces | Steam; 2 | 122 | 149 |
| Steam-boilers | Steam; 2 | 122 | 148 |
| Steam-blast for furnaces | Steam; 2 | 122 | 12 |
| Steam carving-tables | Stoves and Furnaces | 126 | 154 |
| Steam-chests of engines | Steam; 1 | 121 | 150 |
| Steam-cocks | Valves | 136 | 148 |
| Steam-cooking apparatus | Stoves and Furnaces | 126 | 154 |
| Steam-cultivators | Plows | 97 | 114 |
| Steam-cylinder heads | Steam; 1 | 121 | 150 |
| Steam-cylinders | Steam; 1 | 121 | 150 |
| Steam-engines | Steam; 1 | 121 | 150 |
| Steam-engines for car-brakes | Steam; 1 | 121 | 150 |
| Steam-engines for hammers | Steam; 1 | 121 | 150 |
| Steam-engines for hoisting | Steam; 1 | 121 | 150 |
| Steam-engines for pumps | Steam; 1 | 121 | 150 |
| Steam-engine indicators | Measuring-Instruments | 73 | 148 |
| Steam-engines, Lubricators for | Journals and Bearings | 64 | 87 |
| Steam-engines, Valve-gear for | Steam; 1 | 121 | 150 |
| Steam-finishing presses for cloth | Presses | 100 | 37 |
| Steam fire-engines | Pumps | 103 | 118 |
| Steam fire-engines, (Engines of) | Steam; 1 | 121 | 118 |
| Steam fire-proof safes | Safes | 109 | 128 |
| Steam-gages | Measuring-Instruments | 73 | 148 |
| Steam-generators | Steam; 2 | 122 | 148 |
| Steam-governors | Governors | 50 | 69 |
| Steam-hammers | Metal-Working; 3 | 78 | 100 |
| Steam-hammers, Engines of | Steam; 1 | 121 | 150 |
| Steam-heaters | Stoves and Furnaces | 126 | 147 |
| Steam-heaters, Boilers of | Steam; 2 | 122 | 147 |
| Steam-heating tables | Stoves and Furnaces | 126 | 147 |
| Steam in furnaces, Use of | Steam; 2 | 122 | 149 |
| Steaming-apparatus, Boilers of | Steam; 2 | 122 | 148 |
| Steaming cloth | Cloth | 26 | 37 |
| Steam-packing | Steam; 1 | 121 | 151 |
| Steam-pipes | Steam; 2 | 122 | 149 |
| Steam-pipes, Non conducting covering for | Steam; 2 | 122 | 149 |
| Steam-piston packing | Steam; 1 | 121 | 151 |
| Steam-pistons | Steam; 1 | 121 | 151 |
| Steam-plates for cloth-presses | Presses | 100 | 7 |
| Steam-plows | Plows | 97 | 114 |
| Steam-pressure gages | Measuring-Instruments | 73 | 148 |
| Steam-pressure indicators | Measuring-Instruments | 73 | 148 |
| Steam-pressure regulators | Steam; 2 | 122 | 148 |
| Steam-pumps | Pumps | 103 | 118 |
| Steam-pumps, Engines of | Steam; 1 | 122 | 118 |
| Steam-regulators | Steam; 2 | 122 | 148 |
| Steam spading-machines | Plows | 97 | 114 |
| Steam-traps | Steam; 1 | 121 | 150 |
| Steam vacuum-pumps | Steam; 1 | 121 | 150 |
| Steam-valves | Valves | 136 | 152 |
| Steam-whistles | Steam; 2 | 122 | 148 |
| Stearine | Oil, Fat, and Glue | 87 | 26 |
| Stearine-presses | Presses | 100 | 26 |
| Steel, Compositions in manufacturing | Metallurgy | 75 | 85 |
| Steel, Machinery in manufacturing | Metallurgy | 75 | 85 |
| Steel pens, Manufacture of | Hardware Manufactures | 53I | 136 |

| SUBJECT. | Name of class—official classification. | No. of class—official classification. | No. of class—subscription classification. |
|---|---|---|---|
| Steel, Processes in manufacturing | Metallurgy | 75 | 85 |
| Steels | Grinding and Polishing | 51 | 70 |
| Steering-apparatus | Ships; 1 | 114 | 137 |
| Stem-winders | Horology | 58 | 36 |
| Stench-traps | Baths and Closets | 4 | 8 |
| Stencil-plates | Fine Arts | 41 | 54 |
| Stencil-printing machines | Printing | 101 | 54 |
| Step-ladders | Carpentry | 20 | 30 |
| Stereoscopes | Optics | 88 | 111 |
| Stereoscopic-picture holders | Fine Arts | 41 | 111 |
| Stereotype-molds, Machines for producing | Printing | 101 | 164 |
| Stereotyping | Printing | 101 | 164 |
| Stereotyping, Compositions for | Printing | 101 | 164 |
| Stereotyping, Processes for | Printing | 101 | 164 |
| Stethometers | Measuring-Instruments | 73 | 156 |
| Stethoscopes | Surgery | 128 | 156 |
| Stills | Stills | 124 | 47 |
| Stilts | Games and Toys | 46 | 143 |
| Stirrups | Harness | 54 | 72 |
| Stirrups, Manufacture of | Hardware Manufactures | 53F | 72 |
| Stirrups, Safety | Harness | 54 | 72 |
| Stitches for sewing-machines | Sewing-Machines | 112 | 134 |
| Stitching horses | Harness | 54 | 72 |
| Stock-cars | Railways; 2 | 105 | 121 |
| Stock-feeders | Stabling | 119 | 144 |
| Stocking and sock machines | Knitting and Netting | 66 | 89 |
| Stocking-supporters | Apparel | 2 | 38 |
| Stockings, Toe and heel protectors for | Apparel | 2 | 38 |
| Stocks and dies | Bolts, Nuts, and Rivets | 10 | 14 |
| Stocks, Bit | Wood-Working; 4 | 145 | 29 |
| Stocks for fire-arms | Fire-Arms | 42 | 55 |
| Stocks for refractory animals | Stabling | 119 | 144 |
| Stocks for shoeing animals | Stabling | 119 | 60 |
| Stocks, Neck | Apparel | 2 | 38 |
| Stomach-pumps | Surgery | 128 | 118 |
| Stone and cement, Indurating | Stone, Lime, and Cement | 125 | 153 |
| Stone and slate tanks, Construction of | Masonry | 72 | 51 |
| Stone, Artificial | Stone, Lime, and Cement | 125 | 153 |
| Stone-breakers | Ore | 90 | 3 |
| Stone bridges | Bridges | 14 | 21 |
| Stone, Channeling | Stone, Lime, and Cement | 125 | 153 |
| Stone, Compositions for porous | Stone, Lime, and Cement | 125 | 153 |
| Stone-crushers | Ore | 90 | 3 |
| Stone-dressing | Stone, Lime, and Cement | 125 | 153 |
| Stone-drills | Stone, Lime, and Cement | 125 | 153 |
| Stone facing for houses | Masonry | 72 | 4 |
| Stone fronts for houses | Masonry | 72 | 4 |
| Stone-gatherers | Harrows | 55 | 73 |
| Stone, Indurating | Stone, Lime, and Cement | 125 | 153 |
| Stone, Manufacture of paving | Stone, Lime, and Cement | 125 | 110 |
| Stone-polishing | Grinding and Polishing | 51 | 153 |
| Stone-saws | Stone, Lime, and Cement | 125 | 153 |
| Stone-staves for millers | Stone, Lime, and Cement | 125 | 153 |
| Stone, Tools for working | Stone, Lime, and Cement | 125 | 153 |
| Stone-ware | Clay | 25 | 115 |
| Stone-working machines | Stone, Lime, and Cement | 125 | 153 |
| Stone-working tools, Manufacture of | Metal-Working; 6 | 81B | 153 |
| Stone-work of buildings | Masonry | 72 | 4 |
| Stools | Furniture | 45 | 62 |

| SUBJECT. | Name of class—official classification. | No. of class—official classification. | No. of class—subscription classification. |
|---|---|---|---|
| Stools, Camp | Furniture | 45 | 62 |
| Stools, Folding | Furniture | 45 | 62 |
| Stools, Foot | Furniture | 45 | 62 |
| Stools, Piano | Furniture | 45 | 62 |
| Stop-cocks | Valves | 136 | 152 |
| Stop-mechanisms for looms | Weaving | 139 | 95 |
| Stop-motions for carding-machines | Carding | 19 | 28 |
| Stop-motions for knitting-machines | Knitting and Netting | 66 | 89 |
| Stop-motions for spinning-machines | Spinning | 118 | 142 |
| Stop-motions for steam-engines | Steam; 1 | 121 | 150 |
| Stopping crevasses | Hydraulic Engineering | 61 | 82 |
| Stops, Door | Builders' Hardware | 16 | 24 |
| Stop-valves | Valves | 136 | 152 |
| Store-houses | Carpentry | 20 | 30 |
| Store shelving | Furniture | 45 | 62 |
| Storm-signals, Electrical | Electricity | 36 | 50 |
| Stove-aprons | Stoves and Furnaces | 126 | 154 |
| Stove-boilers | Stoves and Furnaces | 126 | 154 |
| Stove-cover holders | Stoves and Furnaces | 126 | 154 |
| Stove-covers | Stoves and Furnaces | 126 | 154 |
| Stove-door handles | Stoves and Furnaces | 126 | 154 |
| Stove-door linings | Stoves and Furnaces | 126 | 154 |
| Stove-doors | Stoves and Furnaces | 126 | 154 |
| Stove-drums | Stoves and Furnaces | 126 | 154 |
| Stove-fasteners | Stoves and Furnaces | 126 | 154 |
| Stove-flues | Stoves and Furnaces | 126 | 154 |
| Stove-furniture, Making sheet-metal | Sheet-Metal | 113B | 136 |
| Stove-illuminators | Stoves and Furnaces | 126 | 154 |
| Stove-irons | Stoves and Furnaces | 126 | 154 |
| Stove-leg casters | Stoves and Furnaces | 126 | 154 |
| Stove-legs | Stoves and Furnaces | 126 | 154 |
| Stove-lid holders | Stoves and Furnaces | 126 | 154 |
| Stove-lids | Stoves and Furnaces | 126 | 154 |
| Stove-linings | Stoves and Furnaces | 126 | 154 |
| Stove-pipe cleaners | Stoves and Furnaces | 126 | 154 |
| Stove-pipe drums | Stoves and Furnaces | 126 | 154 |
| Stove-pipe shelves | Kitchen Utensils | 65 | 154 |
| Stove-pipes | Stoves and Furnaces | 126 | 154 |
| Stove-pipes, Making | Sheet-Metal | 113B | 136 |
| Stove-pipe thimbles | Stoves and Furnaces | 126 | 136 |
| Stove-plates | Stoves and Furnaces | 126 | 154 |
| Stove-polish | Grinding and Polishing | 51 | 70 |
| Stoves | Stoves and Furnaces | 126 | 154 |
| Stoves, Base-burning | Stoves and Furnaces | 126 | 154 |
| Stoves, Camp | Stoves and Furnaces | 126 | 154 |
| Stoves, Coal | Stoves and Furnaces | 126 | 154 |
| Stoves, Coal-oil | Stoves and Furnaces | 126 | 154 |
| Stoves, Cooking | Stoves and Furnaces | 126 | 154 |
| Stoves, Foot | Stoves and Furnaces | 126 | 154 |
| Stoves, Gas | Stoves and Furnaces | 126 | 65 |
| Stoves, Heating | Stoves and Furnaces | 126 | 154 |
| Stoves, Lamp | Stoves and Furnaces | 126 | 154 |
| Stoves, Soapstone | Stoves and Furnaces | 126 | 65 |
| Stoves, Utensils | Stoves and Furnaces | 126 | 154 |
| Stoves, Window | Stoves and Furnaces | 126 | 154 |
| Stoves, Wood | Stoves and Furnaces | 126 | 154 |
| Strabismus-goggles | Surgery | 128 | 156 |
| Straightening bars and plates | Metal-Working; 1 | 76 | 125 |
| Straightening gun-barrels | Metal-Working; 1 | 76 | 163 |
| Straightening rails | Metal-Working; 1 | 76 | 125 |

| SUBJECT. | Name of class—official classification. | No. of class—official classification. | No. of class—subscription classification. |
|---|---|---|---|
| Straightening rods and shafts | Metal-Working; 1 | 76 | 125 |
| Strainer-presses | Presses | 100 | 7 |
| Strainers | Wire-Working | 140 | 170 |
| Strainers, Milk | Dairy | 31 | 45 |
| Straw-boards | Paper-Making | 92 | 108 |
| Straw-carriers | Thrashing | 130 | 159 |
| Straw-cutters | Stabling | 119 | 59 |
| Straw for paper, Treating | Paper-Making | 92 | 108 |
| Straw-scatterers | Harvesters | 56 | 77 |
| Straw-stackers | Thrashing | 130 | 159 |
| Street-car heaters | Stoves and Furnaces | 126 | 154 |
| Street-cars | Railways; 2 | 105 | 121 |
| Street-indicators | Stationery | 120 | 146 |
| Street-levelers | Paving | 94 | 110 |
| Street-paving | Paving | 94 | 110 |
| Street-railways | Railways; 1 | 104 | 120 |
| Street-rammers | Paving | 94 | 110 |
| Street-scrapers | Paving | 94 | 110 |
| Street-sprinkling carts | Water-Distribution | 137 | 167 |
| Street-sweepers | Paving | 94 | 110 |
| Street-washers | Water-Distribution | 137 | 167 |
| Stretchers | Carriages and Wagons | 21 | 31 |
| Stretchers for pictures | Fine Arts | 41 | 54 |
| Stretching cloth | Cloth | 26 | 37 |
| Stretching machines, Leather | Leather | 69 | 92 |
| Stretching silk | Silk | 117 | 140 |
| Stringed instruments | Music | 84 | 102 |
| Strings for musical instruments | Music | 84 | 102 |
| Strops | Grinding and Polishing | 51 | 70 |
| Studs | Clasps and Buckles | 24 | 23 |
| Stuffers, Collar | Harness | 54 | 72 |
| Stuffing-boxes | Steam; 1 | 121 | 151 |
| Stuffing-box packing | Steam; 1 | 121 | 151 |
| Stuffing for mattresses, pillows, &c | Beds | 5 | 62 |
| Stuffing for upholstering purposes | Beds | 5 | 62 |
| Stump-extractors | Hoisting | 57 | 79 |
| Stumps, Machines for cutting | Wood-Working; 3 | 144 | 175 |
| Sub-aqueous foundations | Hydraulic Engineering | 61 | 82 |
| Sub-aqueous railways | Hydraulic Engineering | 61 | 82 |
| Sub-aqueous tunneling | Hydraulic Engineering | 61 | 82 |
| Sub-fluvian tunnels | Hydraulic Engineering | 61 | 82 |
| Submarine armor | Hydraulic Engineering | 61 | 82 |
| Submarine cables, Laying | Hydraulic Engineering | 61 | 82 |
| Submarine excavating | Hydraulic Engineering | 61 | 82 |
| Submarine excavators | Excavators | 37 | 82 |
| Submarine helmets | Hydraulic Engineering | 61 | 82 |
| Submarine lamps | Lamps and Gas-Fittings | 67 | 92 |
| Submarine operators | Hydraulic Engineering | 61 | 85 |
| Submarine ordnance | Ordnance | 89 | 107 |
| Submarine vessels | Ships; 1 | 114 | 134 |
| Subsoil-plows | Plows | 97 | 115 |
| Sugar-breakers | Sugar | 127 | 155 |
| Sugar | Sugar | 127 | 155 |
| Sugar-cane cutters | Harvesters | 56 | 75 |
| Sugar-cane mills | Mills | 83 | 155 |
| Sugar-crushers | Mills | 83 | 155 |
| Sugar-filters | Sugar | 127 | 155 |
| Sugar from grain, Obtaining | Sugar | 127 | 155 |
| Sugar-furnaces | Sugar | 127 | 155 |
| Sugar-furnaces | Sugar | 127 | 155 |

| SUBJECT. | Name of class—official classification. | No. of class—official classification. | No. of class—subscription classification. |
|---|---|---|---|
| Sugar-making processes | Sugar | 127 | 155 |
| Sugar-mills | Mills | 83 | 155 |
| Sugar-mixing machine | Sugar | 127 | 157 |
| Sugar-packers | Presses | 100 | 5 |
| Sugar-pans | Sugar | 127 | 155 |
| Sugar-pan skimmers | Sugar | 127 | 157 |
| Sugar-presses | Presses | 100 | 5 |
| Sugar-refining | Sugar | 127 | 155 |
| Sugar-strainers | Sugar | 127 | 151 |
| Sulkies | Carriages | 21 | 31 |
| Sulky-seats | Carriages | 21 | 31 |
| Sulphuric acid, Recovering | Stills | 124 | 47 |
| Sun-shades | Umbrellas and Fans | 135 | 165 |
| Sun-shades for horses | Harness | 54 | 72 |
| Super-heaters for steam-boilers | Steam; 2 | 122 | 149 |
| Supporters, Abdominal | Surgery | 128 | 156 |
| Supporters, Menstrual | Surgery | 128 | 156 |
| Supporters, Obstetrical | Surgery | 128 | 156 |
| Supporters, Pile | Surgery | 128 | 156 |
| Supporters, Uterine | Surgery | 128 | 156 |
| Supporters, Vaginal | Surgery | 128 | 156 |
| Supporters, Window | Builders' Hardware | 16 | 24 |
| Supporting vessels | Ships; 1 | 114 | 82 |
| Suppositories | Surgery | 128 | 156 |
| Surcingles | Harness | 54 | 72 |
| Surf-boats | Boats | 9 | 13 |
| Surge-relievers | Ships; 1 | 114 | 138 |
| Surveying-instruments | Optics | 88 | 157 |
| Surveyors' chains | Optics | 88 | 157 |
| Surveyors' compasses | Optics | 88 | 157 |
| Surveyors' marks | Optics | 88 | 157 |
| Suspenders | Apparel | 2 | 38 |
| Suspending ships' boats | Boats | 9 | 13 |
| Suspension-bridge cables | Wire-Working | 140 | 170 |
| Suspension-bridges | Bridges | 14 | 21 |
| Suspension-cable anchors | Bridges | 14 | 138 |
| Suspension-cable saddles | Bridges | 14 | 21 |
| Suspension-cable wire-carriers | Bridges | 14 | 21 |
| Suspensory bandages | Surgery | 128 | 156 |
| Suture-instruments | Surgery | 128 | 156 |
| Swaging metals | Metal-Working; 3 | 78 | 100 |
| Swaging sheet-metal | Sheet-Metal | 113A | 136 |
| Swarm-indicators | Bee-Hives | 6 | 9 |
| Sward-cutters | Harvesters | 56 | 75 |
| Sweat-leathers for hats | Felting and Hats | 38 | 76 |
| Sweet-potato cultivators | Plows | 97 | 114 |
| Swells for musical instruments | Music | 84 | 106 |
| Swifts | Spinning | 118 | 142 |
| Swimming, Appliances for | Games and Toys | 46 | 143 |
| Swing-bridges | Bridges | 14 | 21 |
| Swings | Games and Toys | 46 | 143 |
| Switch-boards | Electricity | 36 | 50 |
| Switches, Electrical | Electricity | 36 | 50 |
| Switches, Hair | Toilet | 132 | 161 |
| Switches, Railway | Railways; 1 | 104 | 120 |
| Switch-indicators | Signals | 116 | 120 |
| Switch-signals | Signals | 116 | 120 |
| Swivel-hooks | Clasps and Buckles | 24 | 23 |
| Swivel-hooks, Manufacture of | Hardware Manufactures | 53F | 23 |
| Swords, Manufacture of | Hardware Manufactures | 53A | 44 |

| SUBJECT. | Name of class—official classification. | No. of class—official classification. | No. of class—subscription classification. |
|---|---|---|---|
| Sympiesometers | Measuring-Instruments | 73 | 101 |
| Syringes | Pumps | 103 | 118 |
| Syringes, Catarrhal | Surgery | 128 | 156 |
| Syringes, Elastic-bulb | Surgery | 128 | 156 |
| Syringes, Enema | Surgery | 128 | 156 |
| Syringes, Eye | Surgery | 128 | 156 |
| Syringes, Hemorrhoidal | Surgery | 128 | 156 |
| Syringes, Penis | Surgery | 128 | 156 |
| Syringes, Vaginal | Surgery | 128 | 156 |
| T. | | | |
| Table-bells, Manufacture of | Hardware Manufactures | 53B | 24 |
| Table-cars | Kitchen Utensils | 65 | 62 |
| Table-covers | Furniture | 45 | 62 |
| Table-forks | Cutlery | 30 | 44 |
| Table-furniture, Making sheet-metal | Sheet-Metal | 113B | 136 |
| Table-knives | Cutlery | 30 | 44 |
| Table-leaf supporters | Furniture | 45 | 62 |
| Table-leaves, Jointing | Wood-Working; 2 | 143 | 62 |
| Table-mats | Kitchen Utensils | 65 | 88 |
| Tables | Furniture | 45 | 62 |
| Tables, Arithmetical | Measuring-Instruments | 73 | 146 |
| Tables, Card | Furniture | 45 | 62 |
| Tables, Computing | Measuring-Instruments | 73 | 146 |
| Tables, Drawing | Furniture | 45 | 62 |
| Tables, Extension | Furniture | 45 | 62 |
| Tables for glass-manufacture | Glass | 49 | 68 |
| Tables for molders | Casting | 22 | 33 |
| Tables, Ironing | Furniture | 45 | 91 |
| Tables, Operating | Furniture | 45 | 62 |
| Tables, Self-waiting | Furniture | 45 | 62 |
| Table, Sewing-machine | Sewing-Machines | 45 | 134 |
| Tables, Ships' cabin | Ships; 1 | 114 | 137 |
| Tables, Writing | Furniture | 45 | 62 |
| Table-trays | Kitchen Utensils | 65 | 88 |
| Tablets, Compositions for | Stationery | 120 | 146 |
| Tablets, Writing | Stationery | 120 | 146 |
| Tack-extractors | Wood-Working; 4 | 145 | 29 |
| Tackle and blocks | Hoisting | 57 | 79 |
| Tackle-blocks | Hoisting | 57 | 79 |
| Tackle, Hoisting | Ships; 1 | 114 | 79 |
| Tackle, Safety | Ships; 1 | 114 | 79 |
| Tacks | Nails | 85 | 103 |
| Tacks, Leathering | Nails | 85 | 103 |
| Tacks, Machines for making | Nails | 85 | 103 |
| Tag-holders | Stationery | 120 | 146 |
| Tag-machines, Paper | Paper Manufacture | 93 | 108 |
| Tags | Stationery | 120 | 146 |
| Tags, Machines for cutting shoe | Boots and Shoes | 12 | 16 |
| Tags, Printing | Printing | 101 | 117 |
| Tail-boards for wagons | Carriages and Wagons | 21 | 31 |
| Tailoring | Apparel | 2 | 38 |
| Tailors' ironing-machines | Apparel | 2 | 38 |
| Tail-supporters | Harness | 54 | 72 |
| Take-up motions for looms | Weaving | 139 | 95 |
| Tallies for grain, lumber, &c | Measuring-Instruments | 73 | 42 |
| Tallow | Oil, Fat, and Glue | 87 | 26 |
| Tallow, Treatment of | Oil, Fat, and Glue | 87 | 26 |
| Tank-cars | Railways; 2 | 105 | 121 |

| SUBJECT. | Name of class—official classification. | No. of class—official classification. | No. of class—sub-scription classification. |
|---|---|---|---|
| Tanks for storage, Excavated | Masonry | 72 | 51 |
| Tanks for tenders | Steam; 3 | 123 | 94 |
| Tanning | Tanning | 129 | 92 |
| Tanning-machines | Tanning | 129 | 92 |
| Tanning-materials | Tanning | 129 | 92 |
| Tanning-processes | Tanning | 129 | 92 |
| Tape | Weaving | 139 | 17 |
| Tape-fuses | Projectiles | 102 | 105 |
| Tape-primers | Projectiles | 102 | 105 |
| Tapes for skirts | Crinoline and Corsets | 29 | 80 |
| Tape-trimming, Machines for making | Sewing-Machines | 112 | 17 |
| Tape-worm, Apparatus for removing | Surgery | 128 | 156 |
| Taps | Bolts, Nuts, and Rivets | 10 | 14 |
| Taps for liquors | Valves | 136 | 167 |
| Tar, Extracting | Stills | 124 | 47 |
| Targets | Games and Toys | 46 | 143 |
| Tassels | Cordage | 28 | 127 |
| Tatting-machines | Knitting and Netting | 66 | 89 |
| Tatting-shuttles | Apparel | 2 | 38 |
| Tawing leather | Tanning | 129 | 92 |
| Tea-kettles | Stoves and Furnaces | 126 | 136 |
| Tea-pots | Stoves and Furnaces | 126 | 154 |
| Teazel-grading machines | Cloth | 26 | 37 |
| Teazeling cloth | Cloth | 26 | 37 |
| Teazel-trimmers | Garden and Orchard | 47 | 63 |
| Tedders, Hay | Harvesters | 56 | 77 |
| Teeth, Artificial | Dental | 32 | 46 |
| Teeth, artificial, Compositions for | Dental | 32 | 46 |
| Teeth, Insertable saw | Saws | 110 | 136 |
| Telegraphic codes | Electricity | 36 | 50 |
| Telegraphic signals | Electricity | 36 | 50 |
| Telegraph-instruments | Electricity | 36 | 50 |
| Telegraphs, Acoustic | Signals | 116 | 50 |
| Telegraphs, Electric | Electricity | 36 | 50 |
| Telegraphs, Electro-chemical | Electricity | 36 | 50 |
| Telegraphs, Electro-magnetic | Electricity | 36 | 50 |
| Telegraphs, Magneto-electric | Electricity | 36 | 50 |
| Telegraphs, Optical | Signals | 116 | 158 |
| Telegraphs, Pneumatic | Signals | 116 | 158 |
| Telegraph-posts | Fences | 39 | 52 |
| Telegraphs, Printing | Electricity | 36 | 50 |
| Telegraphs, Semaphoric | Signals | 116 | 158 |
| Telegraphs, Submarine | Electricity | 36 | 50 |
| Telephones | Electricity | 36 | 50 |
| Telescopes | Optics | 88 | 157 |
| Tellurians | Educational | 35 | 146 |
| Temperature-alarms | Signals | 116 | 101 |
| Tempering glass | Glass | 49 | 68 |
| Tempering steel | Metallurgy | 75 | 85 |
| Tempering tools | Metallurgy | 75 | 85 |
| Temples, Loom | Weaving | 139 | 96 |
| Temporary binders | Stationery | 120 | 15 |
| Tenaculums | Surgery | 128 | 156 |
| Tender-loaders | Hoisting | 57 | 79 |
| Tenders | Steam; 3 | 123 | 173 |
| Tenoning-machines | Wood-Working; 2 | 143 | 173 |
| Tenon-saws | Saws | 110 | 131 |
| Tension-devices for sewing-machines | Sewing-Machines | 112 | 134 |
| Tenter-bars for cloth | Cloth | 26 | 38 |
| Tents | Trunks | 133 | 162 |

| SUBJECT. | Name of class—official classification. | No. of class—official classification. | No. of class—subscription classification. |
|---|---|---|---|
| Terra cotta | Clay | 25 | 115 |
| Terrets, Manufacture of | Hardware Manufactures | 53F | 24 |
| Tethers | Harness | 54 | 72 |
| Theatrical scenery and effects | Fine Arts | 41 | 54 |
| Theodolites | Optics | 88 | 157 |
| Therapeutic applications of heat, cold, air, vacuum | Surgery | 128 | 156 |
| Thermometers | Measuring-Instruments | 73 | 101 |
| Thermostats | Measuring-Instruments | 73 | 101 |
| Thill-couplings | Carriages and Wagons | 21 | 31 |
| Thill-couplings, Manufacture of | Hardware Manufactures | 53C | 31 |
| Thills | Carriages and Wagons | 21 | 31 |
| Thill-tugs | Carriages and Wagons | 21 | 31 |
| Thimble-machines for vessels | Ships; 1 | 114 | 138 |
| Thimbles, Nautical | Ships; 1 | 114 | 138 |
| Thimbles | Apparel | 2 | 38 |
| Thrashing-floors | Thrashing | 130 | 159 |
| Thrashing-machines | Thrashing | 130 | 159 |
| Thread and spool boxes | Apparel | 2 | 38 |
| Thread assorting, balling, dressing, gassing, measuring, packaging, polishing, reeling, sizing, spooling, tying, winding | Cordage | 28 | 17 |
| Thread-controllers for sewing-machines | Sewing-Machines | 112 | 134 |
| Thread-cutters for sewing-machines | Sewing-Machines | 112 | 134 |
| Thread-cutters | Apparel | 2 | 38 |
| Thread-envelopes | Apparel | 2 | 38 |
| Threaders | Apparel | 2 | 38 |
| Thread-waxers for sewing-machines | Sewing-Machines | 112 | 134 |
| Thresholds | Carpentry | 20 | 30 |
| Throstles | Spinning | 118 | 142 |
| Throttle-valves | Valves | 136 | 152 |
| Throwing horses, Apparatus for | Harness | 54 | 72 |
| Throwing silk | Silk | 117 | 141 |
| Ticket-boxes | Stationery | 120 | 146 |
| Ticket-holders | Stationery | 120 | 146 |
| Ticket-punches | Clasps and Buckles | 24 | 146 |
| Ticket-racks | Stationery | 120 | 146 |
| Tickets, Railway | Stationery | 120 | 146 |
| Tidal docks | Hydraulic Engineering | 61 | 82 |
| Tidal valves | Hydraulic Engineering | 61 | 82 |
| Tide-basins | Hydraulic Engineering | 61 | 82 |
| Tide-gates | Hydraulic Engineering | 61 | 82 |
| Tide-powers | Hydraulic Engineering | 61 | 82 |
| Tide-powers | Mechanical Powers | 74 | 82 |
| Tide-wheels | Water-Wheels | 138 | 167 |
| Ties for bales | Presses | 100 | 7 |
| Ties, Railway | Railways; 1 | 104 | 120 |
| Tightening standing rigging | Ships; 1 | 114 | 138 |
| Tile-laying plows | Plows | 97 | 114 |
| Tile-machines | Clay | 25 | 20 |
| Tiles, Drain | Clay | 25 | 20 |
| Tiles, Fire | Clay | 25 | 20 |
| Tiles, Floor | Clay | 25 | 20 |
| Tiles, Garden | Clay | 25 | 20 |
| Tiles, Mosaic | Clay | 25 | 20 |
| Tiles, Paving | Clay | 25 | 20 |
| Tiles, Roofing | Clay | 25 | 20 |
| Tiling for roofs | Roofing | 108 | 126 |
| Timber, Preservation of | Chemical Miscellaneous | 23 | 35 |

| SUBJECT. | Name of class—official classification. | No. of class—official classification. | No. of class—subscription classification. |
|---|---|---|---|
| Time-balls, Electrical | Electricity | 36 | 50 |
| Time-indicators, Automatic | Horology | 58 | 36 |
| Time-indicators, (*non-automatic*) | Stationery | 120 | 146 |
| Time-fuses | Projectiles | 102 | 105 |
| Time-locks | Locks and Latches | 70 | 93 |
| Time-registers | Horology | 58 | 36 |
| Time-reporters | Horology | 58 | 36 |
| Time-tables | Stationery | 120 | 146 |
| Timothy-headers | Harvesters | 56 | 75 |
| Tin cans and boxes, Making | Sheet-Metal | 113B | 136 |
| Tinmen's shears | Cutlery | 30 | 44 |
| Tinners' tools | Metal-Working; 6 | 81A | 136 |
| Tinning | Plating | 96 | 113 |
| Tin-plate working | Sheet-Metal | 113A | 136 |
| Tin-scrap, Utilizing | Sheet-Metal | 113A | 136 |
| Tippets | Apparel | 2 | 38 |
| Tips, Shoe | Boots and Shoes | 12 | 16 |
| Tire-coolers | Metal-Working; 3 | 78 | 60 |
| Tire-making | Metal-Working; 3 | 78 | 60 |
| Tires, Bending and upsetting | Metal-Working; 3 | 78 | 60 |
| Tire-setters | Wood-Working; 3 | 144 | 60 |
| Tires, Making | Metal-Working; 3 | 78 | 60 |
| Tires, Securing | Carriages and Wagons | 21 | 31 |
| Tires, Tightening | Carriages and Wagons | 21 | 31 |
| Tire-tighteners | Carriages and Wagons | 21 | 31 |
| Tire-upsetting | Metal Working; 3 | 78 | 60 |
| Toasters | Stoves and Furnaces | 126 | 154 |
| Tobacco-cartridges | Tobacco | 131 | 160 |
| Tobacco-curing | Garden and Orchard | 47 | 160 |
| Tobacco-curing, (*except Drying, Class 34*) | Tobacco | 131 | 160 |
| Tobacco-cutting | Tobacco | 131 | 160 |
| Tobacco-dryers | Dryers and Kilns | 34 | 48 |
| Tobacco-flavoring | Tobacco | 131 | 160 |
| Tobacco, Flavoring | Chemical Miscellaneous | 23 | 160 |
| Tobacco-granulating machines | Tobacco | 131 | 160 |
| Tobacco-hooks | Tobacco | 131 | 160 |
| Tobacco packing | Tobacco | 131 | 160 |
| Tobacco-paper | Paper-Making | 92 | 108 |
| Tobacco-pipes | Tobacco | 131 | 160 |
| Tobacco-pipes, Machines for making wooden | Wood-Working; 1 | 142 | 175 |
| Tobacco-pipe stems | Tobacco | 131 | 160 |
| Tobacco-pouches | Tobacco | 131 | 160 |
| Tobacco-preparing | Tobacco | 131 | 160 |
| Tobacco-presses | Presses | 100 | 7 |
| Tobacco-sheeting | Tobacco | 131 | 160 |
| Tobacco-stemming | Tobacco | 131 | 160 |
| Tobacco-stoppers | Tobacco | 131 | 160 |
| Toe-caps | Boots and Shoes | 12 | 16 |
| Toilet-mirrors | Furniture | 45 | 62 |
| Tolling devices for grain-mills | Mills | 83 | 57 |
| Tombstones | Fine Arts | 41 | 54 |
| Tompions | Ordnance | 89 | 105 |
| Tongues, Carriage and wagon | Carriages | 21 | 31 |
| Tongs, Blacksmiths' | Metal-Working; 6 | 81A | 60 |
| Tongs, Oyster | Fishing | 43 | 143 |
| Tongs, Pipe | Metal-Working; 6 | 81A | 100 |
| Tonsils, Instruments for operating on the | Surgery | 128 | 156 |

| SUBJECT. | Name of class—official classification. | No. of class—official classification. | No. of class—subscription classification. |
|---|---|---|---|
| Tool-grinding | Grinding and Polishing | 51 | 70 |
| Tool-handles, (*except Turning, Class* 142) | Wood-Working; 4 | 145 | 175 |
| Tool-holders | Wood-Working; 4 | 145 | 172 |
| Tool-holders for grindstones | Grinding and Polishing | 51 | 70 |
| Tool-holders for lathes | Metal-Working; 7 | 82 | 100 |
| Tool-rests for lathes | Metal-Working; 7 | 82 | 100 |
| Tool-rests for wood-lathes | Wood-Working; 1 | 142 | 172 |
| Tools for glass manufacture | Glass | 49 | 68 |
| Tools for metal-lathes, Turning | Metal-Working; 7 | 82 | 100 |
| Tools for molding clay | Clay | 25 | 115 |
| Tools for scaling boilers | Steam; 2 | 122 | 149 |
| Tools for wood-lathes | Wood-Working; 1 | 142 | 172 |
| Tools for wood-working | Wood-Working; 4 | 145 | 174 |
| Tools for working sheet-metal | Sheet-Metal | 113A | 136 |
| Tools for working wire | Wire-Working | 140 | 170 |
| Tool-stocks | Wood-Working; 4 | 145 | 172 |
| Tools to handles, Attaching | Wood-Working; 4 | 145 | 175 |
| Tooth-brushes | Brushes and Brooms | 15 | 161 |
| Tooth-picks | Toilet | 132 | 161 |
| Tooth-picks, Machines for making | Wood-Working; 3 | 144 | 174 |
| Tooth-powder | Chemical Miscellaneous | 23 | 161 |
| Topical applications, Appliances for | Surgery | 128 | 156 |
| Tops, Carriage | Carriages and Wagons | 21 | 31 |
| Tops, Spinning | Games and Toys | 46 | 143 |
| Torpedoes | Projectiles | 102 | 105 |
| Torpedoes for oil-wells | Projectiles | 102 | 105 |
| Torpedoes, Submarine | Projectiles | 102 | 105 |
| Torpedo-vessels | Ships; 1 | 114 | 137 |
| Tortoise-shell, Machines for working in | Wood-Working; 3 | 144 | 176 |
| Tourniquets | Surgery | 128 | 156 |
| Tournures | Crinoline and Corsets | 29 | 80 |
| Tow-hooks | Ships; 1 | 114 | 137 |
| Tow-lines | Ships; 1 | 114 | 127 |
| Towel-racks | Furniture | 45 | 62 |
| Towel-stands | Furniture | 45 | 62 |
| Toy-blocks | Games and Toys | 46 | 143 |
| Toys | Games and Toys | 46 | 143 |
| Trace-couplings | Harness | 54 | 72 |
| Traces | Harness | 54 | 72 |
| Traces, Safety-attachment of | Carriages and Wagons | 21 | 31 |
| Tracing-table | Fine Arts | 41 | 54 |
| Track-clearers | Railways; 3 | 106 | 120 |
| Track-clearers for harvesters | Harvesters | 56 | 74 |
| Tracking vessels | Ships; 1 | 114 | 137 |
| Track-layers' tools, Manufacture of | Metal-Working; 6 | 81B | 100 |
| Track-laying railways | Railways; 1 | 104 | 120 |
| Track raisers, Railway | Railways; 1 | 104 | 120 |
| Track raisers, Railway | Hoisting | 57 | 120 |
| Traction-engines | Steam; 3 | 123 | 94 |
| Traction-wheels | Carriages and Wagons | 21 | 67 |
| Trainways for ferry-boats | Hydraulic Engineering | 61 | 82 |
| Trammels | Drafting | 33 | 97 |
| Tramming millstones | Mills | 83 | 57 |
| Tram-staffs for mills | Mills | 83 | 57 |
| Transfer, Methods of pictorial | Fine Arts | 41 | 54 |
| Transit-instruments | Optics | 88 | 157 |
| Transmitting motion, Apparatus for | Mechanical Powers | 74 | 67 |
| Transplanting-machines | Garden and Orchard | 47 | 63 |
| Transplanting-tools | Garden and Orchard | 47 | 63 |
| Trap-doors | Carpentry | 20 | 30 |

| SUBJECT. | Name of class—official classification. | No. of class—official classification. | No. of class—subscription classification. |
|---|---|---|---|
| Traps, Animal | Games and Toys | 46 | 143 |
| Traps, Bird | Games and Toys | 46 | 143 |
| Traps, Curculio | Garden and Orchard | 47 | 63 |
| Traps, Earth-worm | Garden and Orchard | 47 | 63 |
| Traps, Fish | Fishing | 43 | 143 |
| Traps for gutters | Paving | 94 | 110 |
| Traps, Household insect | Furniture | 45 | 143 |
| Traps in pavements, Coal | Paving | 94 | 110 |
| Traps, Steam | Steam; 1 | 121 | 150 |
| Traps, Stench | Baths and Closets | 4 | 8 |
| Traps, Water | Baths and Closets | 4 | 8 |
| Travelers for masts and booms | Ships; 1 | 114 | 138 |
| Traveling-bags | Trunks | 133 | 162 |
| Traveling-flasks | Trunks | 133 | 162 |
| Traverses of guns | Ordnance | 89 | 105 |
| Trays, Dish | Kitchen Utensils | 65 | 88 |
| Trays, Food | Kitchen Utensils | 65 | 88 |
| Trays, Fruit | Kitchen Utensils | 65 | 88 |
| Trays, Machines for making wooden | Wood-Working; 3 | 144 | 174 |
| Trays, Tea | Kitchen Utensils | 65 | 88 |
| Treadle-machines | Mechanical Powers | 74 | 67 |
| Treadles | Mechanical Powers | 74 | 67 |
| Tree-boxes | Garden and Orchard | 47 | 63 |
| Tree-diggers | Garden and Orchard | 47 | 63 |
| Tree-dwarfers | Garden and Orchard | 47 | 63 |
| Treenails | Ships; 1 | 114 | 138 |
| Treenails, Machines for making | Wood-Working; 1 | 142 | 174 |
| Treenails, Machines for turning | Wood-Working; 1 | 142 | 172 |
| Tree-planter | Garden and Orchard | 47 | 63 |
| Tree-protectors | Garden and Orchard | 47 | 63 |
| Trees, Boot | Boots and Shoes | 12 | 16 |
| Trees, Saddle | Harness | 54 | 72 |
| Tree-training | Garden and Orchard | 47 | 63 |
| Trellises | Garden and Orchard | 47 | 63 |
| Trellises, Grape | Garden and Orchard | 47 | 63 |
| Trellises, Hop | Garden and Orchard | 47 | 63 |
| Tremolo-attachments for musical instruments | Music | 84 | 106 |
| Trenches | Excavators | 37 | 51 |
| Trephines | Surgery | 128 | 156 |
| Trimmings for garments | Apparel | 2 | 38 |
| Trimming vessels | Ships; 1 | 114 | 138 |
| Tripods, Surveyors' | Optics | 88 | 157 |
| Trolling-devices | Fishing | 43 | 143 |
| Troughs, Feed | Stabling | 119 | 144 |
| Troughs, Machines for making wooden | Wood-Working; 3 | 144 | 174 |
| Trowels, Bricklayers' | Masonry | 72 | 4 |
| Trowels, Garden | Garden and Orchard | 47 | 63 |
| Trowels, Manufacture of | Metal-Working; 6 | 81B | 100 |
| Trowels, Plasterers' | Masonry | 72 | 4 |
| Trucks | Carriages and Wagons | 21 | 31 |
| Trucks, Car | Railways; 3 | 106 | 121 |
| Trumpets | Music | 84 | 102 |
| Trunk-corners | Trunks | 133 | 162 |
| Trunk-frames | Trunks | 133 | 162 |
| Trunk-guards | Trunks | 133 | 162 |
| Trunk-handles | Trunks | 133 | 162 |
| Trunk-labels | Stationery | 120 | 146 |
| Trunk-lid guides | Trunks | 133 | 162 |
| Trunk-lid supporters | Trunks | 133 | 162 |

| SUBJECT. | Name of class—official classification. | No. of class—official classification. | No. of class—sub-scription classification. |
|---|---|---|---|
| Trunk-locks | Locks and Latches | 70 | 93 |
| Trunk-rivets, Manufacture of | Hardware Manufactures | 53J | 24 |
| Trunk-rollers | Trunk-Rollers | 16 | 162 |
| Trunks | Trunks | 133 | 162 |
| Trunks and traveling-bags, Special tools for | Trunks | 133 | 162 |
| Trunk-stays | Trunks | 133 | 162 |
| Trunnions for steam-engines | Steam; 1 | 121 | 150 |
| Truss-bridges | Bridges | 14 | 21 |
| Trusses | Surgery | 128 | 156 |
| Trusses for bridges, floors, and roofs | Bridges | 14 | 21 |
| Trusses for vessels | Ships; 1 | 114 | 137 |
| Trusses of iron or wood | Bridges | 14 | 21 |
| Truss-hoops | Ships; 1 | 114 | 40 |
| Tube-cleaners for steam-boilers | Steam; 2 | 122 | 149 |
| Tube-couplings | Water-Distribution | 137 | 167 |
| Tube-expanders | Metal-Working; 6 | 81A | 100 |
| Tube-filters | Water-Distribution | 137 | 167 |
| Tube-plugs for steam-boilers | Steam; 2 | 122 | 149 |
| Tubes | Tubing and Wire | 134 | 163 |
| Tubes, Bushing | Tubing and Wire | 134 | 163 |
| Tubes, Casing | Tubing and Wire | 134 | 163 |
| Tubes for steam-boilers | Steam; 2 | 122 | 149 |
| Tube-sheets | Steam; 2 | 122 | 125 |
| Tube-sheets, Punching | Metal-Working; 4 | 79 | 119 |
| Tubes, Ornamenting | Tubing and Wire | 134 | 163 |
| Tube-wells | Excavators | 37 | 51 |
| Tubing | Tubing and Wire | 134 | 51 |
| Tubing, Coating of elastic | Hose and Belting | 60 | 64 |
| Tubing, Compositions for elastic | Hose and Belting | 60 | 64 |
| Tubing, Elastic | Hose and Belting | 60 | 64 |
| Tubing for gas, Flexible | Hose and Belting | 60 | 64 |
| Tubing, Lining of elastic | Hose and Belting | 60 | 64 |
| Tubing, Well | Excavators | 37 | 51 |
| Tubs, Machines for making leather | Leather | 69 | 92 |
| Tubular boilers | Steam; 2 | 122 | 149 |
| Tubular bridges | Bridges | 14 | 21 |
| Tuckers for sewing-machines | Sewing-Machines | 112 | 134 |
| Tug-carriers | Harness | 54 | 72 |
| Tumblers for small castings | Casting | 22 | 33 |
| Tumbler-washers | Kitchen Utensils | 65 | 88 |
| Tumbling-boxes | Casting | 22 | 33 |
| Tuning instruments | Music | 84 | 102 |
| Tunneling, Rock | Stone, Lime, and Cement | 125 | 153 |
| Tunnels, Iron | Hydraulic Engineering | 61 | 82 |
| Tunnels, Sub-aqueous | Hydraulic Engineering | 61 | 82 |
| Turbines | Water-Wheels | 138 | 167 |
| Turf-cutters | Garden and Orchard | 47 | 114 |
| Turf-cutting machines | Plows | 97 | 114 |
| Turning cannon | Metal-Working; 7 | 82 | 100 |
| Turning balls, bullets, and shells | Metal-Working; 7 | 82 | 100 |
| Turning gun-barrels | Metal-Working; 7 | 82 | 100 |
| Turning irregular forms | Wood-Working; 1 | 142 | 172 |
| Turning lathes, Metal | Metal-Working; 7 | 82 | 100 |
| Turning stone | Stone, Lime, and Cement | 125 | 153 |
| Turnip-peelers | Kitchen Utensils | 65 | 88 |
| Turnip-toppers | Garden and Orchard | 47 | 63 |
| Turn-out, Railway | Railways; 1 | 104 | 120 |
| Turn-tables, Railway | Railways; 1 | 104 | 120 |
| Turpentine-stills | Stills | 124 | 47 |

| SUBJECT. | Name of class—official classification. | No. of class—official classification. | No. of class—subscription classification. |
|---|---|---|---|
| Turrets | Ordnance | 89 | 105 |
| Turtles | Printing | 101 | 117 |
| Tuyeres | Pneumatics | 98 | 12 |
| Tuyeres, Manufacture of | Metal-Working; 6 | 81B | 12 |
| Twine-cutters | Cutlery | 30 | 44 |
| Twine-holders | Furniture | 45 | 24 |
| Twine-machines | Cordage | 28 | 127 |
| Twist-drills | Metal-Working; 2 | 77 | 100 |
| Twisting bars, rods, and fagots | Metal-Working; 1 | 76 | 125 |
| Twisting silk | Silk | 117 | 140 |
| Twisting wire | Wire-Working | 140 | 170 |
| Twist-lace machines | Knitting and Netting | 66 | 89 |
| Tying yarn | Cordage | 28 | 17 |
| Type | Printing | 101 | 117 |
| Type, Casting | Printing | 101 | 164 |
| Type, Composing | Printing | 101 | 117 |
| Type-cutting machines | Metal-Working; 7 | 82 | 164 |
| Type, Distributing | Printing | 101 | 117 |
| Type-finishing | Printing | 101 | 164 |
| Type, Flexible | Printing | 101 | 164 |
| Type, Founding | Printing | 101 | 164 |
| Type-scouring | Printing | 101 | 164 |
| Type-setting | Printing | 101 | 117 |
| Type-writing machines | Printing | 101 | 117 |
| Typography | Printing | 101 | 117 |
| U. | | | |
| Umbrella and cane combined | Umbrellas and Fans | 135 | 165 |
| Umbrella-cases | Umbrellas and Fans | 135 | 165 |
| Umbrella-covers | Umbrellas and Fans | 135 | 165 |
| Umbrella-fasteners | Umbrellas and Fans | 135 | 165 |
| Umbrella-ferrules | Umbrellas and Fans | 135 | 165 |
| Umbrella-frames | Umbrellas and Fans | 135 | 165 |
| Umbrella-joints | Umbrellas and Fans | 135 | 165 |
| Umbrella ribs and stretchers | Umbrellas and Fans | 135 | 165 |
| Umbrella-runners | Umbrellas and Fans | 135 | 165 |
| Umbrellas | Umbrellas and Fans | 135 | 165 |
| Umbrella-supporters | Umbrellas and Fans | 135 | 165 |
| Umbrella-supporters on carriages | Carriages and Wagons | 21 | 31 |
| Umbrella-tips | Umbrellas and Fans | 135 | 165 |
| Umbrella-trimmings, Manufacture of | Hardware Manufactures | 53I | 165 |
| Under-garments, (*except Knitted, Class* 66) | Apparel | 2 | 38 |
| Under-sleeves | Apparel | 2 | 38 |
| Universal joints | Mechanical Powers | 74 | 67 |
| Unloading-machines | Hoisting | 57 | 79 |
| Untwisting rope, Machines for | Cordage | 28 | 127 |
| Upholstering-stuffing | Beds | 5 | 62 |
| Upholstery | Furniture | 45 | 62 |
| Uppers, Machines for cutting | Boots and Shoes | 12 | 16 |
| Upsetting machines, Tire | Metal-Working; 3 | 78 | 60 |
| Upsetting metals | Metal-Working; 3 | 78 | 60 |
| Urinals | Baths and Closets | 4 | 8 |
| Urinals, Portable | Baths and Closets | 4 | 8 |
| Urinometers | Measuring-Instruments | 73 | 156 |
| Urn-stands | Kitchen Utensils | 65 | 88 |
| Urn-stands and heaters | Stoves and Furnaces | 126 | 154 |
| Urns for coffee, &c | Stoves and Furnaces | 126 | 154 |

| SUBJECT. | Name of class—official classification. | No. of class—official classification. | No. of class—subscription classification. |
|---|---|---|---|
| Uterine sounds | Surgery | 128 | 156 |
| Uterine suppositories | Surgery | 128 | 156 |
| V. | | | |
| Vaccinators | Surgery | 128 | 156 |
| Vacuum-cups | Surgery | 128 | 156 |
| Vacuum-gauges for steam-engines | Measuring-Instruments | 73 | 150 |
| Vacuum-pans | Sugar | 127 | 155 |
| Vacuum-pumps | Pumps | 103 | 118 |
| Valances | Beds | 5 | 62 |
| Valise-locks | Locks and Latches | 70 | 162 |
| Valises | Trunks | 133 | 162 |
| Valve-gear for steam-engines | Steam; 1 | 121 | 152 |
| Valve-motions for locomotives | Steam; 3 | 123 | 152 |
| Valve-regulators | Valves | 136 | 152 |
| Valves, Arrangement of | Valves | 136 | 152 |
| Valves, Automatic | Valves | 136 | 152 |
| Valves, Back-pressure | Valves | 136 | 152 |
| Valves, Balance | Valves | 136 | 152 |
| Valves, Butterfly | Valves | 136 | 152 |
| Valves, Check | Valves | 136 | 152 |
| Valves, Clack | Valves | 136 | 152 |
| Valves, Core | Valves | 136 | 152 |
| Valves, Cut-off | Valves | 136 | 152 |
| Valves, Disk | Valves | 136 | 152 |
| Valve-seats | Valves | 136 | 152 |
| Valves, Escape | Valves | 136 | 152 |
| Valves for draining tidal lands | Hydraulic Engineering | 61 | 82 |
| Valves for gas-engines | Valves | 136 | 2 |
| Valves for submarine ordnance | Ordnance | 89 | 105 |
| Valves for syringes | Valves | 136 | 118 |
| Valves, Globe | Valves | 136 | 152 |
| Valves, Governor | Valves | 136 | 152 |
| Valves, Oscillating | Valves | 136 | 152 |
| Valves, Piston | Valves | 136 | 152 |
| Valves, Pump | Valves | 136 | 118 |
| Valves, Puppet | Valves | 136 | 152 |
| Valves, Relief | Valves | 136 | 152 |
| Valves, Rotary | Valves | 136 | 152 |
| Valves, Safety | Valves | 136 | 152 |
| Valves, Slide | Valves | 136 | 152 |
| Valves, Spring-balance | Valves | 136 | 152 |
| Valves, Steam | Valves | 136 | 152 |
| Valves, Stop | Valves | 136 | 152 |
| Valves, Throttle | Valves | 136 | 152 |
| Valves, Trimmer | Valves | 136 | 152 |
| Valves, Water-closet | Valves | 136 | 8 |
| Valve-tubes | Valves | 136 | 152 |
| Vanes | Measuring-Instruments | 73 | 118 |
| Vapor-burners | Lamps and Gas-Fittings | 67 | 90 |
| Vapor-engines, (*except steam*) | Pneumatics | 98 | 2 |
| Vapor-inhalers | Surgery | 128 | 156 |
| Vapor-lamps | Lamps and Gas-Fittings | 67 | 90 |
| Variable exhaust for steam-engines | Steam; 3 | 123 | 150 |
| Varicose veins, Bandages for | Surgery | 128 | 156 |
| Varnishes | Paint | 91 | 107 |
| Varnishing-machines | Paint | 91 | 107 |
| Vats, Tanning | Tanning | 129 | 92 |
| Vault-covers | Paving | 94 | 110 |
| Vault-lights | Paving | 94 | 110 |

| SUBJECT. | Name of class—official classification. | No. of class—official classification. | No. of class—sub-scription classification. |
|---|---|---|---|
| Vaults | Masonry | 72 | 4 |
| Vaults, Safe | Safes | 109 | 128 |
| Vegetable-assorters | Garden and Orchard | 47 | 59 |
| Vegetable-boilers | Stoves and Furnaces | 126 | 154 |
| Vegetable-cutters | Stabling | 119 | 59 |
| Vegetable parchment | Paper-Making | 92 | 108 |
| Vegetable-peelers | Kitchen Utensils | 65 | 88 |
| Vegetable-slicers | Stabling | 119 | 59 |
| Vegetables, Preserving | Preserving Food | 99 | 58 |
| Vegetable-washers | Stabling | 119 | 59 |
| Vegetable-washers | Kitchen Utensils | 65 | 88 |
| Vehicles | Carriages and Wagons | 21 | 31 |
| Veils | Apparel | 2 | 38 |
| Velinches | Water-Distribution | 137 | 167 |
| Velocimeters | Measuring-Instruments | 73 | 42 |
| Velocipedes | Carriages and Wagons | 21 | 31 |
| Veneering | Wood-Working; 3 | 144 | 176 |
| Veneering-tools | Wood-Working; 3 | 144 | 176 |
| Veneer-saws | Saws | 110 | 130 |
| Veneers, Machines for cutting | Wood-Working; 3 | 144 | 176 |
| Veneers, Machines for planing | Wood-Working; 3 | 144 | 176 |
| Veneers, Machines for polishing | Wood-Working; 3 | 144 | 176 |
| Veneers, Machines for pressing | Wood-Working; 3 | 144 | 176 |
| Veneers, Machines for straightening | Wood-Working; 3 | 144 | 176 |
| Ventilating-cores | Casting | 22 | 33 |
| Ventilating hay and grain | Stabling | 119 | 144 |
| Ventilating millstones | Mills | 83 | 57 |
| Ventilation by heated air | Pneumatics | 98 | 166 |
| Ventilation by hot water | Pneumatics | 98 | 166 |
| Ventilation by steam | Pneumatics | 98 | 166 |
| Ventilation of cars | Pneumatics | 98 | 166 |
| Ventilation of grain | Pneumatics | 98 | 166 |
| Ventilation of halls | Pneumatics | 98 | 166 |
| Ventilation of houses | Pneumatics | 98 | 166 |
| Ventilation of mines | Pneumatics | 98 | 166 |
| Ventilation of ships | Pneumatics | 98 | 166 |
| Ventilators | Pneumatics | 98 | 166 |
| Ventilators for boots | Boots and Shoes | 12 | 16 |
| Ventilators, Hat | Felting and Hats | 38 | 76 |
| Ventilators, Plenum | Pneumatics | 98 | 166 |
| Ventilators, Vacuum | Pneumatics | 98 | 166 |
| Vents for barrels | Aëration and Bottling | 1 | 40 |
| Vents of ordnance | Ordnance | 89 | 105 |
| Vent-stoppers | Ordnance | 89 | 105 |
| Vermin-destroying compounds | Chemical Miscellaneous | 23 | 98 |
| Verniers | Measuring-Instruments | 73 | 97 |
| Vessels | Ships; 1 | 114 | 137 |
| Vessels, Armor for | Ships; 1 | 114 | 137 |
| Vessels, Construction of | Ships; 1 | 114 | 137 |
| Vessels, Grain-levelers for | Ships; 1 | 114 | 137 |
| Vessels, Iron | Ships; 1 | 114 | 137 |
| Vessels, Iron-clad | Ships; 1 | 114 | 137 |
| Vessels, Launching | Hydraulic Engineering | 61 | 137 |
| Vessels, Loading and unloading, Machinery for | Hoisting | 57 | 79 |
| Vessels, Propulsion of | Ships; 2 | 115 | 139 |
| Vessels, Steam | Ships; 1 | 114 | 137 |
| Vessels, Submarine | Ships; 1 | 114 | 137 |
| Vessels, Torpedo | Ships; 1 | 114 | 137 |
| Vessels, Trimming | Ships; 1 | 114 | 137 |

| SUBJECT. | Name of class—official classification. | No. of class—official classification. | No. of class—sub-scription classification. |
|---|---|---|---|
| Vests | Apparel | 2 | 38 |
| Viaducts | Bridges | 14 | 21 |
| Vine cutter, Potato | Plows | 97 | 114 |
| Vinegar-apparatus | Beer and Wine | 7 | 19 |
| Vinegar processes | Beer and Wine | 7 | 19 |
| Vine-locks | Garden and Orchard | 47 | 63 |
| Vine-supports | Garden and Orchard | 47 | 63 |
| Vine-trellises | Garden and Orchard | 47 | 63 |
| Violin-attachments | Music | 84 | 102 |
| Violin-bows | Music | 84 | 102 |
| Violins | Music | 84 | 102 |
| Violin-screws | Music | 84 | 102 |
| Violin sound-boards | Music | 84 | 102 |
| Violoncellos | Music | 84 | 102 |
| Vise-making, Machinists' | Metal-Working; 3 | 78 | 100 |
| Vises, Carpenters' | Wood-Working; 4 | 145 | 29 |
| Vises, Machinists' | Metal-Working; 6 | 81A | 100 |
| Visiting-cards | Stationery | 120 | 146 |
| Vitreography | Fine Arts | 41 | 54 |
| Vitreous enamels | Glass | 49 | 68 |
| Voltaic apparatus | Electricity | 36 | 50 |
| Voltaic baths | Electricity | 36 | 50 |
| Voltaic batteries | Electricity | 36 | 50 |
| Voltaic instruments | Electricity | 36 | 50 |
| Voltameters | Electricity | 36 | 50 |
| Votes, Instruments for recording | Measuring-Instruments | 73 | 42 |
| Vulcanite | Caoutchouc | 18 | 84 |
| Vulcanizing-flasks for dentists' uses | Caoutchouc | 18 | 84 |
| Vulcanizing-furnaces | Caoutchouc | 18 | 84 |
| Vulcanizing-processes | Caoutchouc | 18 | 84 |
| W. | | | |
| Wadding | Carding | 19 | 28 |
| Wadding-machines | Carding | 19 | 28 |
| Wad-punches | Clasps and Buckles | 24 | 23 |
| Wads, Machines for making leather | Leather | 69 | 72 |
| Waffle-irons | Stoves and Furnaces | 126 | 154 |
| Wagon-bolsters | Carriages and Wagons | 21 | 31 |
| Wagons | Carriages and Wagons | 21 | 31 |
| Wagons, Dumping | Carriages and Wagons | 21 | 31 |
| Wagon-seats | Carriages and Wagons | 21 | 31 |
| Wagons, Unloading-attachment for | Carriages and Wagons | 21 | 31 |
| Wagon-tongue supporters | Carriages and Wagons | 21 | 31 |
| Wainscoting, Making | Wood-Working; 3 | 144 | 30 |
| Wall-covering, Ornamental | Fine Arts | 41 | 54 |
| Wall-paper hanging | Fine Arts | 41 | 107 |
| Wall-paper printing | Printing | 101 | 117 |
| Wall-paper trimmers | Paper Manufacture | 93 | 107 |
| Wall-protectors | Furniture | 45 | 62 |
| Wall-shields | Furniture | 45 | 62 |
| Walls, (other than wooden) | Masonry | 72 | 4 |
| Wardrobes | Furniture | 45 | 62 |
| Warp-beams | Weaving | 139 | 96 |
| Warp-dressing | Cordage | 28 | 127 |
| Warping-bitts | Ships; 1 | 114 | 137 |
| Warping-checks | Ships; 1 | 114 | 137 |
| Warping-machines | Cordage | 28 | 127 |
| Wash-basins | Baths and Closets | 4 | 8 |
| Wash-benches | Laundry | 68 | 90 |

| SUBJECT. | Name of class—official classification. | No. of class—official classification. | No. of class—subscription classification. |
|---|---|---|---|
| Wash-boards | Laundry | 68 | 90 |
| Wash-boards, Machines for making | Wood-Working; 2 | 143 | 174 |
| Wash-boilers | Laundry | 68 | 136 |
| Washer-making machines, Metallic | Bolts, Nuts, and Rivets | 10 | 14 |
| Washer-punches | Clasps and Buckles | 24 | 23 |
| Washers, Gold | Ore | 90 | 3 |
| Washers, Machines for making leather | Leather | 69 | 72 |
| Washers, Metallic | Bolts, Nuts, and Rivets | 10 | 14 |
| Washing-compositions | Oil, Fat, and Glue | 87 | 26 |
| Washing-compounds | Chemical Miscellaneous | 23 | 91 |
| Washing-machines | Laundry | 68 | 91 |
| Washing-tables | Ore | 90 | 91 |
| Wash-stands | Baths and Closets | 4 | 8 |
| Watch and fob chains, Manufacture of | Hardware Manufactures | 53E | 86 |
| Watch-cases | Horology | 58 | 36 |
| Watch-chains | Jewelry | 63 | 86 |
| Watch-clocks, Electrical | Electricity | 36 | 36 |
| Watch-dials | Horology | 58 | 36 |
| Watches, Manufacture of | Hardware Manufactures | 53E | 36 |
| Watches | Horology | 58 | 36 |
| Watch-keys | Horology | 58 | 36 |
| Watchmakers' lathes | Metal-Working; 7 | 82 | 36 |
| Watchmakers' tools | Metal-Working; 6 | 81A | 36 |
| Watchman's time-registers | Horology | 58 | 36 |
| Watch-motions | Horology | 58 | 36 |
| Watch-springs | Horology | 58 | 36 |
| Water-backs for ranges and stoves | Stoves and Furnaces | 126 | 154 |
| Water-closets | Baths and Closets | 4 | 8 |
| Water-closets, Appliances of | Baths and Closets | 4 | 8 |
| Water-closet valves | Valves | 136 | 8 |
| Water-coolers | Ice | 62 | 58 |
| Water-elevators | Pumps | 103 | 118 |
| Waterfall head-dresses | Toilet | 132 | 161 |
| Water-gauges for steam-boilers | Steam; 2 | 122 | 148 |
| Water-gauges for steam-boilers, Registering-devices of | Measuring-Instruments | 73 | 148 |
| Water-gauges, Registering-devices of | Measuring-Instruments | 73 | 148 |
| Water-heaters for steam-boilers | Steam; 2 | 122 | 147 |
| Water-indicators for steam-boilers | Steam; 2 | 122 | 148 |
| Water-marks in paper | Paper-Making | 92 | 108 |
| Water-meters, Registering-devices of | Measuring-Instruments | 73 | 167 |
| Water-pressure regulating-valves | Valves | 136 | 167 |
| Water-proofing compounds | Paint | 91 | 107 |
| Water-proofing leather | Tanning | 129 | 92 |
| Water-proofing paper | Paper-Making | 92 | 108 |
| Water-proofing processes | Paint | 91 | 107 |
| Water-proof safes | Safes | 109 | 128 |
| Water pumps, Feed | Steam; 2 | 122 | 118 |
| Water-purifiers for steam-boilers | Steam; 2 | 122 | 148 |
| Water-rams | Pumps | 103 | 118 |
| Water-tanks for railways, &c | Water-Distribution | 137 | 118 |
| Water-traps | Baths and Closets | 4 | 8 |
| Water-valves | Valves | 136 | 167 |
| Water-wheels | Water-Wheels | 138 | 167 |
| Wave-powers | Hydraulic Engineering | 61 | 82 |
| Wax-cutting machines | Oil, Fat, and Glue | 87 | 26 |
| Waxed-ends, Machines for twisting | Boots and Shoes | 12 | 16 |
| Waxing thread, Machines for | Boots and Shoes | 12 | 16 |
| Wax | Oil, Fat, and Glue | 87 | 26 |

| SUBJECT. | Name of class—official classification. | No. of class—official classification. | No. of class—subscription classification. |
|---|---|---|---|
| Wax-thread heaters for sewing-machines | Sewing-Machines | 112 | 134 |
| Wax, Treatment of | Oil, Fat, and Glue | 87 | 26 |
| Weather-boarding | Carpentry | 20 | 30 |
| Weather-guides | Measuring-Instruments | 73 | 101 |
| Weather-strips | Carpentry | 20 | 30 |
| Weather-strips, (metallic,) Manufacture of | Hardware Manufactures | 53B | 136 |
| Weavers' harness, Machines for making | Weaving | 139 | 72 |
| Weaving and working human hair | Toilet | 132 | 161 |
| Weaving baskets | Weaving | 139 | 169 |
| Web-saws | Saws | 110 | 131 |
| Wedges, Machines for making | Wood-Working; 3 | 144 | 174 |
| Weeding-implements | Garden and Orchard | 47 | 63 |
| Weed-pullers | Garden and Orchard | 47 | 63 |
| Weft-forks | Weaving | 139 | 96 |
| Weighing-apparatus | Measuring-Instruments | 73 | 132 |
| Weighing-apparatus, Electrical | Electricity | 36 | 132 |
| Weighing-buckets | Measuring-Instruments | 73 | 132 |
| Weighing-scales | Measuring-Instruments | 73 | 132 |
| Weights, Sash | Builders' Hardware | 16 | 24 |
| Weirs | Hydraulic Engineering | 61 | 82 |
| Welded-pipe making | Tubing and Wire | 134 | 163 |
| Welding-apparatus | Metal-Working; 3 | 78 | 60 |
| Welding-compounds | Metallurgy | 75 | 100 |
| Well-boring, (*except Rock, Class* 125) | Stone, Lime, and Cement | 125 | 51 |
| Well-curbs | Water-Distribution | 137 | 118 |
| Well-digging | Excavators | 37 | 51 |
| Wells, Artesian | Stone, Lime and Cement | 125 | 124 |
| Wells, Driven | Water-Distribution | 137 | 51 |
| Well-tubes | Water-Distribution | 137 | 51 |
| Well-tubes, Filters for | Water-Distribution | 137 | 51 |
| Well-tubes, Packing for | Water-Distribution | 137 | 51 |
| Well-tubes, Strainers for | Water-Distribution | 137 | 51 |
| Well-tubing | Water-Distribution | 137 | 51 |
| Welting cord | Cordage | 28 | 127 |
| Welts | Boots and Shoes | 12 | 16 |
| Welts, Machines for forming | Boots and Shoes | 12 | 16 |
| Welts, Machines for trimming | Boots and Shoes | 12 | 16 |
| Wet-docks | Hydraulic Engineering | 61 | 82 |
| Whalebone, Machines for dressing | Wood-Working; 3 | 144 | 176 |
| Whalebone, Machines for splitting | Wood-Working; 3 | 144 | 176 |
| Whaling-apparatus, Electrical | Electricity | 36 | 50 |
| Wharf-boats | Ships; 1 | 114 | 137 |
| Wharfs | Hydraulic Engineering | 61 | 82 |
| Wharfs, Floating | Ships; 1 | 114 | 137 |
| Wheelbarrows | Carriages and Wagons | 21 | 31 |
| Wheel-heads | Spinning | 118 | 142 |
| Wheel-presses, (*not Hydraulic, Class* 103) | Presses | 100 | 7 |
| Wheels, Car | Railways; 3 | 106 | 122 |
| Wheels, Fly | Mechanical Powers | 74 | 67 |
| Wheels for vehicles | Carriages and Wagons | 21 | 31 |
| Wheels, (metallic,) Manufacture of | Hardware Manufactures | 53C | 33 |
| Wheel-spokes | Carriages and Wagons | 21 | 31 |
| Wheels to axles, Securing | Carriages and Wagons | 21 | 31 |
| Wheelwrights' machines | Wood-Working; 3 | 144 | 175 |
| Wheelwrights' tools of metal, Manufacture of | Metal-Working; 6 | 81B | 100 |
| Wheelwrights' wood-working tools | Wood-Working; 4 | 145 | 100 |
| Whetstones | Grinding and Polishing | 51 | 70 |

| SUBJECT. | Name of class—official classification. | No. of class—official classification. | No. of class—subscription classification. |
|---|---|---|---|
| Whiffletrees | Carriages and Wagons | 21 | 31 |
| Whip-hangers | Harness | 54 | 72 |
| Whipping cloth | Cloth | 26 | 37 |
| Whip-racks | Harness | 54 | 72 |
| Whips | Harness | 54 | 72 |
| Whip-saws | Saws | 110 | 131 |
| Whips, Machines for making leather | Leather | 69 | 72 |
| Whip-sockets | Carriages and Wagons | 21 | 31 |
| Whip-stocks | Harness | 54 | 72 |
| Whistles, Electrical | Electricity | 36 | 50 |
| Whistles for steam-boilers | Steam; 2 | 122 | 149 |
| White-lead machinery | Paint | 91 | 107 |
| White-lead processes | Paint | 91 | 107 |
| Whitening leather | Tanning | 129 | 92 |
| White rubber | Caoutchouc | 18 | 84 |
| Wicks, Compositions for treating | Lamps and Gas-Fittings | 67 | 90 |
| Wick-trimmers | Lamps and Gas-Fittings | 67 | 90 |
| Wigs | Toilet | 132 | 161 |
| Willow-peeling | Wood-Working; 3 | 144 | 169 |
| Willow-shaving machines | Wood-Working; 3 | 144 | 169 |
| Winches | Hoisting | 57 | 79 |
| Wind-chests for organs | Music | 84 | 106 |
| Winding-bobbins | Cordage | 28 | 17 |
| Winding cord, cordage, fishing-lines, rope, thread, warp, yarn | Cordage | 28 | 17 |
| Winding silk | Silk | 117 | 140 |
| Winding-spools | Cordage | 28 | 17 |
| Wind-instruments | Music | 84 | 106 |
| Windlasses | Hoisting | 57 | 79 |
| Wind-mills | Pneumatics | 98 | 118 |
| Window-alarms | Signals | 116 | 24 |
| Window-blinds | Carpentry | 20 | 30 |
| Window-blinds, Movable | Furniture | 45 | 62 |
| Window-buttons | Builders' Hardware | 16 | 24 |
| Window-fasteners | Builders' Hardware | 16 | 24 |
| Window-glass | Glass | 49 | 68 |
| Window-illuminators | Lamps and Gas-Fittings | 67 | 90 |
| Window-locks | Builders' Hardware | 16 | 93 |
| Window-mirrors | Furniture | 45 | 62 |
| Window-operators | Builders' Hardware | 16 | 24 |
| Window-reflectors | Lamps and Gas-Fittings | 67 | 90 |
| Windows | Carpentry | 20 | 30 |
| Window-sash, (metallic,) Manufacture of | Hardware Manufactures | 53B | 4 |
| Window-screens | Carpentry | 20 | 30 |
| Windows for railway-cars | Railways; 2 | 105 | 30 |
| Window-shades | Furniture | 45 | 62 |
| Window-shutters | Carpentry | 20 | 30 |
| Window-supporters | Builders' Hardware | 16 | 24 |
| Window-washers | Furniture | 45 | 62 |
| Wind-wheels | Pneumatics | 98 | 118 |
| Wine-ageing | Beer and Wine | 7 | 19 |
| Wine-making | Beer and Wine | 7 | 19 |
| Wine-presses | Presses | 100 | 19 |
| Wines, Treatment of | Beer and Wine | 7 | 19 |
| Winnowing-machines | Thrashing | 130 | 159 |
| Wire-annealing | Metallurgy | 75 | 163 |
| Wire-bending | Wire-Working | 140 | 170 |
| Wire, Braided thread-covering | Cordage | 28 | 127 |
| Wire brushes | Wire-Working | 140 | 170 |

| SUBJECT. | Name of class—official classification. | No. of class—official classification. | No. of class—sub-scription classification. |
|---|---|---|---|
| Wire cables | Wire-Working | 140 | 170 |
| Wire-cloth crimping, Machines for | Wire-Working | 140 | 170 |
| Wire coiling | Wire-Working | 140 | 163 |
| Wire covering | Wire-Working | 140 | 163 |
| Wire dish-covers, Making | Wire-Working | 140 | 163 |
| Wire dish-stands, Making | Wire-Working | 140 | 170 |
| Wire-drawing | Wire-Working | 140 | 163 |
| Wire fabrics | Wire-Working | 140 | 170 |
| Wire fences | Fences | 39 | 52 |
| Wire fences, Panels for | Wire-Working | 140 | 52 |
| Wire ferriage | Hydraulic Engineering | 61 | 82 |
| Wire for hoops of skirts | Crinoline and Corsets | 29 | 80 |
| Wire gauze | Wire-Working | 140 | 170 |
| Wire heddles for weavers' harness, Making | Wire-Working | 140 | 170 |
| Wire, Looms for weaving | Weaving | 139 | 170 |
| Wire-making | Tubing and Wire | 134 | 163 |
| Wire-pointing | Wire-Working | 140 | 163 |
| Wire-rolling | Tubing and Wire | 134 | 163 |
| Wire ropes | Wire-Working | 140 | 163 |
| Wire screens | Wire-Working | 140 | 170 |
| Wire sifters | Wire-Working | 140 | 170 |
| Wire springs, Making | Wire-Working | 140 | 170 |
| Wire straightening | Wire-Working | 140 | 163 |
| Wire strainers | Wire-Working | 140 | 170 |
| Wire-stretchers for fences | Fences | 39 | 52 |
| Wire-twisting | Wire-Working | 140 | 163 |
| Wire-ways | Railways; 1 | 104 | 120 |
| Wire with wire for skirt-hoops, &c., Covering | Wire-Working | 140 | 80 |
| Wire-working for ornamental fences | Wire-Working | 140 | 52 |
| Wiring sheet-metal ware | Sheet-Metal | 113A | 136 |
| Withes, Machines for twisting | Wood-Working; 3 | 144 | 174 |
| Wood and asphalt pavements combined | Paving | 94 | 110 |
| Wood and iron pavements combined | Paving | 94 | 110 |
| Wood and stone pavements combined | Paving | 94 | 110 |
| Wood, Artificial | Chemical Miscellaneous | 23 | 35 |
| Wood-bending machines | Wood-Working; 3 | 144 | 176 |
| Wood-bundling machines | Wood-Working; 3 | 144 | 174 |
| Wooden bowls, Machines for making | Wood-Working; 3 | 144 | 172 |
| Wooden boxes, Machines for turning | Wood-Working; 1 | 142 | 172 |
| Wooden bridges | Bridges | 14 | 21 |
| Wood-engraving | Fine Arts | 41 | 54 |
| Wooden pavements | Paving | 94 | 110 |
| Wooden pins, Machines for turning | Wood-Working; 1 | 142 | 172 |
| Wooden screws, Machines for turning | Wood-Working; 1 | 142 | 172 |
| Wooden trays, Machines for making | Wood-Working; 3 | 144 | 172 |
| Wooden ware, Machines for making | Wood-Working; 3 | 144 | 172 |
| Wood for paper, Treating | Paper-Making | 92 | 108 |
| Wood-gear cutting | Wood-Working; 3 | 144 | 172 |
| Wood-lathes | Wood-Working; 1 | 142 | 172 |
| Wood-polishing machines | Wood-Working; 3 | 144 | 173 |
| Wood, Preservation of | Chemical Miscellaneous | 23 | 35 |
| Wood-screws | Wood-Screws | 141 | 171 |
| Wood-screws, Machines for making | Wood-Screws | 141 | 171 |
| Wood-splitting machines | Wood-Working; 3 | 144 | 174 |
| Wood-work of buildings | Carpentry | 20 | 30 |
| Wool-burring machines | Carding | 19 | 28 |
| Wool-cleaning machines | Carding | 19 | 28 |
| Wool, Compositions for treating | Chemical Miscellaneous | 23 | 35 |

| SUBJECT. | Name of class—official classification. | No. of class—official classification. | No. of class—subscription classification. |
|---|---|---|---|
| Wool-dryers | Dryers and Kilns | 34 | 48 |
| Wool-oiling machines | Carding | 19 | 28 |
| Wool-packers | Presses | 100 | 7 |
| Wool-presses | Presses | 100 | 7 |
| Wool, Removing suint and grease from | Oil, Fat, and Glue | 87 | 26 |
| Wool-tyers | Presses | 100 | 135 |
| Wool-washing machines | Laundry | 68 | 135 |
| Work-baskets | Apparel | 2 | 38 |
| Work boxes, Sewing | Apparel | 2 | 38 |
| Work holders, Sewing | Apparel | 2 | 38 |
| Work-table, Appliances for the | Apparel | 2 | 38 |
| Worming rope | Cordage | 28 | 127 |
| Worms for fire-arms | Fire-Arms | 42 | 55 |
| Wrappers | Apparel | 2 | 38 |
| Wrecking-cars | Railways; 2 | 105 | 121 |
| Wrenches, Monkey | Metal-Working; 6 | 81A | 100 |
| Wrenches, Pipe | Metal-Working; 6 | 81A | 100 |
| Wringers | Laundry | 68 | 91 |
| Wringers, Clamps for | Laundry | 68 | 91 |
| Wringers, Rolls for | Laundry | 68 | 91 |
| Writing-apparatus for the blind | Educational | 35 | 146 |
| Writing-cases | Stationery | 120 | 146 |
| Writing-ink | Chemical Miscellaneous | 23 | 146 |
| Writing-tables | Furniture | 45 | 62 |
| Wrought-metal railway-wheels | Railways; 4 | 107 | 122 |
| Wrought nails | Nails | 85 | 103 |
| Y. | | | |
| Yarn assorting, balling, dressing, drying, folding, gassing, measuring, packaging, polishing, reeling, sizing, spooling, tying, winding | Cordage | 28 | 17 |
| Yarn coating or covering | Carding | 19 | 28 |
| Yarn-hooks | Cordage | 28 | 17 |
| Yarn-presses | Presses | 100 | 17 |
| Yards to masts, Attaching | Ships; 1 | 114 | 138 |
| Yeast and its substitutes | Chemical Miscellaneous | 23 | 35 |
| Yokes for breechy animals | Harness | 54 | 72 |
| Yokes, Neck | Carriages and Wagons | 21 | 31 |
| Yokes, Ox | Carriages and Wagons | 21 | 31 |
| Z. | | | |
| Zincography | Fine Arts | 41 | 54 |
| Zinc roofing | Roofing | 108 | 126 |
| Zinc, white, processes and machinery | Paint | 91 | 107 |

O

www.ingramcontent.com/pod-product-compliance
Lightning Source LLC
LaVergne TN
LVHW020224110826
845151LV00003B/818

* 9 7 8 1 4 2 5 5 3 3 0 8 3 *